Handbook of Energy Engineering

Handbook of Energy Engineering

Edited by Lana Roche

SYRAWOOD
PUBLISHING HOUSE

New York

Published by Syrawood Publishing House,
750 Third Avenue, 9th Floor,
New York, NY 10017, USA
www.syrawoodpublishinghouse.com

Handbook of Energy Engineering
Edited by Lana Roche

International Standard Book Number: 978-1-64740-121-4 (Hardback)

Cataloging-in-Publication Data

Handbook of energy engineering / edited by Lana Roche.
    p. cm.
Includes bibliographical references and index.
ISBN 978-1-64740-121-4
1. Power (Mechanics). 2. Energy conservation. 3. Energy development.
4. Sustainable engineering. I. Roche, Lana.
TJ163.9 .H36 2022
621.042--dc23

TABLE OF CONTENTS

Permissions

List of Contributors

Index

PREFACE

Energy engineering is a multidisciplinary field which merges knowledge from the fields of mathematics, chemistry and physics with environmental and economic engineering practices. There are various aspects which are dealt with under this discipline such as energy efficiency, energy services, plant engineering and alternative energy technologies. The main focus of energy engineering is to develop the most efficient and sustainable methods to operate buildings and manufacturing processes. This is generally accomplished by observing the use of energy and then suggesting approaches to improve the efficiency. Power engineering is a sub-discipline of energy engineering where mathematics and physics are applied to the movement and transfer of energy to work within a system. The topics included in this book on energy engineering are of utmost significance and bound to provide incredible insights to readers. It is appropriate for students seeking detailed information in this area as well as for experts.

This book is the end result of constructive efforts and intensive research done by experts in this field. The aim of this book is to enlighten the readers with recent information in this area of research. The information provided in this profound book would serve as a valuable reference to students and researchers in this field.

At the end, I would like to thank all the authors for devoting their precious time and providing their valuable contribution to this book. I would also like to express my gratitude to my fellow colleagues who encouraged me throughout the process.

Editor

Creating paradigms for nearly zero energy hotels in South Europe

Theocharis Tsoutsos*, Stavroula Tournaki, Maria Frangou and Marianna Tsitoura

Renewable and Sustainable Energy Systems Lab, School of Environmental Engineering, Technical University of Crete, University Campus, Chania, Crete, Greece

* **Correspondence:** Email: theocharis.tsoutsos@enveng.tuc.gr

Abstract: According to the European Directive on the Energy Performance of Buildings recast, hotels and other buildings of a certain size, frequently visited by the public, should set an example in environmental and energy performance. Moreover, being energy intensive buildings, they are at high priority for becoming nearly Zero Energy Buildings. Even though they represent a specific category, along with restaurants, till today there is a lack of credible data for this type of buildings, especially taking into account the wide range of different typologies (coastal, mountain, urban, rural or business, resort, spa/wellness, bed & breakfast). This paper presents the results of the actual energy performance of six south European countries (Greece, Croatia, France, Romania, Italy and Spain) plus one north (Sweden) for comparison, analyzed in the framework of the nearly Zero Energy Hotels (neZEH) project. The project focused on providing technical assistance to existing pilot hotels for refurbishing into nearly Zero Energy Buildings, demonstrating the sustainability of investments towards zero energy and undertaking training and capacity building activities at regional, national and European level. The results showed that the primary energy use for the hosting functions across all sixteen pilot hotels can decrease dramatically—from an average of 277 kWh/m^2/y to an average of 102 kWh/m^2/y; an average reduction of 63%. At the same time, Renewable Energy Sources share for the hosting functions can be increased from an average of 18% to an average of 46%. The analysis also showed that hotel non-hosting functions, i.e., other facilities that require special indoor environmental conditions, such as spa, kitchen etc. are more critical than the hosting functions; their primary energy use can decrease from an average of 727 kWh/m^2/y to an average of 374 kWh/m^2/y.

Keywords: nearly zero energy; net-zero energy; buildings; sustainable tourism; technology policy; high energy efficiency hotels

Abbreviations: DHW: Domestic Hot Water; EED: Energy Efficiency Directive; EPBD: Energy Performance of Buildings Directive; EU: European Union; GHG: Greenhouse Gases; HVAC: Heating, Ventilation, and Air-Conditioning; LED: Light Emitting Diode; MS: Member States; nZE: nearly Zero Energy; nZEB: nearly Zero Energy Buildings; neZEH: nearly Zero Energy Hotels; PV: Photovoltaic; RES: Renewable Energy Sources; ROI: Return on Investment; UNWTO: World Tourism Organisation

1. Introduction

The existing building stock represents a high potential for energy savings [1,2]. Currently, there are a few successful demonstrations of nearly Zero Energy Buildings (nZEBs) in Europe to motivate and initiate replications in the private sector. Hotels in specific, represent a big challenge, since they are usually complex building systems and at the same time, energy intensive businesses [3,4]. The hotel industry is highly depended on energy, which in some luxury hotels is up to 50% of the total Operation and Maintenance cost [5], therefore their challenge is to minimize energy costs without compromising the quality of their guests' stay [6].

Indeed, the hotel industry is the most energy and resource-intensive branch of the accommodation sector. They are "comfort or service-oriented accommodations" [7] with great heterogeneity in terms of business sizes and other individual features. Major studies investigating energy efficiency in the tourism sector pointed out the high potential of energy savings as well as the lack of a systematic approach for investigating its energy use [8]. The credibility and applicability of results shown in many types of research aiming at improving the understanding of the hotel energy flows [9–11] are always strongly depended on a wide variety of variables. However, based on different studies [12], the idea of a "typical" hotel remains vague and is required to be linked to a bottom-up approach including hosting and non-hosting activities, as will be explained below.

Although several examples of energy saving in non-domestic buildings exist today [13,14], hotels face difficulties when undertaking energy efficiency actions, mainly due to the existing energy policies, as well as due to being unfamiliar with energy technologies [15].

Since hotels place great significance on the comfort of their guests, they can apply a variety of mechanical and passive cooling means, such as natural cooling, ventilation, shading, thermal insulation etc. [16], providing a range of options so that architects and developers select the most sustainable strategy. It is clear that the implementation of renewable applications in the hotel/tourism sectors is case sensitive and follows different routes depending on existing needs and climate changes [17].

Studies suggest to integrate energy in the complete environmental performance of the hotels as a part of their excellence and eco-friendly policy [18]. The significant demand for Domestic Hot Water (DHW), not only for sanitary uses but also for swimming pools or spa facilities, can be covered by Renewable Energy Sources (RES), mostly with solar thermal systems [19,20].

This paper summarizes the approach and methodology of the "nearly Zero Energy Hotels" (neZEH) initiative, which is a response to the Energy Performance of Buildings Directive (EPBD) recast, 2010/31/EU [21], supporting the EU Member States (MS) in their national plans for increasing the number of nZEBs [22]. The paper aims to: (i) describe briefly the methodology for developing neZEH pilot projects; (ii) present the results of energy audits in hotels located in six south European countries (Greece, Croatia, France, Romania, Italy and Spain) plus one north (Sweden) for comparison; (iii) to assess how realistic is the nearly Zero Energy (nZE) vision in the hotel sector.

2. Legislative framework

2.1. European policy

In the EPBD recast (article 2) an nZEB is defined as "a building that has a very high energy performance; the nearly zero or very low amount of energy required should be covered to a very significant extent by energy from renewable sources, including energy from renewable sources produced on-site or nearby". According to the EPBD recast, by 31 December 2020 all new private buildings and after 31 December 2018, all new public buildings should be nZEBs. Even though the EPBD recast provides the framework for nZEBs, the MS should implement it in such a way to reflect their national, regional or local conditions [23].

Furthermore, article 4 of the Energy Efficiency Directive (EED) [24] stipulates that MS shall establish a long-term strategy to trigger investments in the renovation of their existing building stock, both residential and commercial, public or private.

Various reports in the past [25–27] have analyzed the status of nZEB transposition in national legislations. According to most recent data, the status of nZEB definition in the EU countries is as follows: 18 countries have set an official definition, in one country the definition is under approval, whereas in nine countries the definition is under development.

In 13 MS, the definition includes also the proportion of RES production. Although most MS reported a wide variety of policy measures, including financial incentives, strengthening their building regulations, communication activities and demonstrations, not in all cases is well-defined whether these measures specifically target nZEBs. Based on data from Concerted Action EPBD, the numerical definitions do not cover all building typologies in all countries; moreover, they are remarkably different in content, calculation assumptions and ambition level.

2.2. Transposition of EPBD in the selected EU countries

Within the context of the neZEH project, the national framework has been analyzed for the seven studied EU countries [6]. Most of these countries have officially introduced numerical definitions for some categories of new and refurbished buildings (Table 1).

Table 1. nZEB numerical definitions for new and refurbished buildings in the neZEH countries (status February 2016).

Country	New buildings				Refurbished buildings		
	Status of definition	Maximum primary energy (kWh/m^2/y)		Share of RES	Status of definition	Maximum primary energy (kWh/m^2/y)	
		Residential	Non-residential			Residential	Non-residential
Croatia	✓	35–80	25–250	✓ 30%	The same as for new buildings		
France	✓	40–65 [a,b]	70–110	✓ Quantitative [c]	✓	80 [a,b]	60% PE [d]
Greece	Under development	-	-	Minimum share in current requirements for all buildings	Under development	-	-

Continued on next page

| Country | New buildings | | | | Refurbished buildings | | |
| | Status of definition | Maximum primary energy (kWh/m²/y) | | Share of RES | Status of definition | Maximum primary energy (kWh/m²/y) | |
		Residential	Non-residential			Residential	Non-residential
Italy	✓	Class A1		50% RES for DHW + 50% RES for (DHW + heating + cooling)	✓ As for new buildings		
Romania	✓	93–217 [a,b]	50–192 [a,b]	✓ Quantitative	✓	120–230 [a,b]	120–400 [a,b]
Spain	Under development	Buildings will need to comply with class A	Minimum share in current requirements for all buildings	Under development	-	-	-
Sweden	Under development	30–75 [a,b]	30–105 [a,b]	All energy from RES will be excluded from the consumption values	ND*	-	-

*ND = no data; [a] Depending on the reference building; [b] Depending on the location; [c] Requirement depending on the RES measures adopted; [d] Maximum primary energy consumption defined as a percentage of primary energy consumption (PE) of a reference building.

Estonia has been the first MS, which defined numerical targets for the hotel type of buildings (130 kWh/m²/y of primary energy, 27% RES), including heating and cooling, ventilation, DHW, lighting and auxiliary electricity, as well as the use of appliances [28].

3. Methodology

The objective of the European initiative Nearly Zero Energy Hotels [6], involving partners from the tourism and energy efficiency sectors, academia and research and organisations with worldwide impact, like the World Tourism Organisation (UNWTO), was to accelerate the rate of refurbishments of existing hotels into nZEBs. The methodology to achieve this objective consisted of: providing technical assistance to existing pilot hotels for refurbishing into nZEBs, demonstrating the sustainability of investments towards zero energy and undertaking training and capacity building activities at regional, national and EU level (Figure 1).

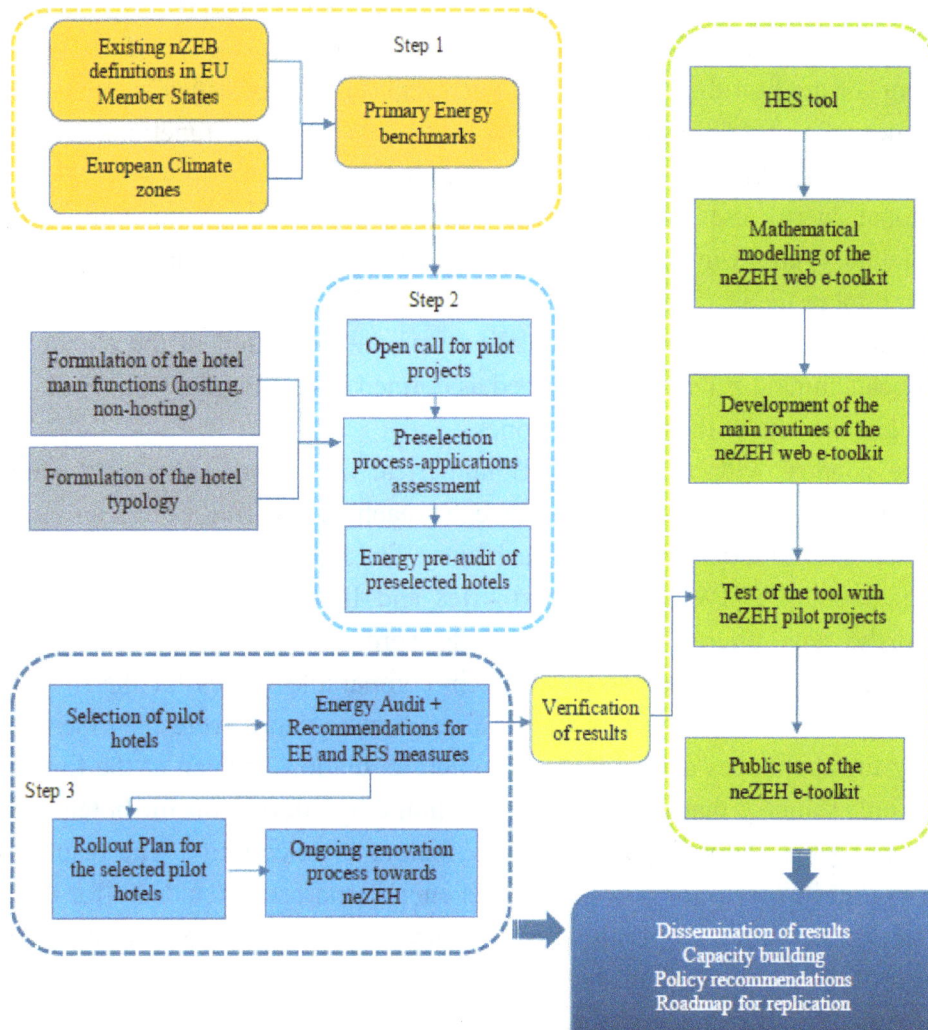

Figure 1. The neZEH benchmarking methodology.

3.1. Benchmarking methodology

Due to the delay of national transposition of the nZEB definition and given the absence of numerical indicators for hotel buildings in the neZEH countries, there was a necessity to introduce certain benchmarks for primary energy use and RES share. The benchmarks were used as targets for the pilot projects, in the cases where the official definition was not in place.

The benchmarking took into account that hotels cannot be considered as typical non-residential buildings; their business model includes numerous energy-intensive operations related to their customers' comfort and expectations, therefore, closely linked to their competitiveness and viability [29]. Data from Spanish hotels estimate that energy use related to the non-hosting function of the hotel i.e., hotel zones where conditions other than the standard indoor environmental conditions may be required, such as swimming pool, kitchen, spa, etc., can reach to an average 35% of the total energy use [30]. Initially, the research team focused on identifying benchmarks for nZE hotels, taking into account only the typical use of the building, as defined in the EPBD.

According to EPBD recast, the typical use (heating, ventilation, cooling, lighting, DHW) must refer to the standard indoor environmental conditions, which in the case of hotels are the comfort conditions for their guests and staff, as recommended in EN15251:2007 Standard ("Indoor environmental input parameters for design and assessment of energy performance of buildings addressing indoor air quality, thermal environment, lighting and acoustics"). With these principles, the standard zones of a hotel to be considered among the hosting functions are guests' rooms, reception hall, offices, meeting rooms, bar and restaurant.

The additional energy uses of non-hosting functions, such as spas, swimming pools, saunas, gym, kitchens, laundry, etc., can be considered in the future in a "modular" benchmarking based on the results of the pilot projects.

The initial step in the benchmarks calculation was the definition of the climate zones; in this work, five European climate zones were assumed as defined by the Ecofys [31]. Each region within a country was assigned to one of these climate zones.

A reference country with existing nZEB numerical definition or relevant minimum energy performance values was chosen as representative for each of the five European climate zones, in order to have reference numerical values for each zone. The first step calculation included the energy uses of heating, cooling, DHW, HVAC auxiliaries and lighting. Since not all reference values corresponded to the same energy uses, adjustments were made to include energy uses that were missing from some reference values, so as to obtain comparable values for the five climatic zones. The primary energy indicators calculated for the five climatic zones were then increased with extra delivered energy for appliances according to Estonian regulation—which at the time, as mentioned before, was the only country that introduced nZEB numerical definition for hotels—converted into primary energy using the national primary energy factors for electricity. The results were primary energy indicators (benchmarks) for new hotels—all energy end uses included—for the seven neZEH countries, as shown in Table 2 [6]. In order to introduce benchmarks for refurbished hotels, a correction factor of 30% was applied, based on review of existing national nZEB definitions and minimum requirements in building codes with different values for new built and renovated buildings.

Table 2. Country-specific neZEH benchmarks.

Country	Primary energy indicator for new hotels (kWh/m^2/y)[a]	Primary energy indicator for refurbished hotels (kWh/m^2/y)[a]
Croatia	77	100
France	115	150
Greece	76	122
Italy	71	93
Romania	80	104
Spain	72	94
Sweden	134	175

[a] including hosting functions + appliances.

Primary energy indicators for existing hotels (2008 level) were also calculated to provide an insight into the energy reduction needed to be achieved to reach benchmarks for new hotels. Energy use data for existing building stock from EU level sources [32–34] were processed, showing that data from residential buildings are more proper to describe the hosting function of hotels than the non-residential building data. The existing residential buildings data by energy source [32] were used in combination with national primary energy parameters in order to calculate primary energy values. Adjustments for additional cooling and ventilation needs were made taking into account EN15251:2007 standard, resulting in primary energy indicators for the hosting function of hotels in each country (2008 level). The comparison of these indicators with benchmarks for new hotels showed that an average 75% energy reduction has to be achieved, a value which is in compliance with previous results [33].

3.2. The pilot projects

In order to demonstrate the feasibility of refurbishment interventions leading to nZEB status, neZEH supported pilot projects in the seven target countries, which were selected following a three-step process (Figure 2):
(a) Public calls per country; this initial step resulted in a pool of 85 applications.
(b) Evaluation of the candidate applications and selection of 35 hotels (five candidates per country) by applying criteria, including the hotel's commitment and maturity to achieve nZE status, the distribution of selected hotels amongst the different climate zones and the alternative typologies.
(c) Energy pre-audits: the selected hotels were pre-audited in order to assess, in an initial phase, their capability and potential to reach the nZE targets. More accurate energy and economic data helped in filtering and ranking the hotels to lead to the selection of at least two pilot hotels per country.

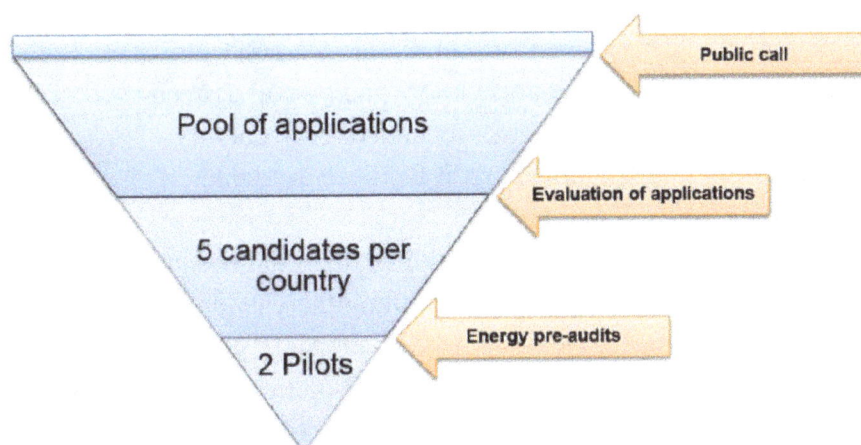

Figure 2. Pilot hotels selection procedure.

3.2.1. Climate zones

As mentioned previously in Section 3.1, five European climatic zones were taken into account in the benchmark calculations. These climatic zones were also used as a reference throughout the implementation of the action, meaning that outcomes should be representative per each zone, to make sure that there are replicable examples for all the regions of Europe.

3.2.2.   Hotel typologies

Four hotel typologies were identified: (i) Coastal; (ii) Mountain; (iii) Urban; (iv) Rural. Each participating country had pre-identified the hotel typologies within its region, so as the five candidates during the pre-audit phase would correspond to these typologies.

3.3. Energy audits and feasibility study

An energy audit was implemented in each pilot hotel, which included a full on-site inspection and examination of energy uses, suggesting the most appropriate solutions to reach nZE levels. In more detail the audit included:

- Data gathering and compilation (drawings, energy statistics, utility bills, etc.).
- On-site inspection (study of HVAC infrastructure, lighting status, windows and insulation standards, etc.).
- Calculation of the current energy flows (inputs and losses).
- Recommendations for proper measures ranked based on savings and future costs.
- A proposed action plan.

Following the energy audit, a detailed energy study, as well as a feasibility study were carried out, which included the most suitable technological solutions combined with a cost scenario and the calculation of the Return on Investment (ROI), to assist in the decision making. The following parameters were taken into account when proposing technological solutions:

- The specific characteristics of each hotel (site and orientation, energy consumption, seasonality, etc.).
- The improvement of the thermal comfort conditions for hotel guests, as well as other critical quality parameters for hotels, such as noise levels and overall convenience of guests.
- The cost of each solution based on real market data of each country.
- The funding opportunities, tax reliefs or credit facilities available.
- The payback period of each solution.

The hotels received advice on available funding opportunities and assistance in the tender preparation and selection of contractors for implementing the energy efficiency and RES measures. Furthermore, training was provided to the hotel owners and staff, guiding them on how to maximize their efficiency and optimize the use of the implemented solutions; training material and guides are available to other hoteliers for upscale [6]. Finally, the hotels received marketing guidelines [6] to support them in their effort to communicate their nZE profile to potential guests. The pilot hotels will continue to receive expanded visibility at national and EU level, inspiring more hoteliers to proceed in similar refurbishments and triggering more investments in nZE projects.

4.   Results and discussion

The results from seven European countries are presented in this paper. The pilot hotels have been audited, gathering a set of technical and qualitative data for each. Per case, the group of experts analysed each function (hosting and non-hosting). The primary energy use was calculated from building plant characteristics and occupation schedules in relation to real consumption from bills in order to validate them. Although the energy demand for non-hosting functions (pools, bars, restaurants, conference room, spa, etc.) varies from country to country, hotel type, operation months and occupancy, in some hotels it reaches to more than three times the energy demand of the hosting functions (Table 3).

Table 3. Primary energy use and RES share before refurbishment for (a) hosting functions and (b) non-hosting functions for the 16 pilot hotels, as resulted from the energy audits.

Hotel	Country	Hotel typology	Operating months	% Average occupancy during opening months	Facilities	Primary energy use (kWh/m^2/y)		RES share (%)	
						Hosting	Non-hosting	Hosting	Non-hosting
1	Croatia	Coastal	May–Oct	35	spa/wellness	130	16	36	N/A
2	Croatia	Coastal	all year	48	spa, sauna, pool, gym	66	60	64	N/A
3	France	Coastal	all year	83	lounge bar, pool, spa	244	N/A	0	N/A
4	France	Rural	Apr–Oct	44	lounge bar, pool, spa	490	N/A	8	N/A
5	Greece	Urban	May–Oct	84	kitchen, restaurant, pools	230	450	30	0
6	Greece	Coastal	Apr–Oct	78	pools, bars, restaurants, conference room	250	293	26	20
7	Greece	Urban	all year	80	Bar	287	0	20	0
8	Italy	Mountain	Jun–Sep	21	Spa/wellness	100	66	74	0
9	Italy	Urban	all year	76	kitchen, gym	266	24	0	0
10	Romania	Urban	all year	70	restaurant, conference room	347	1,451	0	0
11	Romania	Urban	all year	59	restaurant, conference room	227	588	4	0
12	Romania	Mountain	all year	20	spa, pool, adventure park, conference rooms, restaurant, disco, tennis	277	1,460	0	0
13	Spain	Rural	all year	22	spa, pool, shrine room	278	253	9	0
14	Spain	Coastal	all year	80	spa, restaurant, gym, sauna	363	2,273	9	0
15	Sweden	Urban	all year	56	kitchen, restaurant, sauna	722	1,482	0	0
16	Sweden	Urban	all year	35	kitchen, restaurant, sauna, gym	151	1,285	0	0
Averages						277	727	18	2

Generally, in most cases, as can be concluded from the pilot project results:

- the full-year operating hotels are bigger energy consumers than the seasonal ones, when comparing within the same country.
- the rural/mountain ones have higher primary energy use than the coastal and the urban ones.
- the non-hosting activities very often impose an important burden on the energy performance of hotels.
- the use of RES is limited in most cases, which might not be expected at first.

Energy saving and RES measures were proposed per case, focusing on the energy uses with the highest energy consumption and the most inefficient parts and equipment of each hotel. The set of measures proposed for each hotel should satisfy the target of neZEH benchmarks, in terms of the eventual primary energy use for the hosting functions and the eventual RES share, always in conjunction with economic feasibility. The thermal comfort for guests and the overall improvement of indoor environmental conditions, on the basis of international and national standards (usually around 20 °C for winter and 26 °C for summer), was also an important parameter. In addition, guest convenience was taken into account, especially in the case of automatic controls, as well as the maintenance of quality and services standards.

The data gathered from the energy audits were used to optimize the simulation models for the calculations of the energy savings and renewable energy production of the proposed measures. The primary energy reduction percentage and RES share were calculated both for hosting and non-hosting functions, in order to demonstrate the energy efficiency potential of the different services of hotels. Results for the total of 16 hotels are presented in Table 4.

Table 4. Energy savings and RES share in hosting and non-hosting functions in the audited hotels.

	Average primary energy use after suggested interventions (kWh/m^2/y)	Average reduction percentage (%)	Average RES share after (%)
Hosting functions	102	63	46
Non-hosting functions	374	49	11

The overwhelming majority of pilot hotels are able to achieve savings higher than 50% of their initial consumption. That is a substantial saving percentage, taking into account that most of these hotels have been already quite energy efficient. When it came to RES share, the majority of them experienced a coverage increase of at least 20%, while some others have demonstrated up to a remarkable 60%. On average, by applying the measures proposed in the energy audits, the RES share increased from an average of 18% to an average of 46% in the hosting functions.

Primary energy use for the hosting functions across all sixteen pilot hotels can decrease dramatically—from an average of 277 kWh/m^2/y to an average of 102 kWh/m^2/y; a reduction of 63%. At the same time, RES share for the hosting functions can be increased by an average of 18% to an average of 46%.

The non-hosting functions are, on average, much more energy intensive, therefore, extremely important for the overall performance of the hotel energy efficiency towards the nZEB concept. The primary energy use for the non-hosting functions can decrease from an average of 727 kWh/m^2/y to an average of 374 kWh/m^2/y, a reduction of 49%.

Amongst the measures which will improve (directly or indirectly) the energy consumption, simple interventions, offering the most attractive payback (i.e., lighting controls, sun-shading devices, shower head diffusers, etc.), are also included. In contrary, building envelope insulation, although being the first priority in terms of energy efficiency and thermal comfort of guests, remains very often less attractive, with up to 25 years payback period (Table 5).

Table 5. Measures suggested after energy audits.

Measure	Energy savings (%)	Investment (€)	Payback period (yrs)
Building envelope insulation	3–35	30,000–350,000	6.4–25
Installation of solar thermal collectors	5–39	5,600–143,000	10–14
Installation of photovoltaic (PV) panels	4–23	8,000–300,000	4.5–20
Installation of Building Energy Management System	2.6–18.5	12,000–125,000	6.8–21
Replacement of light bulbs with LED	2–14	1,200–32,000	1–9
Lighting controls	1–7	1,300–4,000	0.4–2
Replacement of low efficiency with high efficiency heat pumps	1–36	30,000–300,000	5.4–11.8
Adding ceiling fans and use of control systems for cooling	17	95,000	9.5
Installation of radiant heaters	1	7,000	12
Installation of geothermal heating system	10–58	50,000–500,000	6–16
Installation of sun-shading devices	2–6	8,000–20,000	3
Connection to district heating	1–26	14,000–89,000	5–9
Outdoor redesign for better microclimate	4	25,000	4
Heat recovery in ventilation	5.7–8	20,000–86,000	8–16
Install water saving taps and shower head diffusers	4.9–7.5	350–1,600	0.16–0.3
Balance the DHW network	1	4,000	7
Replacement of the minibars	2.1	32,250	10.7
Balancing and improvements of heating system	1–21	800–38,000	1–5.7
Reduce stand-by consumption	1	400	1.4

5. Energy efficiency in hotels e-toolkit

Based on the overall experience gained from the neZEH initiative and its pilot projects, a practical e-toolkit was developed to help hotel owners benchmark their energy consumption compared to the neZEH levels and to suggest appropriate measures for energy efficiency improvement and RES integration. It is designed to increase the energy efficiency capacity in the accommodation industry and to encourage hotel owners to proceed with energy renovation projects towards nZE targets. The neZEH e-toolkit [35] is based on the UNWTO Hotel Energy Solutions toolkit (also co-funded by the Intelligent Energy Programme), which—since 2011—has engaged 20,000 users worldwide and 1,500 hotels in Europe in better understanding how to become energy efficient and more competitive (Figure 3).

The neZEH e-toolkit uses a ranking methodology at the national level, to rank the identified solutions according to three aspects: potential energy savings, size of the financial investment and profitability of the financial investment, as per the flux diagram shown in Figure 4.

The toolkit is online since 2016 and the target users are the hotel owners or hotel technical staff. It is a user-friendly tool, which requires answers to a short questionnaire and subsequently provides hoteliers with a report assessing their current energy profile, indicating how close the hotel is to from achieving the nearly zero energy levels and providing an indicative set of appropriate renewable energy and energy efficiency measures.

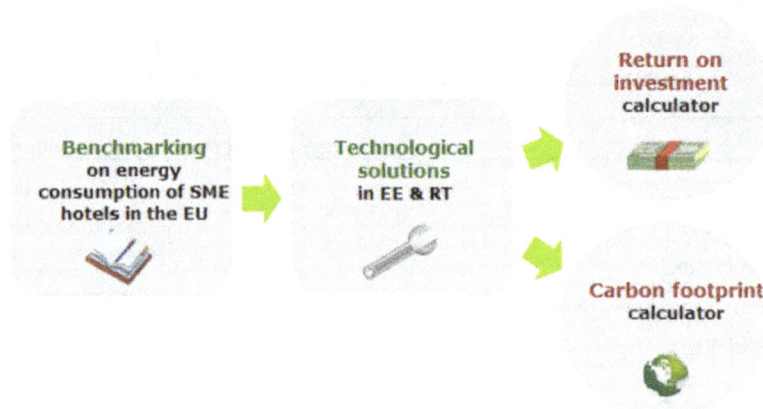

Figure 3. HES e-toolkit methodology diagram [36].

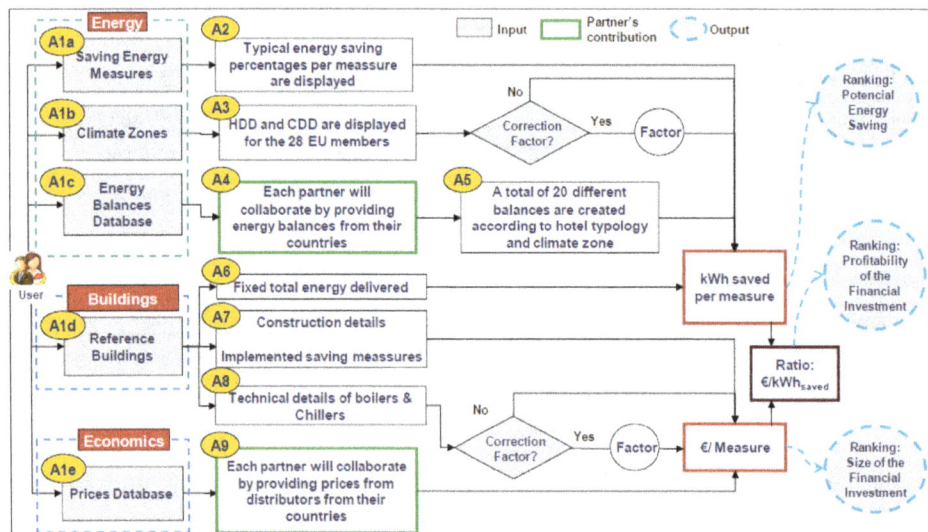

Figure 4. Flux diagram of the technologies/solutions ranking tool.

6.   Annex: Case study hotel

The case study of Hotel No. 6, as appears in Table 3, is presented in more detail in this Annex. The hotel actual status is calculated with the use of energy bills from all the different energy types, on-site visits and measurement of lamps, equipment, heating, cooling and hot water systems, interviews and cooperation with the hotel management and maintenance departments. The energy audit included building information and measurements, occupancy data from the hotel, description of all mechanical and equipment installations, RES, water and energy use of every building of the hotel. Table 6 shows the general data from the hotel located in a coastal location in the island of Crete. The hotel according to all the information gathered had a primary energy consumption of 250 kWh/m^2/y for the hosting functions and primary energy consumption of 293 kWh/m^2/y for the non-hosting functions.

Table 6. General data for the hotel case study, as derived from the energy audit performed.

Hotel typology	Coastal, Resort & Spa hotel
Category	5 stars
Location	Lasithi, Crete
Working season	April–October
Number of beds	1,036
Mean occupancy	78%
Total hotel area (m^2)	20,866
Thermal zones area (m^2)	18,017
Non-thermal zones area (m^2)	2,849
Annual primary energy consumption (MWh)	4,624
Electricity energy consumption (kWh)	1,311,430
Biomass energy consumption (kWh)	428,400
Oil energy consumption (kWh)	98,472
LPG energy consumption (kWh)	270,794

All the information is used for correlations between the occupancy and energy use for different functions of the hotel and also for constructing a simulation model of the hotel. The simulation model is validated from the selected data and is used for the selection and implementation of the proposed energy measures. All the necessary climatic data and thermal comfort boundaries are used for setting points according to the different countries. For this particular example, the simulation model included climatic data from the government database for the location in Lasithi, Crete and the thermal comfort boundaries were set at 20 °C for winter and 26 °C for summer as proposed in the Greek legislation for energy consumption in buildings. The proposed measures as presented in Table 7 are derived from the hotel necessities and installations with large energy demand, like the cooling loads, which reach to a striking 62% as can be seen in Figure 5. The measures resulted in the primary energy use for the hosting functions to drop to 99 kWh/m^2/y (60% energy saving) and RES share to increase to 60%, thus reaching the neZEH benchmarks set for Greece and achieving in this way nZEB status. At the same time, the measures proposed helped the hotel reduce primary energy use for non-hosting functions to 110 kWh/m^2/y (62% energy saving) and RES share to rise to 50%.

The project rollout plan assumed a step by step approach, where the priority measures or more economically attractive measures are installed first. The insulation, which is more expensive and has a longer payback time, was set to be installed from 2020 onwards. Despite the long payback, it was decided to eventually adopt this solution as it will greatly contribute to the improvement of thermal comfort for hotel guests and have a positive impact on the hotel standards.

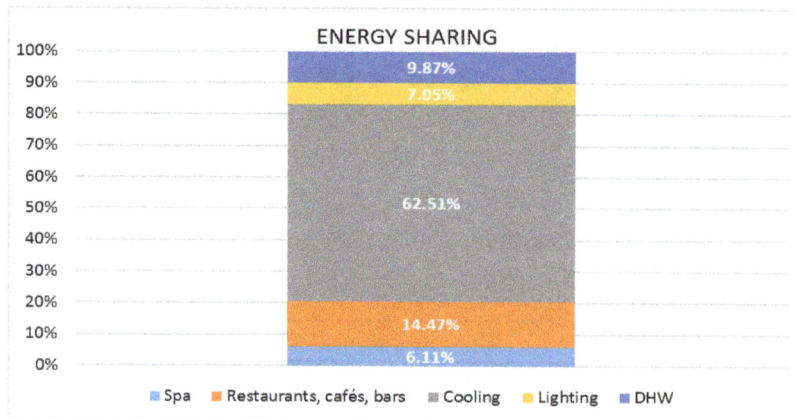

Figure 5. Energy balance for the hotel case study.

Table 7. Results of the proposed energy saving and RES measures for the hotel case study.

Enhancement	Yearly primary energy savings per m² (kWh/m²/y)	Total yearly primary energy savings (kWh/y)	Avoided emissions (kg CO_2/y)	Simple investment payback (y)
Adding heat pumps for cooling and DHW	92	1,667,060	456,118	5.4
Installing PV modules	34	609,000	207,690	6.0
Adding insulation and double glazed windows in the central building	21	371,161	126,578	19.2
Adding solar collectors for DHW and coverage of the spa pool	13	231,184	11,805	10.0
Adding green for better microclimate conditions	11	198,193	67,590	4.1
Total	171	3,076,598	869,781	9.4

7. Conclusions

It is a known fact that buildings consume 40% of total energy and emit 36% of Greenhouse Gases (GHG) in the EU [37], so the existing building stock represents a high potential for energy savings. Accommodation in specific, counting for 21% of total tourism sector GHG emissions [38], represents a bigger challenge, since hotels are more complex building systems and energy intensive businesses.

This work proposes a step by step approach, starting from an energy audit and resulting in a renovation rollout plan and a marketing plan, and provides useful insights regarding the nZEB benchmarks for hotel buildings, in terms of energy indicators. A complete methodology was developed and applied in the hotel sector to support the implementation of a large-scale renovation towards nZE status. The analysis showed that:

(i) The vision of nZE in hotels is close to the reality; economically attractive interventions towards nZE are feasible in existing hotels.

(ii) In most of the cases, the achievement of nZE targets is possible to be implemented in two or more phases.

(iii) The non-hosting functions are more critical than the hosting ones, when targeting for high energy efficiency in hotels; this is due to the fact that hotels are not typical buildings, but resemble small communities.

(iv) In most of the cases, the high-tech solutions are not always the best value for money option; low-cost interventions in selected cases could be very economically efficient.

(v) The hotel industry is very cooperative to become more competitive, but in parallel eco-friendly. Hoteliers acknowledge that they have a lot to gain by implementing energy efficiency measures, including the reduction of operational and maintenance cost, energy security on peak demands, an increase of competitiveness and improved image.

(vi) A key element of the initiative's success is the commitment of the tourism stakeholders to endorse the neZEH concept in hotels, challenging, even more hotels to invest in renovation projects that achieve nZE levels.

In the long-term, the neZEH project plans to support the hospitality sector to reduce operational costs and to enhance competitiveness and sustainability, contributing in parallel to the EU efforts to reach the 2020 and 2030 targets. It is estimated that the investments triggered by 2020 will be 80 M€ and will have as a result: cumulative primary energy savings to reach 42,000 toe/y, RES production of 11,000 toe/y and up to 98,000 t CO_2 eq/y reduction of GHG emissions.

The 16 pilot projects in Greece, Croatia, France, Italy, Romania, Spain and Sweden, will stand out as "real life" lighthouse examples in Europe and globally and inspire other hotel owners to invest in high energy performance refurbishments, including a large share of their energy needs covered by on-site or nearby RES, as requested by the EPBD recast and EED European Directives.

In line with other published reports and papers, this work would like to highlight the significance of capabilities building actions as a crucial part of the policy packages promoting nZEBs, maybe by a potential shift from subsidizing applications [39].

Additionally, the authors suggest future work should consider a "modular" benchmarking, with a specific focus on non-hosting functions, in order to provide recommendations for policy makers for dealing with hotel buildings' complexities. The parameters which affected the results, identified in this work, were the hotel typology, climate zone, offered facilities, operation seasonality. The data results obtained and presented in this paper can be used in a future study to define modular benchmarks for hotels, on the basis of the different parameters that define how much and how energy is being used in European hotels.

Acknowledgments

This work was supported by the European Commission under the Intelligent Energy Europe Programme, within the framework of the project neZEH (Nearly Energy Zero Hotels, www.nezeh.eu) [grant number IEE/12/829/SI2.644758, 01/04/2013 to 31/03/2016]. The sole responsibility for the content of this paper lies with the authors. It does not necessarily reflect the opinion of the European Union. The European Commission is not responsible for any use that may be made of the information contained therein.

The authors would like to thank the following persons for their active involvement in the project neZEH: Rodrigo Morell and Ignacio G. Hernandez (Creara Consultores S.L., Madrid, Spain), Zoritsa Urosevic (World Tourism Organization, Madrid, Spain), Cristina Nunez (Network of European Regions for Competitive and Sustainable Tourism, Brussels, Belgium), Anita Derjanecz (Federation of European Heating, Ventilation and Air-conditioning Associations, Brussels, Belgium), Camelia

Rata (Agency of Braşov for Energy Management and Environment Protection, Braşov, Romania), Marko Bišćan (Energy Institute Hrvoje Požar, Zagreb, Croatia), Stephane Pouffary (ENERGIES 2050, Villeneuve-Loubet, France), Sara L. Sacerdotti (Sistemi Territoriali per l'Innovazione, Torino, Italy), Nigel Claridge (Sustainable Innovation, Stockholm, Sweden).

Conflict of interest

All authors declare no conflicts of interest in this paper.

References

1. Dascalaki E, Balaras C (2004) XENIOS—a methodology for assessing refurbishment scenarios and the potential of application of RES and RUE in hotels. *Energ Buildings* 36: 1091–1105.
2. Chaudhry IS, Das M (2016) Design of optimum reference temperature profiles for energy saving control of indoor temperature in a building. *AIMS Energy* 4: 906–920.
3. Karagiorgas M, Tsoutsos T, Drosou V, et al. (2006) HOTRES: renewable energies in the hotels. An extensive technical tool for the hotel industry. *Renew Sust Energ Rev* 10: 198–224.
4. Tsoutsos T, Tournaki S, Farmaki E, et al. (2017) Benchmarking framework to encourage energy efficiency investments in South Europe. The trust EPC South approach. *Procedia Environ Sci* 38: 413–419.
5. Lai J, Yik F (2008) Benchmarking operation and maintenance costs of luxury hotels. *J Facil Manage* 6: 279–289.
6. Nearly Zero Energy Hotels (neZEH), neZEH results and tools, 2016. Available from: http://www.nezeh.eu/main_menu/library/nezeh_reports/index.html.
7. Becken S, Frampton C, Simmons D (2001) Energy consumption patterns in the accommodation sector—the New Zealand case. *Ecol Econ* 39: 371–386.
8. Soediono B (2011) Analysis on energy use by European hotels: online survey and desk research. *Hotel Energ Solut* 53: 160.
9. Instituto de la Mediana y Pequeña Industria Valenciana (IMPIVA) (1995) Rational use of energy in the Hotel sector. A Thermie Programme Action B-103.
10. Zambrana-Vasquez D, Aranda-Usón A, Zabalza-Bribián I, et al. (2015) Environmental assessment of domestic solar hot water systems: a case study in residential and hotel buildings. *J Clean Prod* 88: 29–42.
11. Beccali M, Gennusa ML, Coco LL, et al. (2009) An empirical approach for ranking environmental and energy saving measures in the hotel sector. *Renew Energ* 34: 82–90.
12. Boemi SN, Slini T, Papadopoulos AM, et al. (2011) A statistical approach to the prediction of the energy performance of hotel stock. *Int J Vent* 10: 163–172.
13. Paolo B, Barbara C (2013) The European Green Building Projects Catalogue July 2011–August 2012. Available from: https://core.ac.uk/download/pdf/38625632.pdf.
14. Paolo B, Barbara C (2012) The European Green Building Projects Catalogue June 2010– October 2011. Available from: https://publications.europa.eu/en/publication-detail/-/publication/275bff78-deba-42ca-bc29-7ad19bbdcb38/language-en.

15. Nearly Zero Energy Hotels (neZEH), Nearly Zero Energy Hotels towards low carbon growth in the European Union—neZEH position paper, 2016. Available from: http://www.nezeh.eu/assets/media/PDF/neZEH_EU_policy_paper__v2_final375.pdf.

16. Samuel D, Nagendra S, Maiya M (2013) Passive alternatives to mechanical air conditioning of building: a review. *Build Environ* 66: 54–64.

17. Michalena E, Hills J, Amat JP (2009) Developing sustainable tourism, using a multicriteria analysis on renewable energy in Mediterranean Islands. *Energy Sustain Dev* 13: 129–136.

18. Oreja-Rodríguez JR, Armas-Cruz Y (2012) Environmental performance in the hotel sector: the case of the Western Canary Islands. *J Clean Prod* 29–30: 64–72.

19. Kyriaki E, Drosou V, Papadopoulos A (2015) Solar thermal systems for low energy hotel buildings: state of the art, perspectives and challenges. *Energy Procedia* 78: 1968–1973.

20. Gabbar AH, Eldessouky A, Runge J (2016) Evaluation of renewable energy deployment scenarios for building energy management. *AIMS Energy* 4: 742–761.

21. European Commission. Directive 2010/31/EU of 19 May 2010 on the energy performance of buildings (recast). Official Journal 2010; L 153/13. Available from: http://edit.eceee.org/policy-areas/buildings/EPBD_Recast/EPBD_recast_19May2010.pdf.

22. Tsoutsos T, Tournaki S, Santos C, et al. (2013) Nearly zero energy buildings application in Mediterranean hotels. *Energy Procedia* 42: 230–238.

23. Kurnitski J, RW Group (2013) REHVA nZEB technical definition and system boundaries for nearly zero energy buildings. Brussels: REHVA; 2013. Available from: https://www.beuth.de/de/publikation/rehva-report-no-4/209769065.

24. European Commission. Directive 2012/27/EU of 25 October 2012 on energy efficiency, amending Directives 2009/125/EC and 2010/30/EU and repealing Directives 2004/8/EC and 2006/32/EC. Official Journal 2012; L315/1. Available from: https://ec.europa.eu/energy/sites/ener/files/documents/article7_en_luxembourg.pdf.

25. D'Agostino D, Zangheri P, Cuniberti B, et al. (2016) Synthesis Report on the National Plans for Nearly Zero Energy Buildings (NZEBs). Joint Research Center (JRC) publications.

26. Buildings Performance Institute Europe (BPIE) Nearly Zero Energy Buildings definitions across Europe. Available from: http://bpie.eu/uploads/lib/document/attachment/128/BPIE_ factsheet_ nZEB_definitions_across_Europe.pdf.

27. EPBD CA participants (2013) Implementing the Energy Performance of Buildings Directive-Featuring country reports 2012.

28. Kurnitski J, Buso T, Corgnati SP, et al. (2014) nZEB Definitions in Europe. *Rehva J* 51: 6–9.

29. Farrou I, Kolokotroni M, Santamouris M (2012) A method for energy classification of hotels: a case-study of Greece. *Energ Buildings* 55: 553–562.

30. Tournaki S, Frangou M, Tsoutsos T, et al. (2014) Nearly Zero Energy Hotels—from European policy to real life examples: the neZEH pilot hotels. Available from: http://www.nezeh.eu/assets/media/PDF/neZEH_EinB2014_Proceedings63.pdf.

31. Pagliano L, Hermelink A, Schimschar S, et al. (2013) Towards nearly zero-energy buildings, Definition of common principles under the EPBD. Available from: http://www.buildup.eu/sites/default/files/Minutes_nZEB_Workshop_0.pdf.

32. ENTRANZE, Data tool, 2014. Available from: http://www.entranze.enerdata.eu/.

33. COHERENO, nZEB criteria for typical single-family home renovations in various countries, 2016. Available from: http://www.cohereno.eu/fileadmin/media/Dateien/D2_1_BPIE_WP2_12092013_3_5_-final.pdf.

34. Buildings Performance Institute Europe (BPIE), Data Hub for the energy performance of buildings, 2015. Available from: http://www.buildingsdata.eu/.

35. Nearly Zero Energy Hotels (neZEH), neZEH e-toolkit, 2016. Available from: http://www.nezeh.eu/etoolkit/index.html.

36. Hotel Energy Solutions (HES), HES e-toolkit, 2011. Available from: http://www.hes-unwto.org/HES_root_asp/index.asp?LangID=1.

37. European Commission, Buildings, 2016. Available from: https://ec.europa.eu/energy/en/topics/energy-efficiency/buildings.

38. World Tourism Organisation (UNWTO), Tourism & Climate Change, Confronting the Common Challenges, 2007, Available from: http://sdt.unwto.org/sites/all/files/docpdf/docuconfrontinge.pdf.

39. Pace L (2016) How do tourism firms innovate for sustainable energy consumption? A capabilities perspective on the adoption of energy efficiency in tourism accommodation establishments. *J Clean Prod* 111: 409–420.

Assess the local electricity consumption: the case of Reunion island through a GIS based method

Fiona Bénard-Sora[1,*], Jean-Philippe Praene[1] and Yatina Calixte[2]

[1] University of La Reunion, PIMENT Laboratory, 117 rue du Général Ailleret 97430 Le Tampon, France
[2] University of La Reunion, Department of Building Sciences and Environment, France

* **Correspondence:** Email: fiona.sora@univ-reunion.fr

Abstract: Succeeding energy transition is the current challenging objective of many remote islands such as Reunion Island to reduce their dependency to fossil resources. To define an efficient energy framework strategy for the territory, it is important to be able to assess the electricity consumption intensity per typology of consumers. A particular attention must be paid on building electricity consumption in energy planning scenarios. This paper proposes to investigate the electricity consumption ratio per square meter per building type which appears as a relevant indicator. The proposed methodology aims at filling the lack of data (ratio kWh/m^2/ type of consumers) when this information doesn't exist for a territory. This type of ratio can be useful in two ways: on the one hand to characterize the building energy demand, and, on the other hand, to understand the consumption mode of the inhabitants. We can therefore provide future energy policy framework in favor of demand-side management what is a key step, a lock to solve for the deployment of sustainable cities. This work calculates electricity consumption ratios per area by using a GIS (Geographic Information System) method, distinguishing the type of building. The case of Reunion Island is studied and four building categories are identified corresponding to the functional characteristics such as industry, administration, companies and residential. The results highlight that residential sector has one of the lowest electricity ratios with a value of 29.84 kWh/m^2, but also the highest part of electricity consumption, 45.2%. The different ratio value has been cross validated by estimating municipalities electricity consumption based on the distribution of consumers and the associated ratio.

Keywords: electricity consumption ratio; building typology; GIS

1. Introduction

Energy efficiency and greenhouse gases reduction of new buildings are key objectives of international, national and local policies that aim to decrease electricity use in buildings. In France, the first regulations in this direction followed the oil shocks of the 1970s and the French government's desire to secure its energy situation. To secure its situation, it sought to reduce demand on the one hand and launched its nuclear program on the other. The first thermal regulation of buildings (RT1974) in France sets a target of reducing the energy consumption of buildings by 25% compared to the standards in force since the end of the 1950s [1]. This first objective was subsequently updated by a fairly rich French legislative package: LAURE, the French Law on Air quality and the Rational Use of Energy [2], POPE, the Energy Program Act [3], ENE, the Grenelle II law [4], TEPCV, the law on energy transition for green growth [5]. This thermal regulation was followed by numerous updates. The RT2012, currently in force, will be soon replaced: a RT2020 is planned. These objectives of energy saving in buildings are well present in the French energy policy and in order to help implement this policy it is important to characterize the consumption of buildings by distinguishing in particular their function: tertiary, industrial and residential. This type of characterization is generally done by energy agencies such as the IEA (International Energy Agency) or ADEME in France.

Buildings are the main consumers of energy in our modern societies and acting on them is unavoidable [6]. In France, the residential-tertiary sector is the main consumer of final energy (45%) [7]. This shows the importance of studying buildings and the interest of differentiating buildings, according to their function [8]. Studying the characteristics of buildings for the analysis of their energy consumption and their carbon emissions is common in the literature [9] and especially in the study of housing [10].

Our work focus on the final energy consumption in buildings, but more particularly on electricity consumption on Reunion Island. This type of data is actually produced at the national level, but most of it does not exist in the ultra-marine territories, which nevertheless benefit from their own thermal regulations with regard to their specific climatic characteristics. The energy question is assessed at different scales. At the regulatory level, coherence must exist between the different levels: international, national and local. For territories such as island territories, the local part is of particular importance, especially when the territory is attached to a distant metropolis. Energy issues are certainly international, at the local level the stakes can be very diversified. Nevertheless, local planners are regularly confronted with a lack of data on these energy issues [11]. By combining electricity consumption data by type of consumer and GIS data produced by a local energy agency (SPL Energies Reunion) [12] and the IGN (National Institute of Geography) respectively, we propose new methodology to define consumption ratios by highlighting the different building functions. The interest of the work is twofold: initially, it is a matter of creating a data still non-existent on the island. On the second stage, this work should be seen as the answer to a broader problem: urban planning in a context of ecological transition where the issue of energy is central. The urban planning represents an effective tool to improve the sustainable development and produce/favor the emergence of energy-efficient cities [13]. A first work to characterize the consumption of electricity has already been carried out on Reunion Island: the work was done at the municipal level and highlighted the determinants of electricity consumption [14]. The objective is now to refine the scale and highlight consumption ratios by type of surface. Instead of considering communal consumption, we consider the consumption by type of consumer, which allows, by crossing with the building databases of the IGN, to produce the ratios. The first work allowed a classification of the municipalities, the present work should make it possible to produce a new visualization of the spatialized consumptions of electricity.

2. Methodology

2.1. Context: Reunion island, a small island facing the challenge of energy transition

The work concerns the island of La Reunion which is characterized by a particular economic situation. The island is a French overseas department widely sustained by the French government. It is the tertiary sector that dominates the economy of Reunion, with 83.6% of GDP in 2014. The agricultural and industrial sectors are thus down with respectively 1.4 and 7.7% of GDP.

Another main characteristic of the island of Reunion is its steep relief: this forces the population and the activities to gather on the littoral and especially on the west coast (see Figure 1).

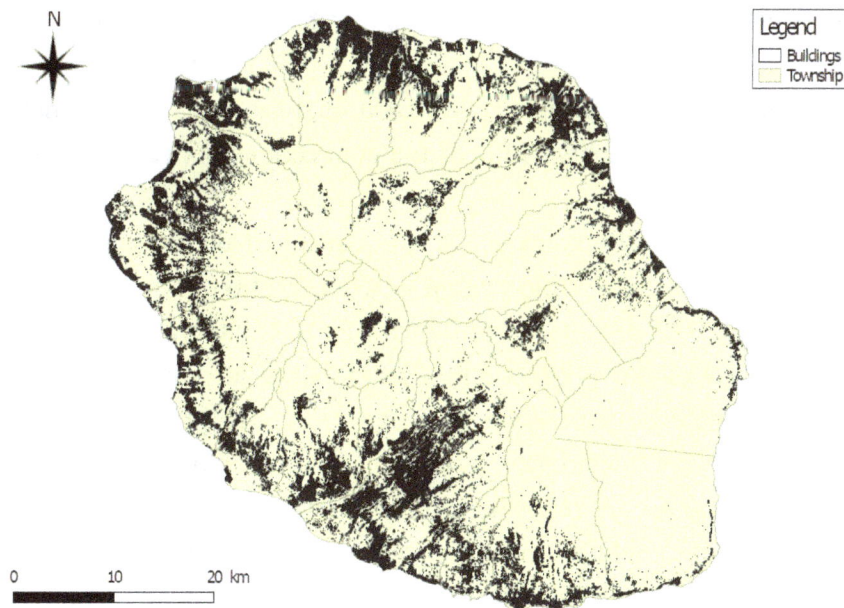

Figure 1. Spatial distribution of buildings.

Reunion Island imports most of its energy resources, including 65% of petroleum products used in transportation, power generation, agriculture and industry. In 2015, the supply of fossil fuels is 1,238.8 ktoe. The local electricity production from renewable resources is 196.4 ktoe, with bagasse and hydraulics as major resources (54% and 22% of local resources respectively). As regards consumption, in 2015, the island shows a primary energy consumption of 1,411.2 ktoe [12].

Like many islands, Reunion Island faces a major problem: the electrification of its rural areas. This type of area is difficult to access, and low density does not provide a cost-effective electricity service for the providers from the global network. Decentralized management of these areas in terms of electricity is often more profitable [15], but the level of poverty of the type of households residing in these areas does not always allow autonomous management [16,17].

2.2. The purpose of a GIS approach

The GIS approach is a useful approach for both quantifying and qualifying a phenomenon. The integration into a single system of analysis of models directly connected to real data facilitates the appropriation of the problem by different stakeholders and consequently permit the development of policy [18]. The GIS approach can build territorial development strategies, develop evolution scenarios and thus improve the process of policy-making.

GIS represents an important analytical potential for the study of energy at the scale of a territory, whether for application to renewable energies or for the modeling of energy consumption [19]. In the scientific literature, it is easy to distinguish several categories of GIS-based models that highlight their potential for urban and energy analysis and policy-making. Indeed, much work has focused on the study of renewable energy potential, whether it is about the solar resource [20–23], the wind resource [23–26] or coupling different resources [27,28]. A number of others studies have focused on urban building energy consumption and microclimate [29–32].

Finally, other work has focused on the spatial study of energy consumption and carbon emissions [33,34].

GIS-based analysis can also be coupled with BIM analysis, as can be demonstrated by recent work on the smart city and urban energy [35] or the connections between a building and its direct urban environment [36].

Since the end of the 1990s, Reunion Island has embarked on an intensive policy to achieve energy autonomy. Promotion of renewable energies, generalization of solar water heaters, coupling of renewable energies with the productive activities of the territory, notably agricultural, energy renovation of buildings, actions to promote energy autonomy are diversified. These efforts have made it possible to achieve a suitable level of green electricity production: in 2015, renewable energy accounted for 36% of electricity production in Reunion, for a total delivered value 2891.3 GWh. Projects for the energy renovation of housing units and the installation of solar water heaters throughout the island to reduce electricity consumption are two axes of a clean policy at the local level: integrating the inhabitants into the energy transition. Indeed, the political authorities raise the importance of integrating the population to make this transition a success. Understanding the consumption of the population is an essential element to better guide them towards change while continuing to meet their needs. An advantage also lies in the ability to locate this consumption. In Reunion, this type of data is not available to the public and must therefore be reconstituted. The method of reconstruction by GIS makes it possible to consider the existing and thus offers a faithful description of the reality.

2.3. Data collection and main assumptions

The objective of this work is to calculate electricity consumption ratios per square meter of built-up area (distinguishing the type of building) by year. In Reunion, there are consumption ratios per inhabitant, but there is no ratio by type of building. However, this type of ratio could make it possible to better simulate urban sprawl or densification by combining electricity needs. The ultimate goal is therefore to forecast the electricity needs of the territory down to the finest possible scale: the building. Two information is needed to calculate this ratio: information on electricity consumption (GWh), and information on the type of building (m^2). For the consumption of electricity, the examination of the BER data was used. For the building surface data (and the typology), it is the INSEE, IGN and Agorah data coupled with the author's knowledge of the territory (see Figure 2).

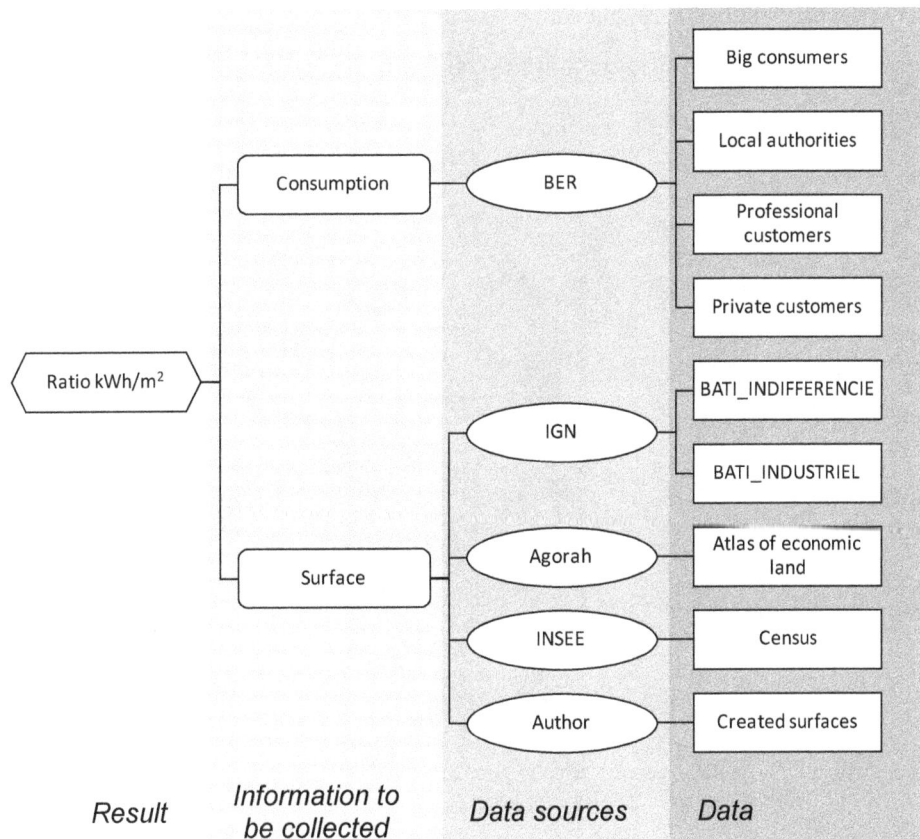

Figure 2. Methodology.

The work is based on a GIS based method and propose a ratio of energy consumption per unit area ($kWh/m^2/yr$). The ratio will also allow distinguishing the type of building. The buildings of the whole territory of Reunion Island are considered. These buildings—which are assimilated to EDF (Electricité de France: Historical electricity supplier in France) "customers"—are segmented into four categories in the Reunion Energy Review (BER) of the Energy Observatory Reunion led by SPL Energie Reunion (see Figure 3):

(1) big consumers, who represent manufacturers, hospitals and airports;

(2) local authorities which means all the premises of the devolved and the decentralized administrations of the French government (Town halls and their annexes, intercommunal bodies, prefectures and sub-prefectures, regional council…), and schools (nursery schools, elementary schools, secondary schools);

(3) professional customers: companies;

(4) private customers: residential customers.

Private customers represent the majority of customers with 329,087 customers out of a total of 370,397, accounting for 45.2% of total consumption (the largest consumer item). The smallest consumer item is local authorities with 16.7% of consumption. It should be noted that the big consumers (hospitals and industry) represent only 1440 customers but 34.6% of consumption. This suggests that the big consumer category should have the highest consumption ratio per unit area, while individual customers should have a much lower ratio.

Thus, the electricity consumption data are taken from the BER and the data on the surfaces must be produced.

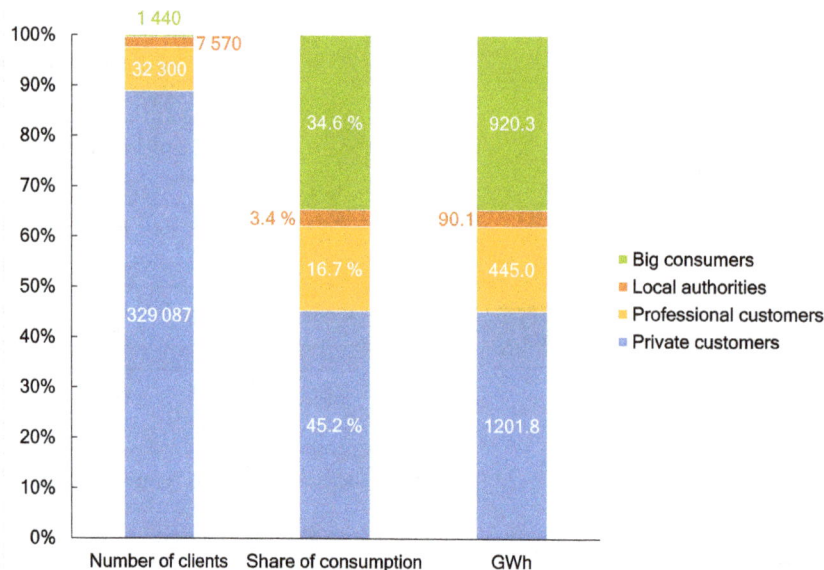

Figure 3. Customers segmentation and share of consumption (2015) Source: EDF-OER.

The methodology proposed is the use of GIS, and more particularly of free and open source software, Quantum GIS (QGIS). It consists of inventorying existing buildings using the IGN's topographic databases: "BATI_INDIFFERENCIE" and "BATI_INDUSTRIEL". To complete the analysis, we use the databases of the census led by the INSEE (National Institute for Statistics and Economic Studies) (RGP on 2013) which supplies data on the number of housing by IRIS zones (statistical division of the French territory).

One of the difficulties of the work rests on the identification in the layers of the IGN of the categories of building by segmenting into four categories (corresponding to the types of customers identified by EDF).

The buildings of the four categories of consumers considered belong to different vector layers:

(1) for the big consumer category: manufacturers correspond to the "BATI_INDUSTRIEL" layer; the hospitals are identified in the "BATI_INDIFFERENCIE" layer and the airports are identified in these two layers. The identification work has been easy here, due to the small number of buildings to be treated and the knowledge of the territory.

(2) for local authorities: the majority of local authorities is in the "BATI_INDIFFERENCIE" layer. The comparison between the total number of local authority customers provided by the BER and the number of customers (buildings) obtained by mapping was necessary for the treatment of this category. The analysis was complemented by the census of the premises on the official websites of the institutions.

(3) professional and private customers correspond to all remaining buildings.

It is therefore on these two latter categories that the work has been more complicated, because belonging to the same layer which is by definition undifferentiated.

For each of the categories of consumers, the buildings missing from the initial layers had to be created.

All the filters and layer mergers that the job required were performed on QGIS. The creation of four vector files corresponding to the four categories of consumers is the basic element for the calculation of the occupied surfaces. For the calculation of the surfaces, it is done by the addition of a new calculation field in the attribute table.

2.4. Special cases

The categories of professional customers and private customers were the most complex to deal with because the vector files of the IGN do not bring any distinction between those two categories. Two possible methods have been identified to separate these two categories. The first is based on the use of information from INSEE and DEAL on the number of dwellings (individual and collective, social and private) to determine the area of residences and then deduct the area of businesses. The second method is based on the use of GIS by considering the total area of firms based on hypotheses based on the work of the Agorah (Regional planning Agency) and the IGN's data and then deduct the area of dwellings. The second method has brought the most satisfaction in terms of precision.

2.4.1. Professional customers

The treatment of this category consists of determining the surface area of buildings of professional clients. To achieve this, the method was to cross two databases:

(1) the basis of the Atlas of economic land produced by the Agorah, which necessitated the creation of 115 zones of activities on QGIS with the help of the IGN Parcel Database.

(2) the base "surfaces_activites" of the IGN Carto database.

The crossing between these two bases and the database including the building made it possible to collect the buildings hosting companies.

The selected buildings then required treatment. Indeed, the hypothesis advanced is that an entity (a building on the vector file) represents a company or a customer. However, the BER informs about the number of professional clients on the island. In addition, NFX35-102 recommends precisely the dimensions of office work spaces and a minimum space of 10 m^2 (Service-public-pro.fr, 2016). All entities whose surface area exceeds 10 m^2 will then be retained. Thus, the area of firms obtained by mapping is thus the sum of the surfaces of the buildings greater than or equal to 10 m^2. The average area per entity was then calculated. The BER data and the number of entities obtained from GIS differ. In order to complete the analysis and obtain a total surface area, this difference was multiplied by the average area per entity. The addition of this latter calculation thus makes it possible to obtain a total area occupied by the businesses.

2.4.2.   Private customers

The processing of this category comes last. It results from the subtraction between the total building and the building corresponding to the categories big consumer, local authorities and professional customers. The problem at this stage is related to the GIS analysis. In fact, only surfaces on the ground are obtained. However, for the housing category it is necessary to consider several levels of floors.

No database is able to obtain this type of information. Therefore, a coupling between the INSEE housing data and the surface data is required. In order to consider a number of floors to be allocated to IRIS, it is necessary to consider the diversity of IRIS and thus to produce classes of IRIS.

To do this, we proceed by learning data: a classification by Natural Ruptures (Jenks), a discretization method which consists of minimizing intra-class variances and maximizing inter-class variances, was carried out under QGIS and 11 classes are highlighted. For the learning, the IRIS-Habitat of Saint-Denis are used. The Habitat type is chosen because the treatment takes place in the category of dwellings, the commune of Saint-Denis is selected because the first 9 IRIS the densest in terms of housing belong to this commune.

Thanks to the classes identified by QGIS, we then allocate a number of floors thanks to a field analysis. This number of floors remains an average and IRIS zoning leads to a bias in the analysis (a density on the island would have been more judicious, but the data does not exist).

The following map summarizes the result of the processing of this category (Figure 4). The classification thus carried out reveals two zones of high density in the North and West of the island and some areas of medium density in the South.

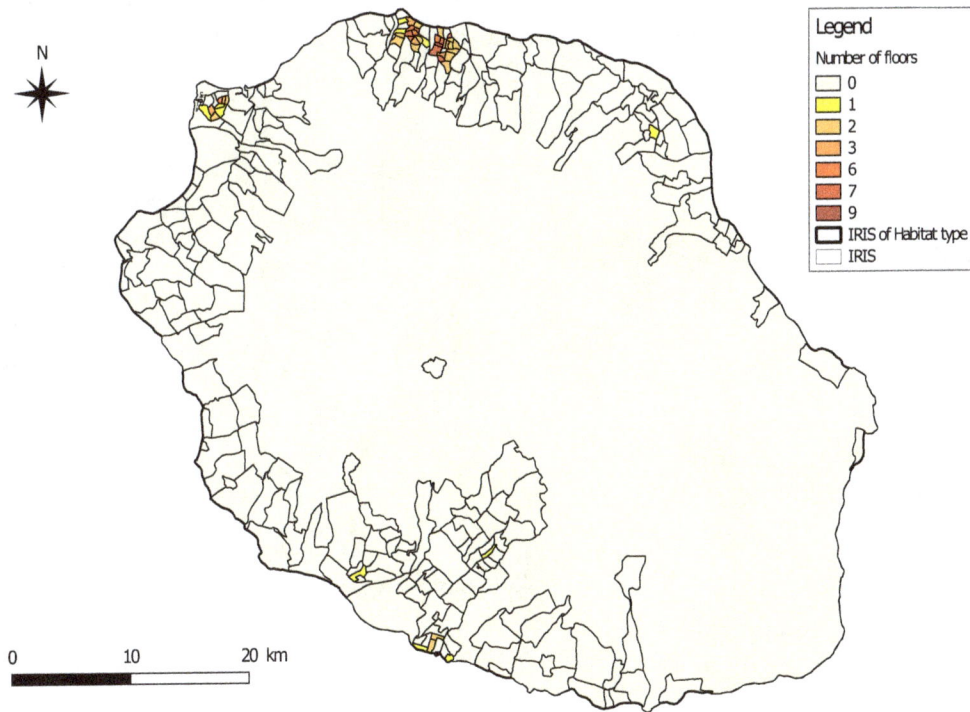

Figure 4. Classification of IRIS by density of residence.

3. Results

An analysis of the entire island structure was carried out and showed a total construction area of 5,515 hectares for a total consumption of 2,657.2 GWh for 2015. This represents an overall consumption of 76.94 $kWh/m^2/yr$ for all sectors combined. This represents 370,397 customers throughout the territory. The results for the four customer categories are presented in the following table (Table 1).

Table 1. Consumption (2015), area and ratio by customers type.

Customers category	Customers type	Electricity consumption (GWh, 2015)[1]	Area (m²)	Ratio (kWh/m²/yr)
Big consumers	industries, hospitals, airports	920.3	5,038,588.000	182.65
Local authorities	Administrations, annexes, schools	90.1	3,260,351.336	27.64
Professional customers	businesses	445	6,578,735.574	67.64
Private customers	residences	1,201.8	40,276,767.000	29.84
Total		2,657.2	55,154,441.910	48.18

[1] source: (Observatoire Energie Réunion, 2016)

Big consumers logically display a higher unit consumption with 182.65 kWh/m^2. The municipalities with the highest consumption are also the most endowed by the area occupied by big consumers (Figure 5).

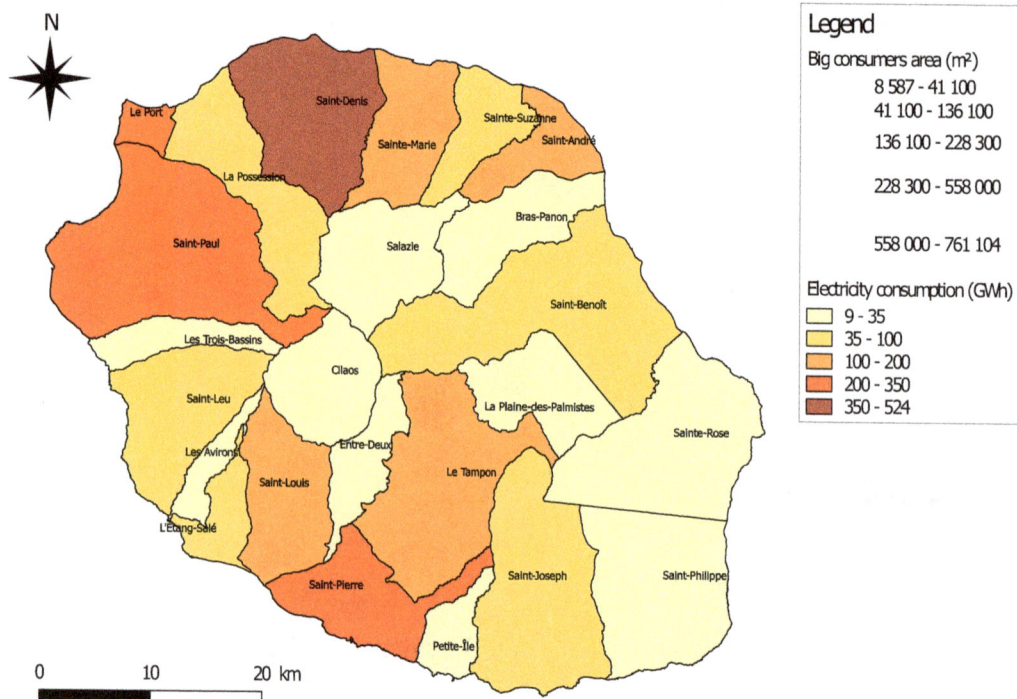

Figure 5. Electricity consumption and big consumers.

Local authorities have the lowest ratio. This is easily explained by the way in which the administrations and schools run: the premises are open only a few hours a day (about 10 hours a day) or even a month (especially for annexes and even schools that are closed weekends and several weeks a year).

Professional customers have a high ratio of 67.64 kWh/m^2. The mode of operation is necessary energy-intensive for this type of consumer, especially because of the use of specific equipment.

Finally, private customers have a ratio of 29.84 kWh/m^2. This ratio is close to the ratio of the BER (with an error of less than 5%), which makes it possible to validate our methodology.

4. Discussion

The "BBC" standard (low energy consumption building standard) imposed by the 2012 thermal regulation in France means a set of measures to be taken on new buildings designed to reduce the consumption of these buildings by 80% compared to a standard building. It then imposes a maximum consumption of 50 kWh/m² per year: this includes ventilation, heating, refreshment, lighting and hot water. In France, in 2011, ADEME stressed that electricity consumption in the residential sector was estimated at 30 kWh/m^2 for buildings located in mainland France, which represents 16.13% of energy consumption [37]. The residential ratio produced for Reunion is consistent with the ratio announced by ADEME for metropolitan France, with an electricity consumption ratio of 29.84 kWh/m^2 produced by our study. In 2015, the BER also produced a residential ratio of 31.37 kWh/m^2 [12], validating the product ratio in our study with a margin of error of less than 5%.

In order to complete our analysis and validate our work, we compared the real municipal consumption (BER) and calculated municipal consumption.

The calculated consumption is the product of the consumption ratios and calculated areas of each consumer category. Figure 6 presents the comparison between the real consumption and the calculated one for all the municipalities of Reunion Island. In the bottom of the plot, a histogram shows the residual for each municipality. The obtained R^2 is 0.97, and the RMSE is 22.8094 for a real value that varies between 9 and 524 GWh. The result is satisfactory with a maximum absolute error of 53.90 GWh for Saint-Paul and a minimum absolute error of 0.11 GWh for Saint-Leu and an average of 16.72 GWh (Figure 6). This corresponds to an error of 560,320 m^2 of residential area or 1.69% of the total residential area.

We can see three types of municipalities: some, for which the estimate is correct (Sainte-Rose, Saint-Philippe, Cilaos for example); others where consumption is underestimated (Saint-Denis, Saint-Pierre, Saint-Paul for example) and others where consumption is overestimated (Le Port, Tampon, Saint-Joseph).

Municipalities whose consumption are underestimated correspond to the major poles of the Island: Saint-Denis, Saint-Pierre and Saint-Paul are the most important urban centers on the island (Figure 7). The extent and complexity of information concerning these poles are therefore probably at stake: these communes deserve a more detailed work with segmentation in city block for a more detailed analysis of the building frame.

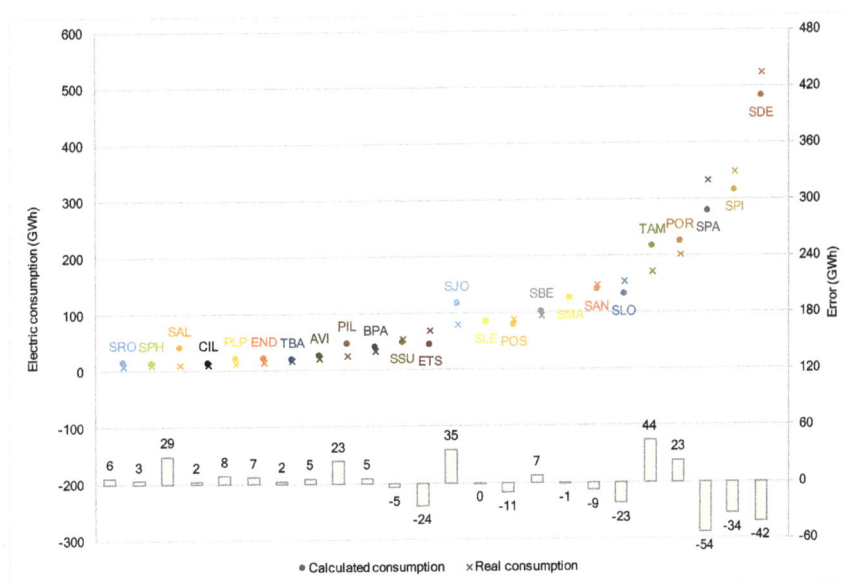

Figure 6. Calculated and real consumption by municipalities.

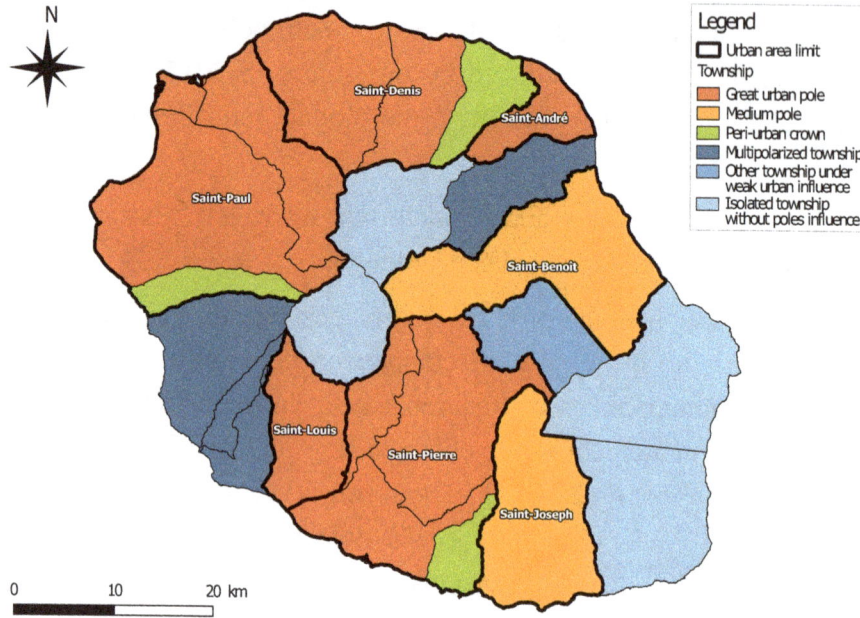

Figure 7. Representation of urban areas.

While the population is the largest share of electricity consumption (45.2 %), it has one of the lowest unit consumption ratios (Figure 8). This confirms that the population is one of the main determinants of the island's electricity consumption [14]. Indeed, the economic organization of the island, with essentially tertiary and a strong economic dependence on France, suggests the weakness of energy-consuming sectors such as industries.

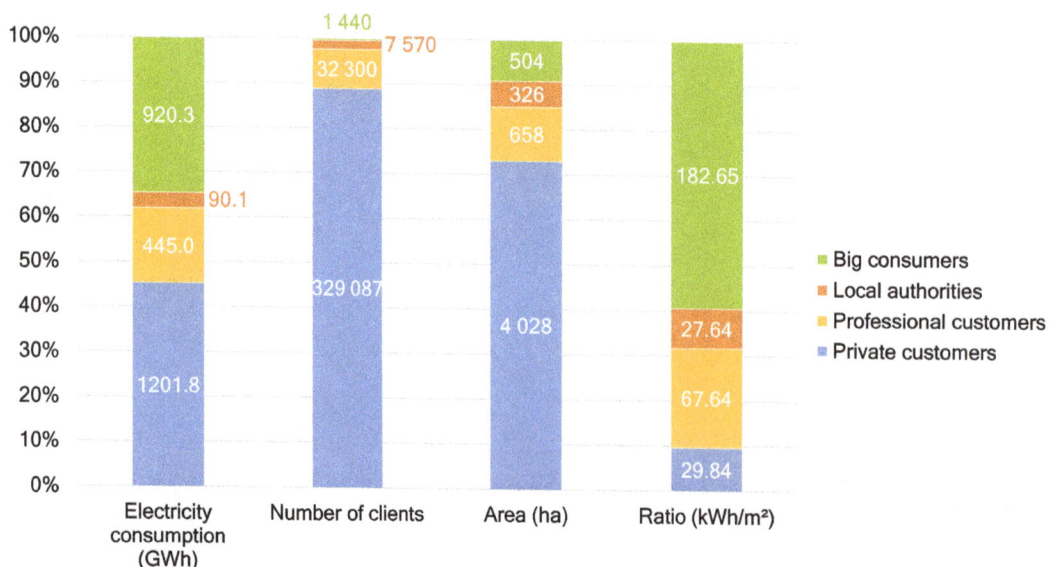

Figure 8. Segmentation of customers: surface, share of consumption and ratio.

The following map shows the spatial distribution of different categories of consumers. The abundance of residential areas is evident in comparison with other types of consumers (Figure 9). The zones corresponding with the large consumers are distributed very unevenly in the territory with a concentration in the cities of Le Port, Saint-Denis and Saint-Pierre. The same applies to the areas occupied by enterprises: the same three zones have been identified in the north, south and west of the island.

Not surprisingly, we can see very clearly that private customers dominate the territory. Our first work to characterize electricity consumption at the municipal level showed that the estimate of electricity needs could indeed focus on the population [14]. This second work is in addition to confirming the importance of the population in the consumption of electricity because of their number, to locate more precisely on the island the electricity needs. Thus, the points of the most consumer and least consumer territories are highlighted. The map shown in Figure 9 could be a great tool for energy planning.

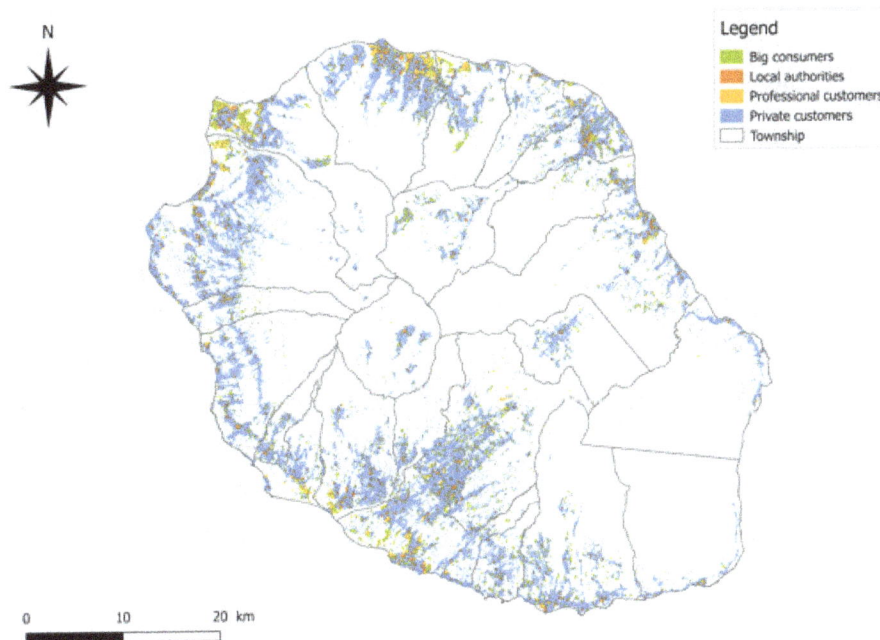

Figure 9. Spatial distribution of the customers.

5. Conclusion

The scope of this work was to produce electricity consumption ratios per square meter of building based on a new GIS approach by segmenting buildings, according to their function: housing, industry, administrations and businesses. These categories correspond to the categories of consumers defined by the electricity supplier: EDF. Their unequal distribution on the territory and their different consumption patterns make it easy to distinguish several consumption ratios. These ratios are based on their distribution on the ground, obtained by mapping, and augmented by hypotheses on the number of floors. The ratios obtained correspond to expectations: the ratio of large consumers is six times higher than that of private customers, the ratio of enterprises is double and the ratio of local authorities is the lowest.

This work is a first attempt to calculate the ratio of electricity consumption per square meter of building. It develops a method using GIS as the basis of calculation. The method allows compensating the lack of information on energy consumption on a territory. It is particularly useful for energy planning scenarios to decrease electricity demand. Most of studies are focused electricity demand per capita, and energy scenarios are mainly based population growth on territory. This article has shown that population is indeed the determinant of electricity consumption. However, it is necessary to consider all economical sector to build a real scenario that takes into account different activities development such in industry or tertiary sector. Our method is very useful for territory that doesn't have any data ratio of electricity consumption by sector and allows in the same to building more accurate scenario.

For further work, it would be useful to define control buildings by consumer categories in order to analyze the actual consumption of these buildings. A second track of deepening would be the analysis of the age of the buildings. Indeed, with the evolution of construction standards, with reference to the various thermal regulations, buildings built today consume less energy than those built ten years ago for example. As we are interested in the way to challenge energy transition, this methodology will be repeated over years, if in the case of a residential or tertiary sector a decrease is actually noticed. The ratio could be used by energy planners as an indicator of success of electricity consumption reduction.

Finally, the most significant point of the ratio is the fact that the information is brought for a typology of the building. This choice was not harmless, because urban development strategy is basically based on urban sprawl or densification, which means most of the time: building area. It seems to us more convenient to define this ratio per sector of activity for better energy planning scenarios modeling.

Electricity consumption represents 37% of the energy consumed in residential and commercial buildings in France in 2012 [37] and 23.1% of total energy consumption in Reunion Island in 2015 [12]. The increasing trend in energy consumption of buildings will continue over the next few years due to the expansion of the built area and associated energy needs, as long as resources and land allow. The private initiative, in conjunction with government intervention through the promotion of energy efficiency, the limitation of energy consumption and social awareness of the rational use of energy, will be essential for a future sustainable energy supply. Thus, the effort to conserve energy in residential, commercial and industrial buildings, while maintaining an optimal indoor environment, is an ongoing struggle to minimize dependence on fossil fuel sources. It also makes it possible to secure the energy balance and reduce the environmental impact of fossil fuels by reducing greenhouse gas emissions.

Acknowledgements

This project is part of the TRANSEETER research project funded by the University of La Reunion through its incentive program on energy transition.

Conflict of interest

The authors declare no conflict of interest in this paper.

References

1. EDF (2013) Il était une fois les réglementations thermiques. Available from: https://travaux.edf.fr/construction-et-renovation/la-reglementation-thermique-2012/il-etait-une-fois-les-reglementations-thermiques.

2. JORF (1996) LOI n° 96-1236 du 30 décembre 1996 sur l'air et l'utilisation rationnelle de l'énergie (1). Available from: https://www.legifrance.gouv.fr/affichTexte.do?cidTexte=JORFTEXT000000381337&categorieLien=id.

3. JORF (2005) Loi n° 2005-781 du 13 juillet 2005 de programme fixant les orientations de la politique énergétique. Available from: https://www.legifrance.gouv.fr/affichTexte.do?cidTexte=JORFTEXT000000813253&categorieLien=id.

4. JORF (2010) LOI n° 2010-788 du 12 juillet 2010 portant engagement national pour l'environnement (1). Available from: https://www.legifrance.gouv.fr/affichTexte.do?cidTexte=JORFTEXT000022470434&categorieLien=id.

5. JORF (2015) LOI n° 2015-992 du 17 août 2015 relative à la transition énergétique pour la croissance verte (1). Available from: https://www.legifrance.gouv.fr/affichTexte.do?cidTexte=JORFTEXT000031044385&categorieLien=id.

6. Cheng MY, Chiu KC, Lien LC, et al. (2016) Economic and energy consumption analysis of smart building–MEGA house. *Build Environ* 100: 215–226.

7. ADEME. Climat, Air et Energie, Chiffres clés; 2013.

8. Durable CGaD. Repères-Ciffres clés de l'environnement; 2015; Paris.

9. Ma JJ, Du G, Zhang ZK, et al. (2017) Life cycle analysis of energy consumption and CO_2 emissions from a typical large office building in Tianjin, China. *Build Environ* 117: 36–48.

10. Blom I, Itard L, Meijer A (2011) Environmental impact of building-related and user-related energy consumption in dwellings. *Build Environ* 46: 1657–1669.

11. Parshall L, Gurney K, Hammer SA, et al. (2010) Modeling energy consumption and CO_2 emissions at the urban scale: Methodological challenges and insights from the United States. *Energ Policy* 38: 4765–4782.

12. Réunion OE. Bilan énergétique Ile de La Réunion 2015; 2016.

13. Amado M, Poggi F, Amado AR (2016) Energy efficient city: A model for urban planning. *Sustain Cities Soc* 26: 476–485.

14. Bénard-Sora F, Praene JP (2016) Territorial analysis of energy consumption of a small remote island: Proposal for classification and highlighting consumption profiles. *Renew Sust Energ Rev* 59: 636–648.

15. Mungwe J, Mandelli S, Colombo E (2016) Community pico and micro hydropower for rural electrification: experiences from the mountain regions of Cameroon. *AIMS Energy* 4: 190–205.

16. Dornan M (2015) Reforms for the expansion of electricity access and rural electrification in small island developing states. *AIMS Energy* 3: 463–479.

17. Wolf F, Surroop D, Singh A, et al. (2016) Energy access and security strategies in small island developing states. *Energ Policy* 98: 663–673.

18. Alhamwi A, Medjroubi W, Vogt T, et al. (2017) GIS-based urban energy systems models and tools: Introducing a model for the optimisation of flexibilisation technologies in urban areas. *Appl Energ* 191: 1–9.

19. Li C (2017) GIS for Urban Energy Analysis. Reference Module in Earth Systems and Environmental Sciences.

20. Man SW, Zhu R, Liu Z, et al. (2016) Estimation of Hong Kong's solar energy potential using GIS and remote sensing technologies. *Renew Energ* 99: 325–335.

21. Nematollahi O, Kim KC (2017) A feasibility study of solar energy in South Korea. *Renew Sust Energ Rev* 77: 566–579.

22. Yushchenko A, Bono AD, Chatenoux B, et al. (2017) GIS-based assessment of photovoltaic (PV) and concentrated solar power (CSP) generation potential in West Africa. *Renew Sust Energ Rev* 81: 2088–2103.

23. Mentis D, Hermann S, Howells M, et al. (2015) Assessing the technical wind energy potential in Africa a GIS-based approach. *Renew Energ* 83: 110–125.

24. Cristea C, Jocea AF (2016) GIS Application for wind energy. *Energ Procedia* 85: 132–140.

25. Cavazzi S, Dutton AG (2016) An offshore wind energy geographic information system (OWE-GIS) for assessment of the UK's offshore wind energy potential. *Renew Energ* 87: 212–228.

26. Jangid J, Bera AK, Joseph M, et al. (2016) Potential zones identification for harvesting wind energy resources in desert region of India? A multi criteria evaluation approach using remote sensing and GIS. *Renew Sust Energ Rev* 65: 1–10.

27. Anwarzai MA, Nagasaka K (2017) Utility-scale implementable potential of wind and solar energies for Afghanistan using GIS multi-criteria decision analysis. *Renew Sust Energ Rev* 71: 150–160.

28. Vasileiou M, Loukogeorgaki E, Vagiona DG (2017) GIS-based multi-criteria decision analysis for site selection of hybrid offshore wind and wave energy systems in Greece. *Renew Sust Energ Rev* 73: 745–757.

29. Quan SJ, Li Q, Augenbroe G, et al. (2015) A GIS-based energy balance modeling system for urban solar buildings. *Energ Procedia* 75: 2946–2952.

30. Mutani G, Delmastro C, Gargiulo M, et al. (2016) Characterization of building thermal energy consumption at the urban scale. *Energ Procedia* 101: 384–391.

31. Tornay N, Schoetter R, Bonhomme M, et al. (2017) GENIUS: A methodology to define a detailed description of buildings for urban climate and building energy consumption simulations. *Urban Climate* 20: 75–93.

32. Palme M, Inostroza L, Villacreses G, et al. (2017) From urban climate to energy consumption. Enhancing building performance simulation by including the urban heat island effect. *Energ Buildings* 145: 107–120.

33. Jiang Y, Gu P, Chen Y, et al. (2017) Modelling household travel energy consumption and CO_2 emissions based on the spatial form of neighborhoods and streets: A case study of Jinan, China. *Computers, Environment and Urban Systems*.

34. Lu H, Liu G (2014) Spatial effects of carbon dioxide emissions from residential energy consumption: A county-level study using enhanced nocturnal lighting. *Appl Energ* 131: 297–306.

35. Yamamura S, Fan L, Suzuki Y (2017) Assessment of urban energy performance through integration of BIM and GIS for smart city planning. *Procedia Eng* 180: 1462–1472.

36. Niu S, Pan W, Zhao Y (2015) A BIM-GIS integrated web-based visualization system for low energy building design. *Procedia Eng* 121: 2184–2192.

37. ADEME. Bâtiment, Edition 2013, Chiffres Clés; 2013.

3

Performance of Zn/Graphite rechargeable cells with 1-ethyl-3-methylimidazolium trifluoromethanesulfonate based gel polymer electrolyte

W. Prasadini, Kumudu S. Perera* and Kamal P. Vidanapathirana

Polymer Electronics Research Laboratory, Department of Electronics, Wayamba University of Sri Lanka, Kuliyapitiya, Sri Lanka

* Correspondence: Email: kumudu31966@gmail.com

Abstract: There is an urgent need to fulfill the global thirst for energy in an efficient and safer way. Since long ago, Li based devices have been employed to cover up the high rising energy demand in the society. But, upon the realization of the hazardous nature and the cost of Li, a substantial attention has been focused on exploiting ecofriendly and naturally abundant materials to be employed for fabricating devices such as rechargeable cells, fuel cells, solar cells and super capacitors. Main aim of this study is to design a low cost, environmental friendly, rechargeable cell. Hence, attempts have been taken to use non Li based electrodes and a gel polymer electrolyte (GPE) with no solvents. The cell consists with natural graphite and Zn electrodes and an ionic liquid (IL) based GPE. ILs are the most recently introduced substitute for solvents in GPEs which possess some attractive features. For the preparation of the GPE, poly(vinylidinefluoride-co-hexafluoropropylene) (PVdF-co-HFP), zinc trifluoromethanesulfonate ($Zn(CF_3SO_3)_2$ and 1-ethyl-3-methylimidazolium trifluoromethanesulfonate (1E3MITF) were used as the polymer, the salt and the ionic liquid (IL) respectively. Configuration of the fabricated cell was Zn/IL based GPE/graphite. Cells had an average open circuit voltage (OCV) value of 1.10 V. Variation of OCV with time was quite low. Cells were characterized by electrochemical impedance spectroscopy (EIS), cyclic voltammetry (CV) as well as continuous galvanostatic charge discharge (GCD) test. Properties of the bulk electrolyte as well as electrode/electrolyte interface were rather stable. Average specific charge was remaining around 4 mAhg^{-1}. Discharge capacity was varying around 2 mAhg^{-1}. These results reflect the possibility of using this cell for applications with some further modifications to increase the performance.

Keywords: rechargeable cells; 1-ethyl-3-methylimidazolium trifluoromethanesulfonate; gel polymer electrolyte; open circuit voltage; natural graphite; poly (vinylidinefluoride-co-hexafluoropropylene); electrochemical impedance spectroscopy

1. Introduction

Due to the unending rapid demand for energy in the modern technological society in various aspects, the whole globe is heading towards an enormous energy crisis. On the other hand, within the next few years, there can be huge challenges to maintain the green environment with the disposal of used power devices which are mainly based on Li. At present, it has been well accepted that Li is significantly a dangerous material even though it possess outstanding performance [1,2]. In addition, presence of liquid electrolytes in many devices available in the present day market is also supposed to create some environmental threats. These issues have laid the foundation to seek materials which are safe and low cost to fabricate devices. In place of Li, materials like Mg, Cu, Zn and Al have been identified as suitable substitutes for anodes with remarkable features [3–5]. On the other hand, natural graphite has a growing attention to be used in place of Li based composite cathodes. Some have carried out investigations using graphite in electrochemical cells and supercapcitors [6–8]. Gel polymer electrolytes (GPEs) have received a major consideration to be employed in various applications to surmount many of the traditional drawbacks of liquid and solid polymer electrolytes [9,10]. In a GPE, a salt/solvent mixture is assumed to be encapsulated inside a polymer host. They possess ionic conductivities similar to liquid electrolytes in the order of 10^{-3} Scm^{-1} [11,12]. Their ionic transference numbers are above 0.9. But, the presence of solvents has given rise to some limitations with respect to safety. This has motivated the interest on replacing solvents by ionic liquid (IL)s which are present as room temperature molten salts. They possess some unique properties such as low toxicity, wide electrochemical stability, high thermal stability, etc. [13]. These ILs are being considered as an innovative approach to avoid many drawbacks of solvents. As such, IL based GPEs are also receiving a considerable attention to be employed for applications [14]. Due to the unrelenting increase in the curiosity towards non Li systems, non Li salts have been incorporated in preparing IL based GPEs. They have been eventually exploit for a diverse range of uses [15,16].

In the present study, cells have been fabricated with the configuration of Zn/IL based GPE/graphite. It is a known fact that Zn is abundant and cheap. Also, it is not toxic as Li. Zn has a high stability as well as satisfactory volumetric and specific density. The graphite electrodes were prepared using Sri Lankan natural graphite. In Sri Lanka, graphite has been renowned as one of the valuable natural resources. But, it has not been properly employed for applications in energy sector. Use of natural graphite provides much safety and also, it is not expensive. The main highlight of the present study is use of Zn and Sri Lankan natural graphite electrodes with an IL based GPE in a rechargeable cell. The average open circuit voltage (OCV) of the cells were 1 V. The stability, charge/discharge characteristics of the cells were determined by electrochemical impedance spectroscopy (EIS), cyclic voltammetry (CV) as well as continuous charge discharge (GCD) test.

2.   Materials and methods

2.1. Preparation of IL based GPE

Electrolyte was prepared using solvent casting technique. Poly (vinylidinefluoride-co-hydrofluoropropylene) (PVdF-co-HFP, average MW 400000), zinc trifluoromethanesulfonate (Zn(CF$_3$SO$_3$)$_2$-ZnTF, 98%) and 1-ethyl-3-methylimidazoliumtrifluoromethanesulfonate (1E3MITF, 98%) were received from Aldrich. ZnTf was vacuum dried at 120 °C for 24 hours. Required amount of PVdF-co-HFP was first magnetically stirred with acetone at room temperature (RT) until PVdF-co-HFP dissolves in acetone. ZnTF and 1E3MITF were added to the resultant mixture and magnetically stirred again overnight. Before mixing, IL was vasuum dried at 120 °C for 12 hours. Homogeneous solution was then poured in to a glass petri dish and it was kept in a vacuum desiccator overnight to evaporate acetone. Film was then vacuum dried at 80 °C for 8 hours. Finally it was possible to obtain a thin, free standing film with no bubbles and pin holes. The composition of the electrolyte was 1 PVdF:1 ZnTf:3 IL (by weight basis).

2.2. Fabrication of rechargeable cells

A circular shaped, well cleaned Zn pellet was used as one electrode (area 1 cm^2). For the other electrode, natural graphite obtained from Bogala Graphite Lanka, Sri Lanka was pre heated to 150 °C and was dissolved in acetone using a homogenizer. Then the resulted homogeneous mixture was coated on the surface of a stainless steel (SS) disk (area 1 cm^2) evenly and dried well in a vacuum oven for 12 hours. Then, a circular shaped (area 1 cm^2) electrolyte was sandwiched between Zn and graphite electrodes. Thicknesses of the Zn electrode, graphite electrode and the electrolyte film were 0.25 mm, 0.30 mm and 0.40 mm respectively. Cell assembly was loaded inside a brass sample holder.

2.3. Performance evaluation of rechargeable cells

Open circuit voltage (OCV) of the cell was measured using a digital multimeter at different time durations. Cyclic voltammetry studies were carried out using a three electrode electrochemical setup using Metrohm AutoLab potetniostat. Graphite electrode was used as the working electrode while Zn electrode was used as the reference and counter electrode. Voltage window of the cyclic voltammogramme was changed to trace the optimum window that results the highest specific charge (C$_s$). Cycling was done at the scan rate of 10 mVs^{-1}. C$_s$ was calculated using the Eq 1,

$$C_s = 2(\int IdV)/S\Delta v \qquad\qquad (1)$$

where $\int IdV$ is the area of CV, Δv is the potential window and S is the scan rate.

Then, cycling was done in that potential window varying the scan rate to determine the corresponding scan rate for the optimum C$_s$. Thereafter, continuous cycling was carried out to check the stability of the cell.

Impedance measurements of the cell were gathered using an Impedance Analyzer (Metrohm AutoLabM101) at different time durations in the frequency range from 0.1 Hz to 400 kHz.

Galvanostatic charge discharge (GCD) test was performed between 0.1 V and 2.1 V for the cell at room temperature under a constant current of 500 μA. Study was done for 1000 GCD cycles and the discharge capacity (C_d) was calculated by the Eq 2:

$$C_d = Idt \qquad (2)$$

where, I is the constant current and dt is the time duration for discharge.

3. Results and discussion

3.1. Open circuit voltage variation

Figure 1 shows the variation of the OCV for the rechargeable cell, Zn/IL based GPE/graphite. As soon as fabrication, the OCV value was 1.10 V. According to the results, the initial value remains constant throughout the investigation. This elucidates the fact that there are no reactions taking place inside the cell. There is a high probability for the occurance of parasitic reactions such as electrode/electrolyte reactions, formation of some interfaces which hinder the efficiency of cell performance [17]. If there were such reactions, they affect OCV and hence, a reduction of OCV with time should be observed. Basically, those reactions are possibly occurring at the electrolyte/electrode interfaces. Thus, the results prove the stability at the interfaces. This would be due to the absence of any reactive material in the cell. Zn and graphite are known to be safer materials with low reactivity. On the other hand, electrolyte does not contain any solvent which is very aggressive with materials. This confirms the user friendly nature of IL based GPEs as well as Zn/graphite electrodes.

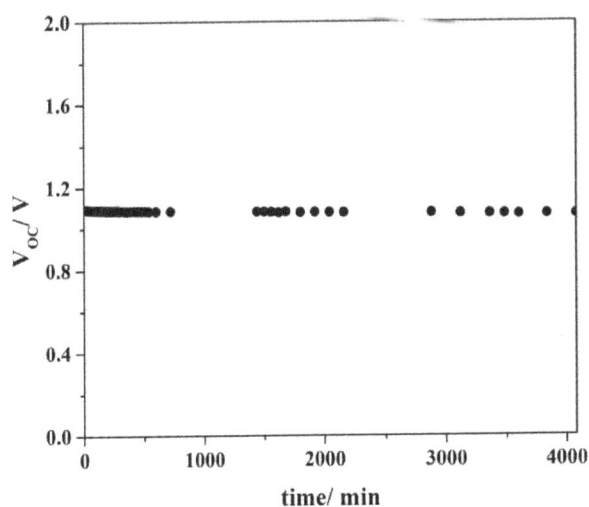

Figure 1. Variation of OCV values of the cells Zn/IL based GPE/graphite with time.

3.2. Cyclic voltammetry

The cyclic voltammogrammes obtained by varying the voltage window are shown in Figure 2.

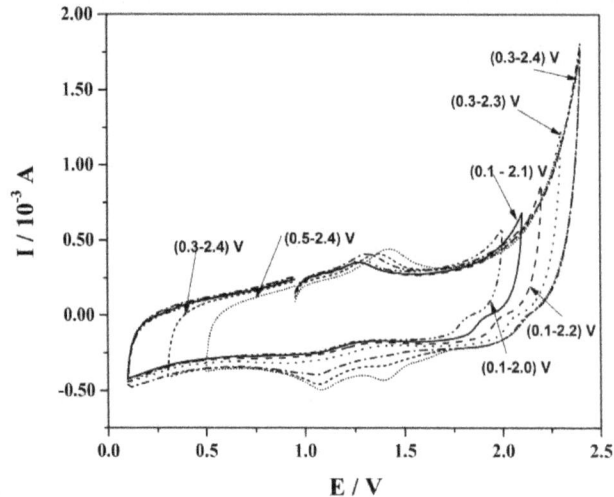

Figure 2. Cyclic voltammograms obtained for different voltage windows of the cell Zn/IL based GPE/graphite at the scan rate of 10 mVs^{-1}.

The widest voltage window was from 0.1 to 2.4 V. According to the results of the present study, the highest value of C_s could be obtained from that voltage window. But when the upper vertex goes to higher values than 2.1 V, cell started to get destroyed as evidenced by the high current. Thus, as the most suitable window, the range from 0.1 to 2.1 V was chosen. Within this window, cycling was done at different scan rates and it is shown in Figure 3.

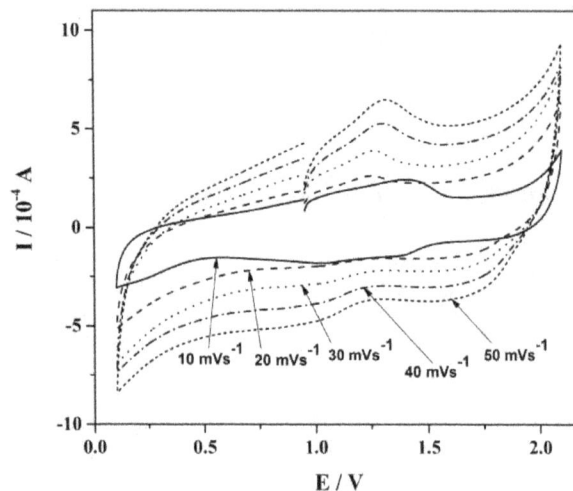

Figure 3. Cyclic voltammograms obtained for different scan rates within the potential window, 0.1 V–2.1 V.

Peaks appearing in the CVs are proving the ion movement that takes place at charge discharge processes of the cell. The peaks at positive side of the current axis are due to charging while negative side are due to discharging. In the present cell, Zn ions are playing a major role in the cell operation. As such, they are getting plated and stripped during charge discharge process [12].

When considering the scan rate variation, higher scan rates resulted lower C_s values. This is illustrated in Figure 4.

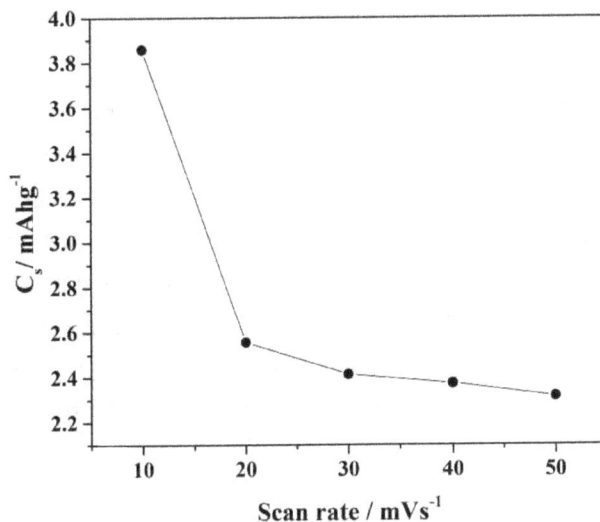

Figure 4. Variation of C_s with the scan rate.

During cycling at a particular scan rate, the ions move between electrodes due to charging and discharging processes. Occurrence of complete cell reactions is highly needed to get high C_s values. This is facilitated by lower scan rates that would allow sufficient time for proper reactions to take place [18]. The scan rate of 10 mVs^{-1} was selected as the optimum scan rate for further studies.

Ability of the cell to withstand continuous cycling was monitored upon performing continuous cycling and it is shown in Figure 5.

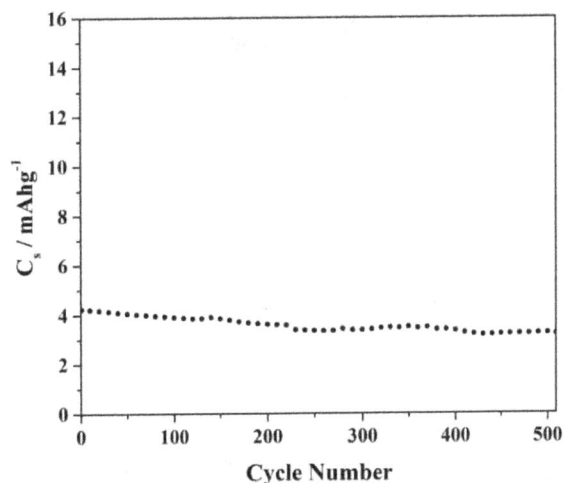

Figure 5. Specific charge variation of the cells Zn/IL based GPE/graphite at scan rate 10 mVs^{-1}.

At the first cycle, the cell has a C_s value of 4.23 mAhg^{-1}. It decreased down to 3.20 mAhg^{-1} during 500 cycles. The reduction of C_s is about 24% which is quite small. This highlights the satisfactory compatibility between GPE and electrodes as evidenced by OCV variation with time [19].

3.3. Electrochemcial Impedance Spectroscopy (EIS)

Figure 6 illustrates several Nyquist plots taken at different time periods. The intercept in the high frequency region represents the bulk electrolyte resistance (R_b) whereas the other intercept in the mid frequency region represents charge transfer resistance (R_{ct}) at the electrolyte/electrode interface [20]. According to the results, it is seen that initial R_{ct} value has dropped with time. Higher R_{ct} at the beginning may be due to the improper contacts at the interface of the electrolyte and the electrodes that lead to high resistance for charge transfer. As time passes, those interfaces may become stable and mature. R_b remains rather constant as shown in the inset. This well proves the stability of the electrolyte. If there were solvents in the system, there is a possibility for solvent evaporation which increases R_b considerably. Salt and IL present in the system may reside in with the polymer without affecting the electrolyte resistance.

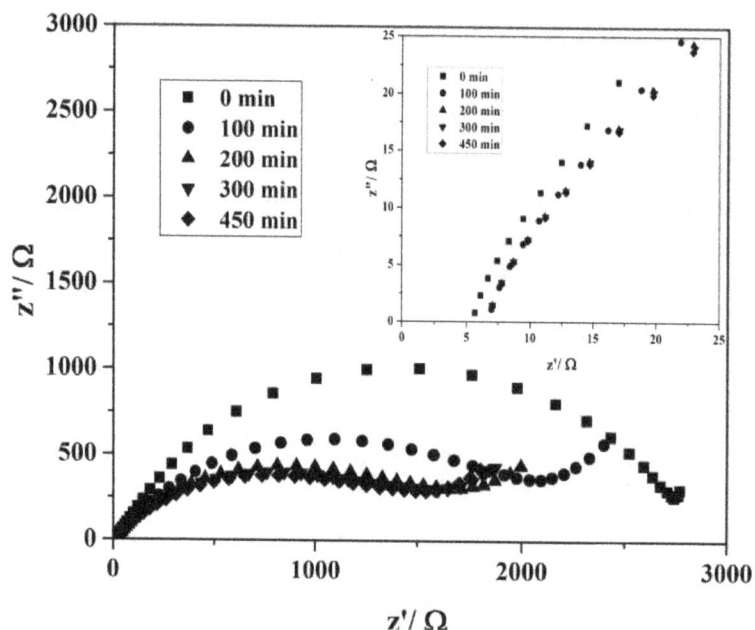

Figure 6. Nyquist plots of the cell taken at different time periods.

3.4. Galvanostatic charge discharge test

Initial galvanostatic charge discharge cycles obtained for the cell under a constant current of 500 μA is shown in Figure 7.

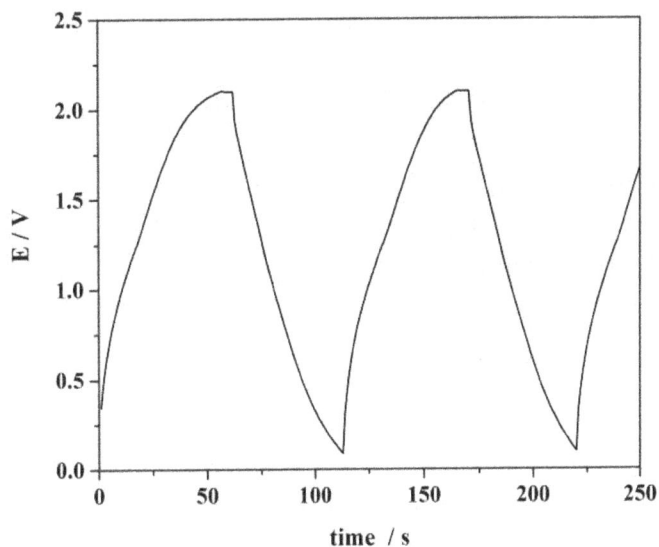

Figure 7. Two galvanostatic charge discharge cycles among the 1000 cycles obtained for the cell.

Variation of the discharge capacity, C_d of the cell with cycle number is shown in Figure 8.

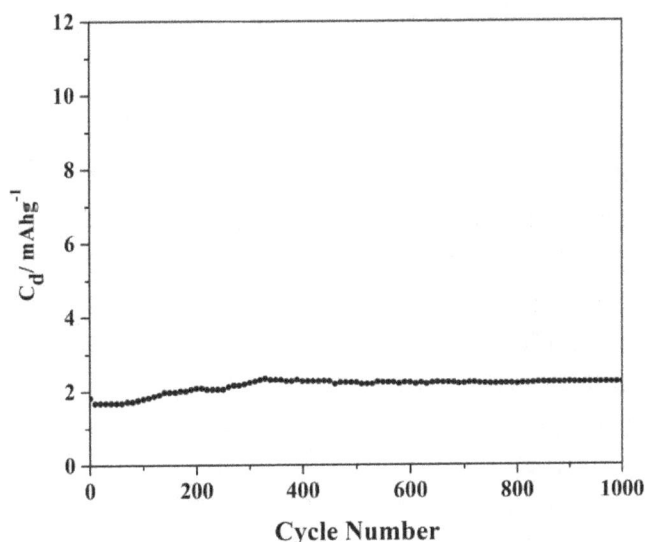

Figure 8. Variation of C_d for the cells Zn/IL based GPE/graphite at constant current of 500 µA.

The initial C_d value is 1.82 mAhg^{-1}. Upon continuous charge-discharge cycling, C_d increases first and thereafter, it reaches to a stable value of 2.2 mAhg^{-1} after about 300 cycles. This suggests that the cell needs some time to reach the maturity [21]. Soon after fabrication, it is not in a state to function in the proper level. The interfaces may not be formed properly to allow respective ion movements initially as evidenced by R_{ct} value obtained from Nyquist plots in Figure 6. As time passes, those interfaces may become stabilized and then, complete reactions may take place. This makes C_d more or less constant.

4. Conclusions

The GPE consisting of PVdF-co-HFP, 1E3MITF and ZnTF is prepared by solvent casting technique with optimized ratio of 1:1:3 by weight basis. OCV value of the cell Zn/IL based GPE/graphite remained constant around 1.10 V. Optimized voltage window of this cell was from 0.1 V to 2.1 V and the scan rate was 10 mVs^{-1}. The C_s value of the cell was 4.23 $mAhg^{-1}$ and its reduction was 24% for 500 cycles. Moreover, it showed a C_d value of 1.82 $mAhg^{-1}$ and it was stable at 2.2 $mAhg^{-1}$ after 300 cycles. Results showcase the suitability of this cell for applications. Further studies are being conducted to improve the performance for moving towards prototype and commercial stages.

Acknowledgements

Authors wish to acknowledge the assistance given by National Science Foundation, Sri Lanka under the grants, RG/2017/BS/02 and RG/2015/EQ/07. In addition, Wayamba University of Sri Lanka is greatly acknowledged for the award of the research grant, SRHDC/RP/04/17/01.

Conflict of interest

All authors declare no conflicts of interest in this paper.

References

1. Agrawal RC, Sahu DK (2013) Magnesium ion conducting polymer electrolytes: Materials characterization and all solid state battery performance studies. *J Phys Sci Appl* 3: 9–17.
2. Osman Z, Samin SM, Othman L, et al. (2012) Ionic transport in PMMA-$NaCF_3SO_3$ gel polymer electrolytes. *Adv Mat Res* 545: 259–163.
3. Tian H, Gao T, Li X, et al. (2017) High power rechargeable Mg/Iodine battery chemistry. *Nat Commun* 14083–14090.
4. Lin MC, Gong M, Lu B (2015) An ultrafast rechargeable Al-Ion battery. *Nature* 520: 325–328.
5. Jayathilaka YMCD, Perera KS, Vidanapathirana KP, et al. (2015) Evaluation of a copper based gel polymer electrolyte and its performance in a primary cell. *Sri Lankan J Phys* 15: 45–51.
6. Kravchyk KV, Wang S, Piveteau L, et al. (2017) Efficient aluminum chloride-natural graphite battery. *Chem Mater* 29: 4484–4492.
7. Zhang M, Song X, Ou X, et al. (2019) Rechargeable batteries based on anion intercalation graphite cathodes. *Energ Stor Mater* 16: 65–84.
8. Jayamaha B, Dissanayake MAKL, Kandasamy V, et al. (2017) Electrochemical double layer capacitors with PEO and Sri Lankan natural graphite. *Adv Ener Res* 5: 219–226.
9. Rosdi A, Zainol NH, Osman Z (2016) Ionic transport and electrochemical stability of PVdF HFP based gel polymer electrolyte. International Symposium on Frontier of Applied Physics. AIP Publishing LLC, 563–572.
10. Sekhon SS (2003) Conductivity behaviour of polymer gel electrolytes: Role of polymer. *B Mater Sci* 2613: 321–328.

11. Kumar GG, Sampath S (2003) Electrochemical characterization of poly(vinylidenefluoride)-zinc triflate gel polymer electrolyte and its application in solid-state zinc batteries. *Solid State Ionics* 160: 289–300.

12. Liu Z, Abedin SZE, Endres F (2014) Electrodeposition and stripping of zinc from an ionic liquid polymer gel electrolyte for rechargeable zinc-based batteries. *J Solid State Electr* 18: 2683–2691.

13. Dileo RA, Marschilok AC, Takeuchi KJ, et al. (2013) Battery electrolytes based on unsaturated ring ionic liquids: conductivity and electrochemical stability. *J Electrochem Soc* 160: A1399–A1405.

14. Armand M, Endres F, Macfarlane DR, et al. (2009) Ionic-liquid materials for the electrochemical challenges of the future. *Nat Mater* 8: 621–629.

15. Liu Z, Li G, Cui T, et al. (2017) A battery supercapacitor hybrid device composed of metallic zinc, a biodegradable ionic liquid electrolyte and graphite. *J Solid State Electr* 22: 1–11.

16. Zhang R, Chen Y, Montazami R (2015) Ionic liquid doped gel polymer electrolyte for flexible lithium ion polymer batteries. *Materials* 8: 2735–2748.

17. Yap SC, Mohamad AA (2007) Proton batteries with hydroponics gel as gel polymer electrolyte. *Electrochem Solid State Lett* 10: 139–141.

18. Vidanapathirana KP, Perera K (2014) Performance variation of Li rechargeable cells having polypyrrole cathodes doped with different anions. *J Natl Sci Found Sri* 42: 143–147.

19. Liu W, Zhang XK, Wu F, et al. (2017) A study no PVDF-HFP gel polymer electrolyte for lithium-ion batteries. *IOP Conf Ser: Mater Sci Eng* 213: 012–036.

20. Larfaillou S, Guy-Bouyssou D, Cras FL, et al. (2016) Comprehensive characterization of all-solid-state thin films commercial microbatteries by electrochemical impedance spectroscopy. *J Power Sources* 319: 139–146.

21. Kim DW, Noh KA, Min HS, et al. (2002) Porous polyacrylonitrile membrane for lithium-ion cells. *Electrochem Solid State Lett* 5: 63–66.

4

Performance evaluation of polyaniline-based redox capacitors with respect to polymerization current density

W.A.D.S.S. Weerasinghe, K.P. Vidanapathirana* and K.S. Perera

Department of Electronics, Faculty of Applied Sciences, Wayamba University of Sri Lanka, Kuliyapitiya, 60200 Sri Lanka

* **Correspondence:** Email: kamalpv41965@gmail.com

Abstract: Supercapacitors (SCs) are promising alternative energy storage devices due to their relatively fast rate of energy storage and delivery. Redox capacitors in the family of SCs are based on conducting polymer (CP) or transition metal oxide electrodes. In this study, symmetric redox capacitors have been fabricated utilizing the CP, polyaniline (PANI) as electrodes and a gel polymer electrolyte (GPE) based on polyvinylidenefluoride (PVdF) as the electrolyte. Investigations have been carried out to study the effect of polymerization current density of PANI electrodes on the performance of redox capacitors. Polymerization current density was varied from 4 mA cm^{-2} to 9 mA cm^{-2} and the performance of redox capacitors was evaluated using electrochemical impedance spectroscopy (EIS), cyclic voltammetry (CV) test and galvanostatic charge discharge (GCD) test. EIS results showed that redox capacitor having electrodes prepared using the current density 7 mA cm^{-2} has the lowest charge transfer resistance and CV test elucidated that the same redox capacitor has maintained around 80% of maximum specific capacity from its initial value after 200 cycles. GCD results exhibited the highest specific discharge capacity of 421.4 F g^{-1}, specific power density of 935.6 W kg^{-1} and specific energy density of 8.0 Wh kg^{-1} after 200 cycles for the same capacitor.

Keywords: polyaniline; redox capacitor; gel polymer electrolyte; cyclic voltammetry; electrochemical impedance spectroscopy

1. Introduction

Supercapacitors (SCs) and batteries are the state-of-the-art electrical energy storage devices. SCs are preferable than batteries when high power density, fast charge-discharge rate and long life cycles are required [1]. Depending on the charge storage mechanism and type of electrodes, SCs can be classified into two categories as electric double layer capacitors (EDLCs) and redox capacitors. EDLCs are made up of carbon-based electrodes whereas redox capacitors are having electrodes based on conducting polymers (CPs) or transition metal oxides [2]. EDLC utilizes mainly the separation of the electronic and ionic charge at the interface between electrode materials and electrolyte solution. Although SCs are having high power density, still they were unable to produce the high energy density requirement especially for the application in electric vehicles. There is a new trend in developing hybrid SCs to find a solution to the low energy density of SCs [3,4]. Different materials such as carbon, oxide, and metal materials, are employed in the negative and positive electrode of hybrid SCs [5,6]. Redox capacitor provides its capacitance from faradic or pseudo redox reactions occurring within the active materials of electrode [7]. CPs usually grow in three-dimensional structures, introducing high porosity and roughness, which generate a large surface area favorable for enhancing electrochemical reactions [8]. They store and release charge through redox processes, resulting in so-called pseudo capacitance. Charging in a CP takes place throughout the bulk volume of the polymer, not just on the surface as is the case with the materials used for EDLCs. With this route the specific capacitance of a redox capacitor increases greatly. The fast charge-discharge processes associated with CPs give rise to high specific power too [9]. Polyaniline (PANI) is one of the most extensively used CPs for engineering redox capacitors [8–10]. PANI has multiple redox states (Leucoemeraldine, Emeraldine and Pernigraniline), high doping-dedoping rates during charging-dischargeing and good environmental stability [10]. Conditions used in the polymerization of PANI have very high contribution to the characteristics of the resulting polymer [8–11]. The surface structure of the electropolymerized film is affected by the conditions used for the electropolymerization such as current density, monomer concentration and types of solvent used [12]. Hence, the conductive property of the film is expected to change due to the morphological difference in the film [13]. In this study, the effect of polymerization current density on the performance of PANI electrodes in a redox capacitor was investigated.

2. Materials and methods

2.1. Preparation of electrodes

The monomer aniline (Aldrich) was distilled and stored under refrigeration prior to use. PANI films were galvanostatically electropolymerized on to commercial grade stainless steel (type 304) electrodes using a three-electrode set up where Ag/AgCl and Pt electrode were served as reference and counter electrodes respectively. The monomer concentration was 0.4 M and 0.5 M concentrated sulfuric acid (H_2SO_4–Aldrich) was used as the oxidizing agent. It has been reported that polymerization of aniline does not take place properly at low polymerization current densities [14]. In this study, polymerization current density was varied from 4 mA cm^{-2} to 9 mA cm^{-2} while maintaining a charge density of 1200 mC cm^{-2}. The area of deposition was 1 cm^2. After polymerization, films were rinsed with distilled water and allowed to dry. The mass of the PANI film was 0.4 mg.

2.2. Preparation of gel polymer electrolyte (GPE)

Polyvinylidene fluoride (PVdF) (Aldrich), ethylene carbonate (EC) (98%, Aldrich), propylene carbonate (PC) (99%, Aldrich) and sodium thiocyanate (NaSCN) (99%, Aldrich) were used as received. EC and PC weight ratio was fixed as 1:1. Firstly, the required amount of PVdF was dissolved in an EC, PC mixture and magnetic stirring was done. Then, NaSCN was added to the mixture and stirring was continued. After that, the mixture was heated at 120 °C for 30 minutes. A very thin gel polymer electrolyte film was obtained from the hot press method and they were left overnight in a vacuum desiccator [15]. The composition of the GPE was 16PVDF/40EC/40PC/3NaSCN in weight basis [16].

2.3. Fabrication of redox capacitors

Sandwich-type symmetric redox capacitors were fabricated using two PANI electrodes. GPE having same the area of the electrode was served the electrolyte. Structure of the redox capacitor was in the form of PANI/PVdF:EC:PC:NaSCN/PANI. Several redox capacitors were fabricated using electrodes prepared with polymerization current densities from 4 to 9 mA cm^{-2}.

2.4. Characterization

2.4.1. Electrochemical impedance spectroscopy (EIS)

EIS measurements of redox capacitors were taken within the frequency range from 0.4 MHz to 0.01 Hz using a frequency response analyzer (Metrohm AUTOLAB 101). Impedance data were analyzed using non–linear least square fitting program [17]. By observing the electrochemical impedance plots or Nyquist plots, variation of the charge transfer resistance (R_{ct}) and bulk electrolyte resistance (R_b) with the polymerization current density of PANI electrode was determined. The high-frequency region plot and the mid frequency region of the Nyquist gives the value of R_b and R_{ct} respectively [18].

Specific capacitance (C_{SC}) of redox capacitors can be calculated using the Nyquist plot as well as the Bode plot (plot of the real part of capacitance (C') vs frequency). When using the Nyquist plot, C_{SC} was calculated using the equation,

$$C_s = 1/(2\pi f z'' m) \tag{1}$$

Here, f is the lowest frequency and Z" is the imaginary part of the impedance at frequency, f and m is the mass of a single electrode. Using the bode, single electrode capacitance C_{SC} was found at the maximum C'. Relaxation time (τ) was determined using the bode plot drawn between the imaginary part of capacitance (C'') vs frequency [19–21] using the equation,

$$\tau_0 = 1/(2\pi f_0) \tag{2}$$

where f_0 corresponds to the frequency at the maximum C''.

2.4.2.   Cyclic voltammetry (CV) test

CV tests were performed for all the redox capacitors having PANI electrodes polymerized using different current densities in the potential window of 0.8 V–(−0.2) V at the scan rate of 5 m Vs^{-1} by means of a computer controlled potentiostat / galvanostat (Metrohm- AUTOLAB 101). One PANI electrode was served as the working electrode and the other as both the counter and the reference electrodes. Using cyclic voltammogram, single electrode (C$_{sc}$) specific capacity was calculated with the equation,

$$C_{SC} = 2 \int I dv / (m \Delta v s)$$ (3)

Here, $\int I dv$ is the integrated area of the cyclic voltammogram, m is the mass of single electrode, Δv is the potential window and s is the scan rate [19–21].

2.4.3.   Galvanostatic charge–discharge (GCD) test

Charge-discharge test of redox capacitors was done in the potential limits of 0.0 V to 0.4 V for 200 cycles. The maximum charge and discharge currents were set to 1.25 mA. Using the charge-discharge curves, specific discharge capacity (C$_d$) of the redox capacitor was calculated using the equation,

$$C_d = \frac{I}{m \left(\frac{dv}{dt} \right)}$$ (4)

where, I is the constant discharge current, dt is the discharge time and dv is the potential drop upon discharging excluding IR drop.

Specific energy density (E$_s$) and specific power density (P$_s$) were calculated using equations

$$E_s = \frac{1}{2} C_d v^2$$ (5)

$$\text{and } P_s = \frac{E_s}{t} \quad [19–20]$$ (6)

here, C$_d$ is the specific discharge capacitance, v is the charging potential and t is the discharging time.

2.4.4.   Morphological study

Scanning electron micrograph (SEM) images were obtained for PANI films prepared on FTO strips at different current densities.

3. Results and discussion

3.1. Polymerization

Changes in the polymerization potential with the time during electro polymerization of the PANI films with changing polymerization current densities are shown in the Figure 1.

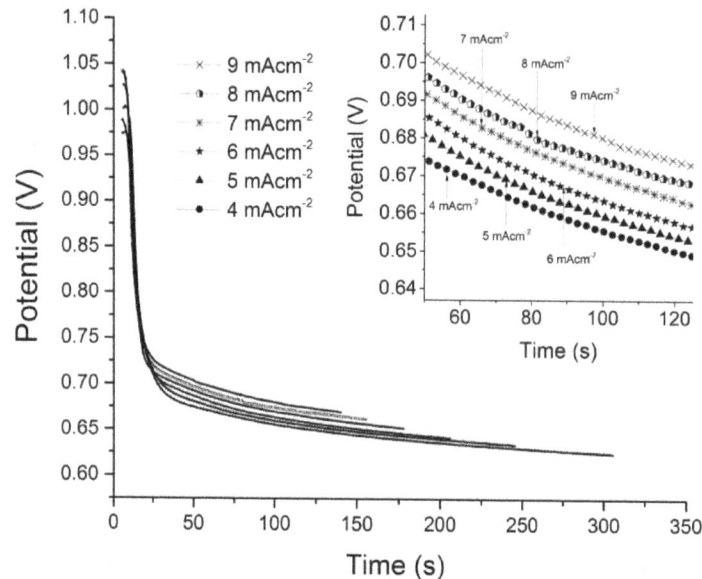

Figure 1. Chronopotentiograms of PANI films polymerized at different polymerization current densities of 4 to 9 mA cm^{-2}.

From Figure 1, it can be seen that the potential of nucleation, where the growth of PANI occurred, is varied with current density. Higher current densities have higher polymerization potentials while lower potentials mark the lower current densities. This may affect the surface structure and morphology of the film. According to Wang, et al., it is expected that roughness of the film increases with the polymerization current density [13]. Also the surface of the film deposited at high current densities seems to be like more "grainy" [13]. Due to this occurrence the electron movement becomes easier in the electrodes deposited at higher polymerization current densities [13].

3.2. Electrochemical impedance spectroscopy

For this type of a system impedance plot should consist of three main regions [22]. They are high frequency region, mid-frequency region and low-frequency region. The characteristics of the electrolyte are shown in the high frequency region by a semi-circle. Mid frequency region represents charge transfer resistance at the electrode/electrolyte interface and the Warburg diffusion. The capacitive properties become dominant at low frequencies and hence, low frequency region in the impedance plot illustrates the capacitive behavior with a spike. If the device shows a proper capacitive behavior, the corresponding spike is parallel to the imaginary axis of the impedance plot. In practical, this does not happen commonly due to several problems such as electrode surface roughness and irregularities of the electrodes. In the resulting impedance plot in Figure 2, the high frequency semi-circle was not present. It may be due to the unavailability of required high-frequency region in the frequency response analyzer used. The steep rising response of impedance measurements in the very low frequency region is an indication for the capacitive behavior of PANI electrodes. The resulting line is not perfectly parallel to Z" as for a typical capacitive behavior. It may be due to the surface roughness as well as non-uniform active layer thickness [23].

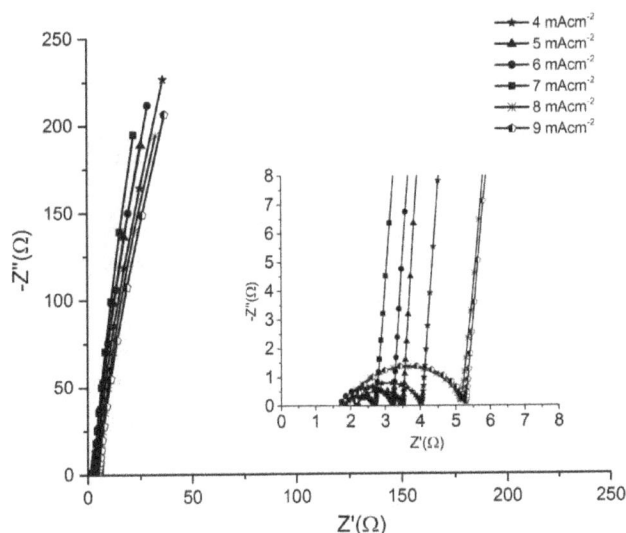

Figure 2. Nyquist plots within the frequency range 0.4 MHz to 0.01 Hz for the PANI/GPE/PANI redox capacitors fabricated using electrodes prepared with different polymerization current densities.

Table 1. Changes in the charge transfer resistance (R_{ct}) of PANI electrodes with the polymerization current density.

Current density/mA·cm^{-2}	R_{ct}/Ω
4	4.0
5	3.5
6	3.2
7	2.7
8	5.7
9	5.9

According to Figure 2 and table 1, it is clear that the R_{ct} has decreased with the increase of the current density up to 7 mA cm^{-2}. After that it has started to increase with the increment of the current density. This contradicts the conclusion of Wang, et al. [13]. According to them, films with higher current densities provide easy pathways to electrons to move. Based on the results obtained in this study, it can be mentioned that up to a certain current density electron movement follows the increase of current density as reported by Wang et al. [13]. However, there is a limiting current density for this behavior and beyond that it would fade.

Figure3 (a) and (b) are the resulting bode plots drawn between the real part of the capacitance

$(C^{'})$ vs frequency and imaginary part of the capacitance $(C^{''})$ with frequency.

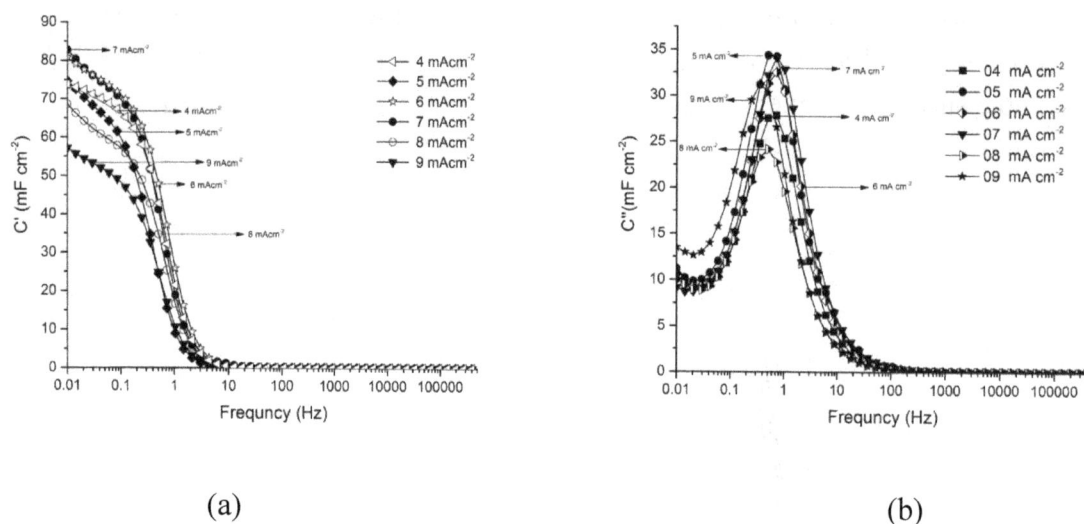

(a)

(b)

Figure 3. (a) and (b) Resulting bode plots of redox capacitors with respect to different current densities used for polymerization of PANI electrodes.

Single electrode capacitance $(C^{'}_{SC})$ values and the calculated relaxation time (τ) with respect to different current densities are given in the table 2.

Table 2. Change in single electrode capacitance and relaxation time with the polymerization current density.

Current density/mA cm^{-2}	$C^{'}$/mF cm^{-2}	τ/s
4	73.8	0.308
5	74.6	0.292
6	80.7	0.218
7	82.7	0.193
8	68.5	0.312
9	57.1	0.447

Low values of τ suggest speedy redox reactions resulting quick ion transfer taking place between electrodes and electrolyte [24].

3.3. Cyclic voltammetry test

Cyclic voltammograms obtained for redox capacitors with PANI electrodes polymerized using different current densities and cycled in the potential window of −0.2 V to 0.8 V are shown in Figure 4.

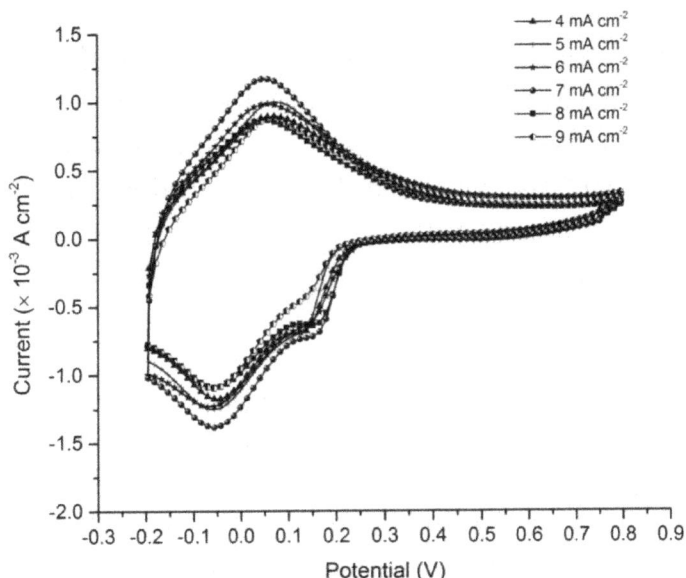

Figure 4. Cyclic voltammograms of PANI/GPE/PANI redox capacitors having PANI electrodes polymerized with different current densities. Scan rate 5 mV s^{-1}.

It shows peaks correspond to both the cathodic and anodic cycles as a result of redox processes that take place in the redox capacitor.

The electrode morphology has a considerable impact on the electrochemical behavior of PANI [6]. This is the reason that the shapes of the cyclic voltammograms for similar PANI electrodes reported in the literature are significantly different. While some exhibit three well-defined redox states [25,26], others show a single state [27,28].

Single electrode specific capacity variation over 200 continuous cycling of redox capacitors having PANI electrodes prepared with different polymerization current densities is given in table 3.

Table 3. Single electrode specific capacity (C_{sc}) of the redox capacitors with respect to polymerization current density of PANI electrodes calculated from the cyclic voltammograms.

Polymerization current density/mA cm^{-2}	C_{sc}/F g^{-1} First cycle	C_{sc}/F g^{-1} After 200 cycles
4	557.9	426.9
5	579.9	445.8
6	620.5	489.0
7	659.2	524.0
8	550.8	400.0
9	532.0	337.1

Results show that specific capacity increases up to a peak value with the increment of polymerization current density and starts to reduce with the further increase of the current density. This may be due to a change in the morphology of the PANI electrode after a certain value of the polymerization current density which affects to the redox reactions that take place between PANI electrode and the GPE. This variation tallies with the results obtained with EIS. Degradation of the capacity was observed with the cycle number. This may be due to lack of full reversibility and less cycling efficiency as a result of side reactions. However, compared to others, redox capacitor with PANI electrodes of 7 mA cm^{-2} current density has maintained around 80% of specific capacity retention from its initial value after 200 cycles.

3.4. Galvanostatic charge-discharge test

Figure 5 shows the continuous charge discharge curves of 7 mA cm^{-2} current density PANI film based redox capacitor and it proves that good stability has been maintained over 200 cycles. Near linear and symmetric charge and discharge curves within the potential window of 0 V and 0.4 V suggest a good capacitive performance with a rapid I–V response [29]. Also, symmetric charge profile suggests a pseudocapacitive characteristic and good capacitive behavior of the electrodes [8].

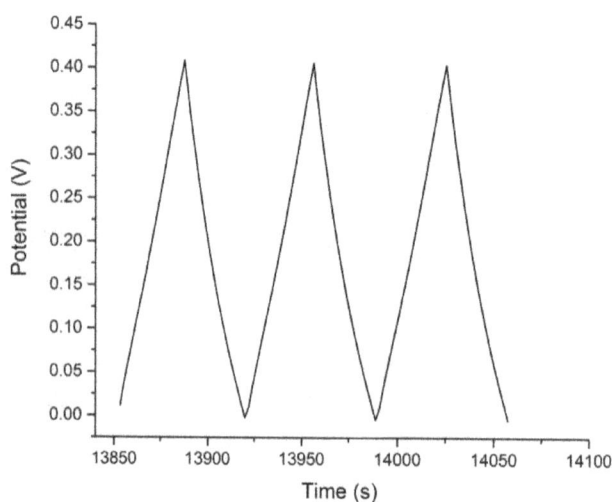

Figure 5. Galvanostatic charge-discharge curves of redox capacitor having PANI electrodes prepared with 7 mA cm^{-2} polymerization current density.

Energy efficiency of the redox capacitors with the change in the polymerization current density of PANI electrodes were calculated using the equation

$$\eta = E_d / E_c \ [30] \tag{7}$$

where E_d is the discharged energy and E_c is charged energy.

Table 4 shows the comparison of the specific discharge capacity, specific energy density, specific power density, and energy efficiency after 200 continuous galvanostatic charge-discharge cycling of redox capacitors having PANI electrodes made with different polymerization current densities. Highest specific discharge capacity of 421.4 F g^{-1}, specific energy density of 8.0 Wh Kg^{-1} and specific power density of 935.6 W kg^{-1} values were recorded for 7 mA cm^{-2} current density

PANI films based redox capacitor. Maximum specific capacity value obtained in this study is higher than the reported values in the range of 80–250 F g^{-1} [10,31,32]. It was observed that when the current density was increased from 4 mA cm^{-2} to 7 mA cm^{-2}, energy efficiency increased by 18.8% and the maximum value of 85.0% was recorded by the capacitor with 7 mA cm^{-2} current density PANI films. With the further increase of current density up to 9 mA cm^{-2} energy efficiency was reduced by 24.8%.

Table 4. Comparison of specific discharge capacity (C_d), specific energy density (E_s), specific power density (P_s), and energy efficiency (η) of redox capacitors with the variation of the polymerization current density of PANI electrodes.

Polymerization current density/mA cm^{-2}	C_d/F g^{-1}	E_s/Wh kg^{-1}	P_s/W kg^{-1}	η%
4	306.0	4.4	699.1	66.2
5	324.4	5.5	775.1	70.0
6	358.8	6.5	840.5	76.3
7	**421.4**	**8.0**	**935.6**	**85.0**
8	281.0	4.0	610.2	63.4
9	270.4	559.3	3.2	60.2

3.5. Morphological studies

Figure 6 shows the SEM images of PANI films fabricated with the polymerization current densities of 4, 7 and 9 mA cm^{-2}.

It can be seen that all samples are having porous structures. Fig 6 (a) and (b) are composed of nanofibers. With the polymerization current density, the quantity of these nanofibers increases, whereas their diameter decreases. This may be due to the fact that aniline is easier to be oxidized into radical cations at higher potential caused by larger current density, resulting in more nuclei, supporting the grow of nanofibers during the next polymerization [33]. Further increase of current density to 9 mA cm^{-2} resulted a film with less porous flake type morphology Figure 6 (c)

(a)   (b)

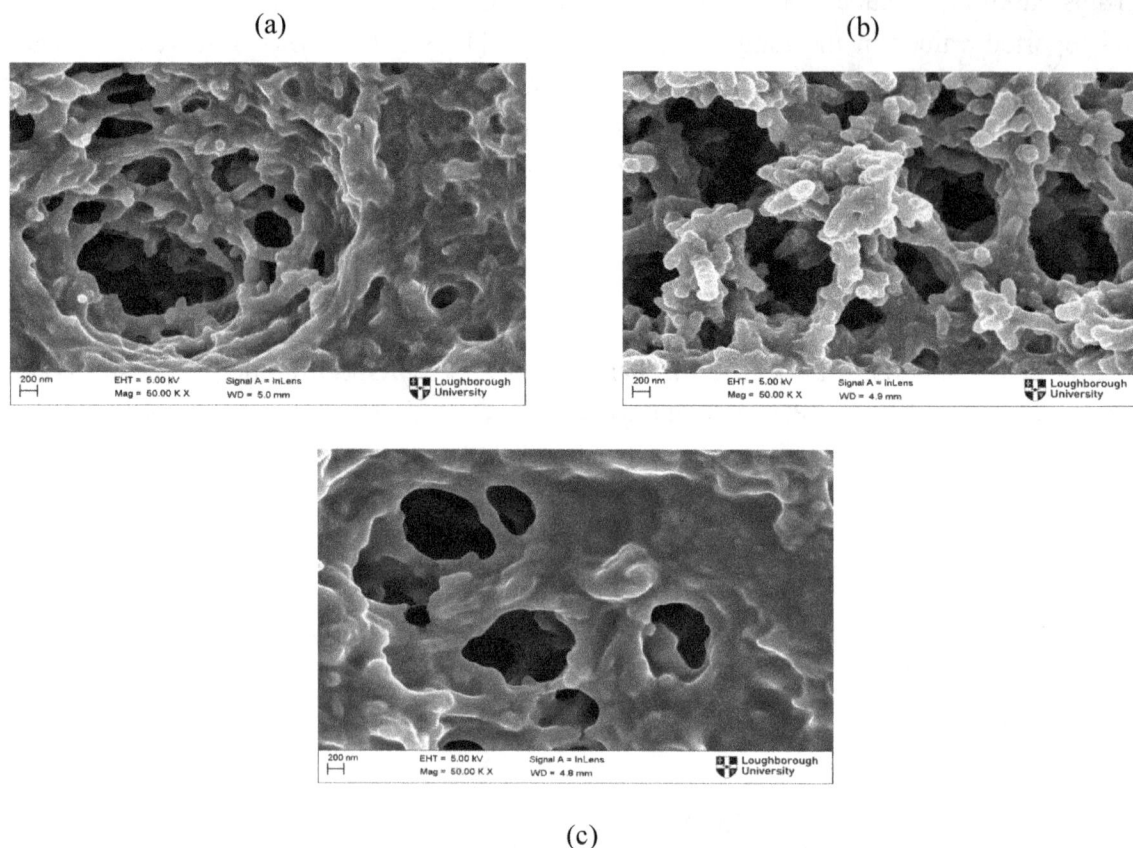

(c)

Figure 6. SEM images of PANI films synthesized at different polymerization current densities: (a) 4 mA cm^{-2}, (b) 7 mA cm^{-2}, (c) 9 mA cm^{-2}.

PANI film prepared with 7 mA cm^{-2} current density has comparably more fibril nature on surface than that of other PANI samples. It is reported that having fibril form morphology is desirable for electrode materials of redox capacitors, as it may provide a large electrochemical surface area that enables the effective and rapid access of electrolyte [34].

Therefore, compared to other current densities, performance enhancement of PANI films prepared with 7 mA cm^{-2} current density can be attributed to this change in morphological, structural and electrochemical properties.

4.   Conclusion

Redox capacitors based on PANI electrodes were fabricated by varying polymerization current density and the effect of the polymerization current density on the performance of the redox capacitors was investigated. Electrochemical impedance spectroscopy showed that redox capacitors have capacitive behavior and low relaxation time which suggests that redox reactions are taking place very quickly. Minimum charge transfer resistance occurs with the redox capacitor fabricated with the electrodes polymerized using a current density of 7 mA cm^{-2}. Cyclic voltammetry studies showed good reversibility of the electrochemical system and 80% of specific charge retention of its initial value was recorded for redox capacitor with 7 mA cm^{-2} polymerization current density. Continuous charge discharge test for 200 cycles also proves that redox capacitor with electrodes polymerized using 7 mA cm^{-2} current density showed the highest performance with specific discharge capacity of 421.4 F g^{-1}, specific energy density of 8.0 Wh kg^{-1}, specific power density of 935.6 W kg^{-1} and energy efficiency of 85.0%. Morphological studies confirmed the results of the other tests. It can be concluded that higher polymerization current densities enhance the performance of the PANI redox capacitor but there exists a peak value beyond that performances fade with the further increment of the current density.

Acknowledgements

Authors wish to acknowledge the assistance given by National Science Foundation, Sri Lanka under the research grant RG/2014/BS/01, RG/2015/EQ/07 and Wayamba University of Sri Lanka (SRHDC/RP/04/16-17(R2)).

Conflict of interest

All authors declare no conflicts of interest in this paper.

References

1. Zhang X, Lin Z, Chen B, et al. (2014) Solid-state flexible polyaniline/silver cellulose nanofibrils aerogel supercapacitors. *J Power Sources* 246: 283–289.

2. Ryu KS, Kim KM, Park NG, et al. (2002) Symmetric redox supercapacitors with conducting polyaniline electrodes. *J Power Sources* 103: 305–309.

3. Lee J, Lee S (2016) Applications of novel Carbon/AlPO4 hybrid-coated $H_2Ti_{12}O_{25}$ as a high-performance anode for cylindrical hybrid supercapacitors. *ACS Appl Matter Interfaces* 8: 28974–28981.

4. Lee B , Lee S (2017) Application of hybrid supercapacitor using granule $Li_4Ti_5O_{12}$ / activated carbon with variation of current density. *J Power Sources* 343: 545–549.

5. Lee J, Kim H, Baek H, et al. (2016) Improved performance of cylindrical hybrid supercapacitor using activated carbon / niobium doped hydrogen titanate. *J Power Sources* 301: 348–354.

6. Lee S, Kim J, Yoon J (2018) Laser scribed Graphene cathode for next generation of high performance hybrid supercapacitors. *Scientific Reports* 8: 8179–8188.

7. Arslan A, Hür E (2012) Supercapacitor applications of polyaniline and poly(N-methylaniline) coated pencil graphite electrode. *Int J Electrochem Sci* 7: 12558–12572.

8. Kulkarni SB, Patil UM, Shackery I, et al. (2014) High-performance supercapacitor electrode based on a polyaniline nanofibers / 3D graphene framework as an efficient charge transporter. *J Mater Chem A* 2: 4989–4998.

9. Liu Q, Nayfeh MH, Yau ST (2010) Supercapacitor electrodes based on polyaniline–silicon nanoparticle composite. *J Power Sources* 195: 3956–3959.

10. Eftekhari A, Li L, Yang Y (2017) Polyaniline super capacitors. *J. Power Sources* 347: 86–107.

11. Patil DS, Pawar SA, Devan RS, et al. (2014) Polyaniline based electrodes for electrochemical supercapacitor: Synergistic effect of silver, activated carbon and polyaniline. *J Electroanal Chem* 724: 21–28.

12. Bhadra J, Al-Thani NJ, Madi NK, et al. (2017) Effects of aniline concentrationson the electrical and mechanical properties of polyaniline polyvinyl alcohol blends. *Arab J Chem* 10: 664–672.

13. Wang G, Hu X, Wong TKS (2001) Effect of deposition current density on the effectiveness of electropolymerized hoe-transport layer in organic electroluminescent device. *Appl Surface Sci* 174: 185–190.

14. Eftekhari A, Jafarkhani P (2014) Galvanodynamic synthesis of polyaniline: a flexible method for the deposition of electroactive materials. *J Electroanal Chem* 717–718: 110–118.

15. Jayathilake YMCD, Perera KS, Vidanapathirana KP (2015) Preparation and characterization of a polyacrylonitrile-based gel polymer electrolyte complexed with 1 methyl-3 propyl immidazolium iodide. *J Solid State Electrochem* 19: 2199–2204.

16. Bandaranayake CM, Jayathilake YMCD, Vidanapathirana KP, et al. (2015) Performance of a sodium thiocyanate based gel polymer electrolyte in redox capacitors. *Sabaragamuwa University J* 14: 149–161.

17. Kumar GG, Munichandraiah N (2000) A gel polymer electrolyte of magnesium triflate. *Solid State Ionics* 128: 203–210.

18. Prasad KR, Munichandraiah N (2002) Potentiodynamically deposited polyaniline on stainless steel inexpensive, high-performance electrodes for electrochemical supercapacitors. *J Electrochem Soc* 149: A1393–A1399.

19. Tey JP, Careem MA, Yarmo MA, et al. (2016) Durian shell-based activated carbon electrode for EDLCs. *Ionics* 22: 1209–1217.

20. Wang W, Guo S, Penchev M, et al. (2013) Three dimensional few layer graphene and carbon nanotube foam architectures for high fidelity supercapacitors. *Nano Energy* 2: 294–303.

21. Harankahawa N, Weerasinghe S, Vidanapathirana K, et al. (2017) Investigation of a pseudo capacitor with polyacrylonitrile based gel polymer electrolyte. *J Electrochem Sci Technol* 8: 107–114.

22. Eftekhari A (2018) The mechanism of ultrafast supercapacitors. *J Mater Chem A* 6: 2866–2877.

23. Prabaharan SRS, Vimala R, Zainal Z (2006) Nanostructured mesoporous carbon as electrodes for supercapacitors. *J Power Sources* 161: 730–736.

24. Ramya R, Sivasubramanian R, Sangaranarayanan MV (2013) Conducting polymers-based electrochemical supercapacitors—progress and prospects. *Electrochim Acta* 101: 109–129.

25. Bandyopadhya P, Kuila T, Balamurugan J, et al. (2017) Facile synthesis of novel sulfonated polyaniline functionalized graphene using m-aminobenzene sulfonic acid for asymmetric supercapacitor application. *Chem Eng J* 308: 1174–1184.

26. Yu T, Zhu P, Xiong Y, et al. (2016) Synthesis of microspherical polyaniline/graphene composites and their application in supercapacitors. *Electrochim Acta* 222: 12–19.

27. Xu J, Ding J, Zhou X, et al. (2017) Enhanced rate performance of flexible and stretchable linear supercapacitors based on polyaniline @ Au @ carbon nanotube with ultrafast axial electron transport. *J Power Sources* 340: 302–308.

28. Chang W, Wang C, Chen C (2016) Plasma-Induced Polyaniline Grafted on Carbon Nanotube-embedded Carbon Nanofibers for High-Performance Supercapacitors. *Electrochim Acta* 212: 30–140.

29. Molapo KM, Ndangili P, MAjayi RF, et al. (2012) Electronics of Conjugated Polymers (I): Polyaniline. *Int J Electrochem Sci* 7: 11859–11875.

30. Laheäär A, Przygocki P, Abbas Q, et al. (2015) Appropriate methods for evaluating the efficiency and capacitive behavior of different types of supercapacitors. *Electrochem Com* 60: 21–25.

31. Prasad KR, Munichandraiah N (2002) Electrochemical Studies of Polyaniline in a Gel Polymer Electrolyte, High Energy and High Power Characteristics of a Solid-State Redox Supercapacitor. *Electrochem Solid-State Lett* 5: A271–A274.

32. Ryu KS, Kim KM, Park YJ, et al. (2002) Redox supercapacitor using polyaniline doped with Li salt as electrode. *Solid State Ionics* 152–153: 861–866.

33. Du X, Xu Y, Xiong L, et al. (2014) Polyaniline with high crystallinity degree: Synthesis, structure, and electrochemical properties. *J Appl Polym Sci* 131: 6–13.

34. Deshmukh PR, Shinde NM, Patil SV, et al. (2013) Supercapacitive behavior of polyaniline thin films deposited on fluorine doped tin oxide (FTO) substrates by microwave-assisted chemical route. *Chem Eng J* 223: 572–577.

In-situ investigation of water distribution in polymer electrolyte membrane fuel cells using high-resolution neutron tomography with 6.5 μm pixel size

Saad S. Alrwashdeh[1,*], Falah M. Alsaraireh[1], Mohammad A. Saraireh[1], Henning Markötter[2], Nikolay Kardjilov[2], Merle Klages[3], Joachim Scholta[3] and Ingo Manke[2]

[1] Mechanical Engineering Department, Faculty of Engineering, Mutah University, P.O Box 7, Al-Karak 61710, Jordan

[2] Helmholtz-Zentrum Berlin, Hahn-Meitner-Platz 1, 14109 Berlin, Germany

[3] Zentrum für Sonnenenergie- und Wasserstoff-Forschung Baden Württemberg (ZSW), Helmholtzstraße 8, 89081 Ulm, Germany

* Correspondence: Email: saad_r1988@yahoo.com

Abstract: In this feasibility study, high-resolution neutron tomography is used to investigate the water distribution in polymer electrolyte membrane fuel cells (PEMFCs). Two PEMFCs were built up with two different gas diffusion layers (GDLs) namely Sigracet® SGL-25BC and Freudenberg H14C10, respectively. High-resolution neutron tomography has the ability to display the water distribution in the flow field channels and the GDLs, with very high accuracy. Here, we found that the water distribution in the cell equipped with the Freudenberg H14C10 material was much more homogenous compared to the cell with the SGL-25BC material.

Keywords: polymer electrolyte membrane fuel cell; water transport; high resolution neutron tomography; gas diffusion layers

---

1. Introduction

The Fuel cell technology plays a major role in offering a pollution free alternative energy supply in contrast to burning fossil fuels [1–6]. Fuel cells are an applicable technology for mobile and stationary applications. A common example of mobile systems is the automotive sector, where polymer electrolyte membrane fuel cells (PEMFC) are considered the most favorable fuel cell type due to their high output power density and flexible operating conditions [7,8]. They are also suitable to provide combined heat and electric power in stationary energy systems to maximize efficiency [9,10].

Especially under critical operating conditions where accumulated product water causes flooding an optimization of the water transport in the PEMFC leads to an improved efficiency. Such conditions include temperatures below 60 °C as well as high current densities that both lead to increased water agglomerations [11–13].

There are many different materials used in PEMFCs, which affect the water transport in the cells. Through the material properties the interaction with liquid water during operation influences the cell performance. Optimization of the water transport in the fuel cell layers leads to enhance the efficiency especially under critical operation conditions that promote flooding. Such as temperatures below 60 °C as well as high currents. A more detailed understanding of the influence of material types was obtained by analyzing several materials [14–17]. So far, X-Ray based imaging methods have been applied to study the liquid water evolution and distribution also during cell operation [18–26].

Neutron imaging can be used as complementary method for fuel cell research because neutrons are strongly affected by hydrogen, imaging methods based on neutrons are very helpful to study hydrogen distributions within the cell materials [23–27]. For the same reason, neutron imaging is used to study the water distribution in operating fuel cells [14, 28–33].

In this study, High-resolution neutron tomography has been used for the investigation of the water distribution in PEMFCs. The detector system was set to a pixel size of ~6.5 µm, which is enough to resolve even smaller water droplets in the GDL of the cells in sufficient detail. Even the catalyst can be observed, which is hardly possible with X-Rays due to the high absorbing platinum [11].

2. Method and experiments

The neutron imaging beam line CONRAD II (COld Neutron RADiography) is used in order to investigate material structure as well as dynamic processes of the water transport in the cell [27–30]. It allows to create radiographs using cold neutrons, which are product of the 10 MW research reactor BER II and moderated in a cold neutron source containing liquid hydrogen. Subsequently they are transported through a neutron guide to the experiment. The neutron beam at CONRAD is polychromatic mainly with wavelengths between 2 and 6 Å and a maximum intensity at about 3.0 Å. The tomography measurements of the fuel cells discussed here are performed 5 m from the end of the neutron guide as shown in Figure 1 [31].

Behind the sample as close as possible a detector system consisting of a scintillator, mirror, lens and CCD camera is placed [31–33]. When neutrons, which are transmitted and scattered by the sample, hit the scintillator, photons in the visible spectrum are emitted. The scintillator used for the tomography measurements is a gadox screen ($Gd_2 O_2 S(Tb)$) with 20 µm thickness.

The photons are projected onto the camera by a mirror/lens combination. The used Andor DW436 camera contains a 16 bit chip with (2048 × 2048) pixels, each corresponding to a size of 6.5 µm. The CCD sensor is continuously cooled to –50 °C to ensure thermal noise being as low as possible. With the used optics an imaging field of view of (13×13) mm² is chosen. Each projection is acquired with an exposure time of 15 s then the projections collected for each tomographic scan.

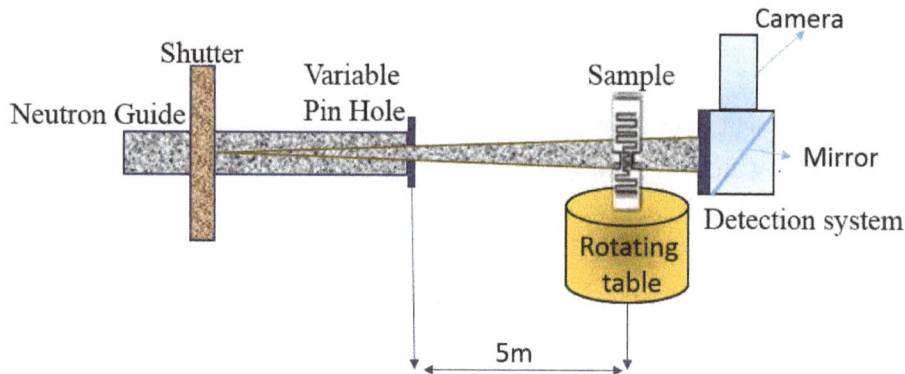

Figure 1. Schematic drawing of the imaging setup at the CONRAD instrument.

2.1 Fuel cell setup

In both cells the regulation of the cell temperature is carried out by using a cooling circuit with deuterium oxide (D_2O). Compared to hydrogen, the attenuation coefficient of deuterium is much smaller [34–36]. As a result, D_2O hardly attenuates the neutron beam and can only be seen faintly in radiographs and tomography. The flow field of the cooling circuit is embedded in the backside of the bipolar plates and is connected with a secondary water coolant circuit via a heat exchanger to maintain a stable temperature distribution.

The two tomography cells have an active area of 5.4 cm². One contains Sigracet® SGL 25BC GDLs and the other Freudenberg H14C10 GDLs.

Each cell is equipped with three flow field channels on anode and cathode side, respectively. 0.5 mm wide meander-shaped channels are milled into the graphite composite material, through which the supply gases, hydrogen on the anode and air on the cathode side, flow from top to bottom. More details of the fuel cell design can be found in previous works [37]. The presented measurements are conducted at a current density of 1 A.cm⁻² and a cell temperature of 65 °C, while the cell is supplied with gases of 120% relative humidity. Via gas flow controllers λ, the ratio of the supplied and consumed gases, is set to 5 for anode (H_2) as well as cathode (air) side.

3.   Results

Figure 2A–C shows tomographic slices of the fuel cell equipped with Sigracet® SGL-25BC material and Figure 2D–F the cell with Freudenberg-H14C10 material. These are parallel to the active layer and positioned in the anode flow field (A and D), in the membrane electrode assembly MEA (B and E) and in the cathode flow field (C and F). It can be seen in the images of the cell with Sigracet® SGL-25BC material that the water accumulates in the corner of the anode channel turnings (see Figure 2A), while it blocks one channel on cathode side with a big droplet due to the high condensation because of the material type. In the MEA, basically in the two gas diffusion layers, water is appearing as small droplets, which are marked in figure 2B with white arrows. The water in the cell with the Freudenberg H14C10 material is distributed all over the cell channels in smaller droplets which aid in a homogeneous distribution of the water inside the cell and enhance the membrane humidity level (Figure 2D) while it is focused on the corners and edges of the cathode channels which will block the channel and reduce the level of the gases supply for a sufficient reaction finally that will aid in a reduction in the cell performance (Figure 2F). The water in Freudenbergs H14C10 appears more in big droplets compared to the Sigracet® SGL-25BC material as in the Figure 2E White arrows. Please note that by using tomography, there is the ability to observe the water in the gas diffusion layers separately from the channels, which is not possible by using radiography.

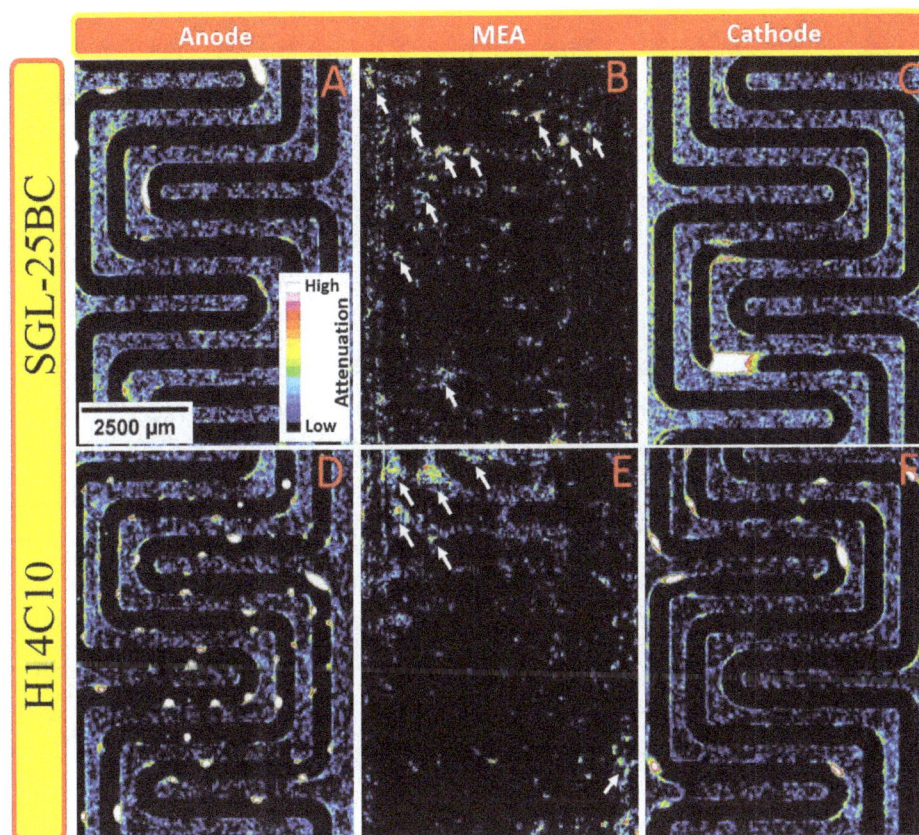

Figure 2. Tomographic slices through the fuel cells with Sigracet® SGL-25BC and Freudenberg H14C10 GDL materials at the position of the anode (A, D), the MEA (B, E) and the Cathode (C, F).

The graphs in Figure 3 show the water volume in both cells. It can be noted that the water volume contained in the Sigracet® SGL 25BC material is less compared to the cell with Freudenberg H1410 material, especially in the region of the MEA. The water in the Freudenberg H1410 material distributed all over the cell in the anode and cathode sides comparing to the water distribution in Sigracet® SGL 25BC material also more water can be found in the MEA region of Freudenberg H1410 material which can play a major role in humidify the membrane and enhance the cell performance comparing to cell containing the Sigracet® SGL 25BC material. Note that the water blocked a complete channel in the cathode side of the cell containing Sigracet® SGL 25BC also concentrated in the corner of the channels.

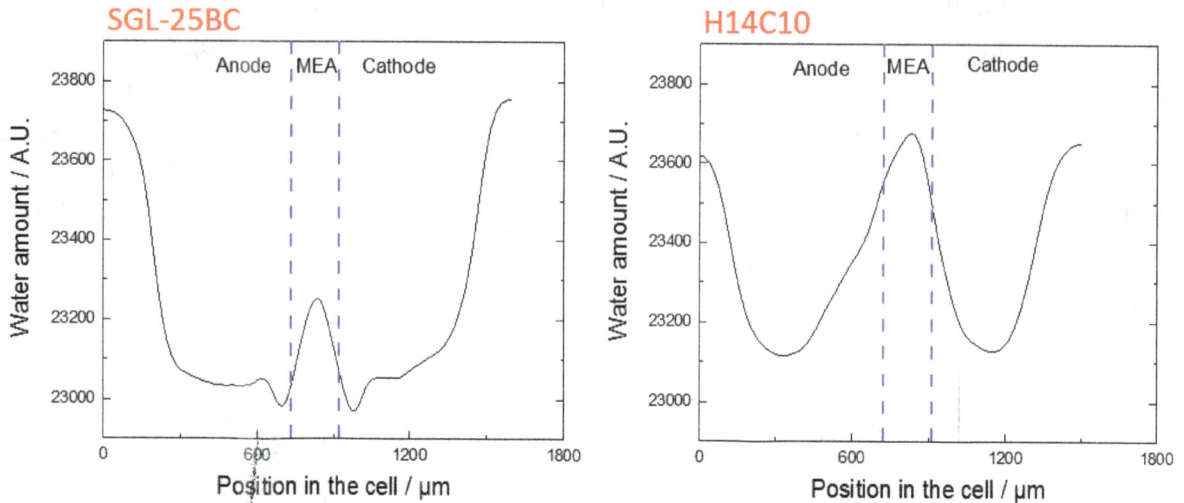

Figure 3. water volume of the fuel cells with Sigracet® SGL 25BC and Freudenberg H1410.

4. Discussion

Despite the same operating parameters more water was found in the MEA of the cell containing the Freudenberg H14C10 material than in the MEA of the cell with Sigracet® SGL 25BC material (see Figure 3). This water is also more homogenously distributed. No larger water agglomerations were found that may yield as bottle necks either affecting the gas stream (e.g., flooding). As there were no bigger areas found without any water, which might affect the membrane humidification (dehydration). In contrast, in the MEA of the cell with Sigracet® SGL 25BC less water is found. It can be assumed that this may reduce humidification of the membrane in the top area and cause a dehydration effect. On the other hand, the larger overall water amount in the MEA of the cell containing Freudenberg H1410 material can also improve membrane humidity and thus ion conductivity at most locations in the cell.

5. Conclusions

Liquid water inside two cells with a specially adapted PEMFC design with a three-channel flow field system containing two different GDLs were studied. High resolution neutron tomography has the ability to detect the water in the MEA even inside the catalyst of the PEMFCs. The technique is a powerful tool to detect liquid water inside the components of PEMFCs. The following points can be concluded in this study:

(1) The spatially resolved water distribution in the flow fields, GDLs and MEA can be detected and studied using high-resolution neutron tomography.

(2) In the presented example we found an overall higher amount of liquid water in the cell containing Freudenbergs H14C10 GDL compared to SGLs 25BC GDL material.

This study provides insights in to the distribution of liquid water in complete operated fuel cells and is not hindered at the catalyst layer, which is still an obstacle when using X-Ray CT. High resolution neutron tomography may contribute to future simulation studies about the water distribution and transport processes in MEAs [38].

Conflict of interest

All authors declare no conflicts of interest in this paper.

References

1. Hoogers G (2003) Fuel Cell Technology Handbook. Boca Raton, FL: CRC Press LLC.
2. Vielstich W, Lamm A, Gasteiger HA (2003) Handbook of fuel cells–fundamentals, technology and applications. Chichester: John Wiley & Sons.
3. Alrwashdeh SS, Manke I, Markötter H, et al. (2017) In operando quantification of three-dimensional water distribution in nanoporous carbon-based layers in polymer electrolyte membrane fuel cells. *ACS Nano* 11: 5944–5949.
4. Alrwashdeh SS, Manke I, Markötter H, et al. (2017) Neutron radiographic in operando investigation of water transport in polymer electrolyte membrane fuel cells with channel barriers. *Energ Convers Manage* 148: 604–610.
5. Alrwashdeh SS, Manke I, Markötter H, et al. (2017) Improved performance of polymer electrolyte membrane fuel cells with modified microporous layer structures. *Energy Technol* 5: 1612–1618.
6. Saad SA, Henning M, Haußmann J, et al. (2017) Investigation of water transport in newly developed micro porous layers for polymer electrolyte membrane fuel cells. *Appl Microscopy* 47: 101–104.
7. Mehta V, Cooper JS (2003) Review and analysis of PEM fuel cell design and manufacturing. *J Power Sources* 114: 32–53.
8. Carrette L, Friedrich KA, Stimming U (2000) Fuel cells: principles, types, fuels, and applications: WILEY-VCH Verlag GmbH, Weinheim.
9. Barelli L, Bidini G, Gallorini F, et al. (2012) Dynamic analysis of PEMFC-based CHP systems for domestic application. *Appl Energ* 91: 13–28.
10. Gigliucci G, Petruzzi L, Cerelli E, et al. (2004) Demonstration of a residential CHP system based on PEM fuel cells. *J Power Sources* 131: 62–68.
11. Krüger P, Markötter H, Haußmann J, et al. (2011) Synchrotron X-ray tomography for investigations of water distribution in polymer electrolyte membrane fuel cells. *J Power Sources* 196: 5250–5255.
12. Manke I, Hartnig C, Kardjilov N, et al. (2009) In-situ investigation of the water distribution in PEM fuel cells by neutron radiography and tomography. *Mater Test* 51: 219–226.
13. Alrwashdeh SS, Markötter H, Haußmann J, et al. (2016) Investigation of water transport dynamics in polymer electrolyte membrane fuel cells based on high porous micro porous layers. *Energy* 102: 161–165.
14. Cindrella L, Kannan AM (2009) Membrane electrode assembly with doped polyaniline interlayer for proton exchange membrane fuel cells under low relative humidity conditions. *J Power Sources* 193: 447–453.
15. Cindrella L, Kannan AM, Ahmad R, et al. (2009) Surface modification of gas diffusion layers by inorganic nanomaterials for performance enhancement of proton exchange membrane fuel cells at low RH conditions. *Int J Hydrogen Energ* 34: 6377–6383.

16. Mohanraju K, Sreejith V, Ananth R, et al. (2015) Enhanced electrocatalytic activity of PANI and CoFe2O4/PANI composite supported on graphene for fuel cell applications. *J Power Sources* 284: 383–391.

17. Cindrella L, Kannan AM, Lin JF, et al. (2009) Gas diffusion layer for proton exchange membrane fuel cells—A review. *J Power Sources* 194: 146–160.

18. Chevalier S, Ge N, George MG, et al. (2017) Synchrotron X-ray radiography as a highly precise and accurate method for measuring the spatial distribution of liquid water in operating polymer electrolyte membrane fuel cells. *J Electrochem Soc* 164: F107–F114.

19. Ge N, Chevalier S, Lee J, et al. (2017) Non-isothermal two-phase transport in a polymer electrolyte membrane fuel cell with crack-free microporous layers. *Int J Heat Mass Transfer* 107: 418–431.

20. Antonacci P, Chevalier S, Lee J, et al. (2015) Feasibility of combining electrochemical impedance spectroscopy and synchrotron X-ray radiography for determining the influence of liquid water on polymer electrolyte membrane fuel cell performance. *Int J Hydrogen Energ* 40: 16494–1502.

21. Lee J, Hinebaugh J, Bazylak A (2013) Synchrotron X-ray radiographic investigations of liquid water transport behavior in a PEMFC with MPL-coated GDLs. *J Power Sources* 227: 123–130.

22. Arlt T, Grothausmann R, Manke I, et al. (2013) Tomographic methods for fuel cell research. *Mater Test* 55: 207–213.

23. Eberhardt SH, Marone F, Stampanoni M, et al. (2016) Operando X-ray tomographic microscopy imaging of HT-PEFC: A comparative study of phosphoric acid electrolyte migration. *J Electrochem Soc* 163: F842–F847.

24. Arlt T, Klages M, Messerschmidt M, et al. (2017) Influence of artificially aged gas diffusion layers on the water management of polymer electrolyte membrane fuel cells analyzed with in-operando synchrotron imaging. *Energy* 118: 502–511.

25. Chevalier S, Ge N, Lee J, et al. (2017) Novel electrospun gas diffusion layers for polymer electrolyte membrane fuel cells: Part II. In operando synchrotron imaging for microscale liquid water transport characterization. *J Power Sources* 352: 281–290.

26. Matsui H, Ishiguro N, Uruga T, et al. (2017) Operando 3D visualization of migration and degradation of a platinum cathode catalyst in a polymer electrolyte fuel cell. *Angew Chem Int Ed Engl* 56: 9371–9375.

27. Kardjilov N, Hilger A, Manke I, et al. (2011) Neutron tomography instrument CONRAD at HZB. *Nucl Instrum Meth A* 651: 47–52.

28. Kardjilov N, Manke I, Hilger A, et al. (2011) Neutron imaging in materials science. *Mater Today* 14: 248–256.

29. Kardjilov N, Hilger A, Manke I, et al. (2015) Imaging with cold neutrons at the CONRAD-2 Facility. In: Lehmann EH, Kaestner AP, Mannes D, editors, Proceedings of the 10th World Conference on Neutron Radiography. Amsterdam: *Elsevier Science Bv*, 60–66.

30. Kardjilov N, Hilger A, Manke I, et al. (2016) CONRAD-2: the new neutron imaging instrument at the Helmholtz-Zentrum Berlin. *J Appl Crystallogr* 49: 195–202.

31. Williams SH, Hilger A, Kardjilov N, et al. (2012) Detection system for microimaging with neutrons. *J Instrum* 7: 1–25.

32. Kardjilov N, Dawson M, Hilger A, et al. (2011) A highly adaptive detector system for high resolution neutron imaging. *Nucl Instrum Meth A* 651: 95–99.

33. Totzke C, Manke I, Hilger A, et al. (2011) Large area high resolution neutron imaging detector for fuel cell research. *J Power Sources* 196: 4631–4637.

34. Banhart J (2008) Advanced tomographic methods in materials research and engineering. Oxford, UK: Oxford University, Press.

35. Manke I, Hartnig C, Kardjilov N, et al. (2008) Characterization of water exchange and two-phase flow in porous gas diffusion materials by hydrogen-deuterium contrast neutron radiography. *Appl Phys Lett* 92: 337–347.

36. Cho KT, Mench MM (2012) Investigation of the role of the micro-porous layer in polymer electrolyte fuel cells with hydrogen deuterium contrast neutron radiography. *Phys Chem Chem Phys* 14: 4296–4302.

37. Haussmann J, Markotter H, Alink R, et al. (2013) Synchrotron radiography and tomography of water transport in perforated gas diffusion media. *J Power Sources* 239: 611–622.

38. Weber AZ, Borup RL, Darling RM, et al. (2014) A critical review of modeling transport phenomena in polymer-electrolyte fuel cells. *J Electrochem Soc* 161: F1254–F1299.

6

Potentiality of biomethane production from slaughtered rumen digesta for reduction of environmental pollution

Muhammad Rashed Al Mamun[1,*], Anika Tasnim[2], Shahidul Bashar[2] and Md. Jasim Uddin[3]

[1] Department of Farm Power and Machinery, Faculty of Agricultural Engineering and Technology, Sylhet Agricultural University, Sylhet, Bangladesh
[2] Department of Farm Power and Machinery, Sylhet Agricultural University, Sylhet, Bangladesh
[3] Department of Animal Nutrition, Faculty of Veterinary, Animal and Biomedical Sciences, Sylhet Agricultural University, Sylhet, Bangladesh

* **Correspondence:** Email: rashedshahi@gmail.com

Abstract: Energy consumes by the world comes from fossil fuel which is non-renewable and pollute environment. Therefore, renewable energy is a possible solution for replacement of fossil fuel. Recycling rumen digesta can be considered as renewable energy source. Anaerobic digestion is one of the best processes for rumen digesta management which will lead to production of biogas, reduction in GHGs emissions and reduce environmental pollution. The purpose of this study was to assess number of slaughterhouse, amount of slaughtered animal, availability of rumen digesta in Sylhet City Corporation and generation of biogas from rumen digesta. The study was conducted in different locations of Sylhet City Corporation The experiment for biogas generation was carried out in 3300 ml digester under mesophilic condition. The mixing ratio of animal digesta and water was 1:1. Hydraulic retention time of the experiment was 40 days. The produced biogas was measured by water displacement method. The result showed that co-digestion of rumen digesta of chicken, cow, goat increased production of biogas. The maximum biogas generation from rumen digesta of chicken, cow, goat and co-digestion of three substrates were 27.2, 3, 39, 7.5 ml/day at the 12th, 28th, 9th, 1st day respectively. The average production of biogas from chicken, cow, goat and co-digestion of rumen digesta were 9.865, 1.32, 6.89, 6.35 ml/day respectively. The average methane production was 58.69%, 58.77%, 57.39% and 56.93% in biogas from chicken, cow, goat and mixed rumen digesta respectively. The study suggests making digester in every slaughterhouse for recycling rumen digesta and produce biogas which can recover future energy crisis and reduce environmental pollution.

Keywords: biogas; renewable energy; anaerobic digestion; slaughterhouse; rumen digesta

---

1. Introduction

Energy is considered as one of the key component for all development activities of human civilization. The source of energy is mainly classified into two categories i.e., renewable energy source and non-renewable energy source. Most of the energy used by world comes from non-renewable sources such as fossil fuel (coal, gas, uranium). But, the scarcity of this energy will be faced by world in the next few decades due to rapid growth of population and uncontrolled urbanization [1]. The sources of fossil fuel are being used in an alarming rate and these resources coming to an end. According to World Coal Association (WCA) and British Petroleum coal and natural gas will run out within 150 years and 52 years respectively. Due to the shortage of fuel, the cost of fossil fuel also increases with time. Also, uses of fossil fuel create some environmental problems. Moreover, greenhouse effect and global warming increase by releasing carbon dioxide when it burns. In this regard, worldwide energy crisis directed the attention to the renewable sources of energy replacement of fossil fuel. Renewable energy sources are an effective and efficient solution to mitigate this problem. Renewable energy source are those sources which are naturally replenished on human timescale such as biomass, wind, solar, hydro, wave. Among the sources rumen digesta as a biomass may be useful sources to produce renewable energy. Rumen digesta is animal waste which store into reticulo rumen, large chamber in where ingested feed is first subjected to microbial digestion. In 2010, the population of ruminant in world was about 3.6 billion, of which 5.38%, 39.59%, 25.19%, and 29.84% were for buffaloes, cattle, goats and sheep, respectively [2]. Although rumen digesta is valuable but it is treated as waste and frequently disposed into open environment and drain. These Disposal system of rumen waste may create environmental pollution, result in health hazard to human due to the presence of millions microorganisms. It also causes water pollution by entering into rivers, streams and other local sources. Moreover, it also responsible for greenhouse effect because of conversion into methane and carbon dioxide. According to Kyoto protocol exploited methane is 23 times more potential greenhouse gas than carbon dioxide [3,4]. Pathogenic organism is also found in slaughterhouse waste including *Clostridium perfringens, vibrio sp.,* [5]. Environmental pollution can be prevented by recycling of rumen digesta. One of the key method to recycle rumen digesta is to produce biogas as a renewable energy source. Due to the present of microorganism (cellulolytic and methanogenic bacteria) in rumen the bio-fertilizer and feeder can also be produced by recycling rumen digesta. The process of microbial fermentation in rumen and fermentation in digester to produce biogas are similar [6]. Biogas is a combustible gas produced by the anaerobic digestion (AD) or fermentation of organic matter. The organic matter can be manure, sewage sludge, municipal solid waste, biodegradable waste or any other biodegradable feedstock. Biogas is mainly methane, carbon dioxide and trace amount of H_2S, CO, NO. Among of the gaseous element, methane is the only combustible.

Anaerobic digestion is a series of biological process in which microorganism break down biodegradable material in the absence of oxygen. Digesters are kept at three different temperatures. They are ambient temperature (less than 25 °C), mesophilic (25–35 °C) or thermophilic (45–60 °C) conditions. It is a step by step process where the organic carbon is mainly converted to carbon dioxide and methane [7]. The process of producing biogas divided into four step. A flow diagram of biogas production is shown in Figure 1.

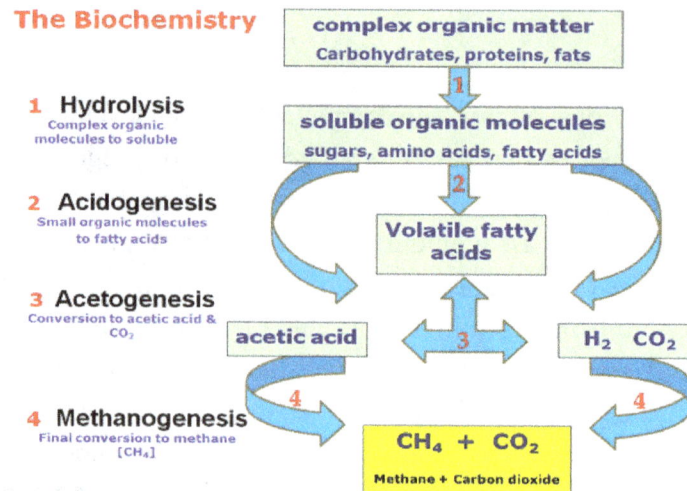

Figure 1. Flow chart of anaerobic digestion [8].

1.1. Hydrolysis

Polymer cannot directly utilize by fermentative micro-organism. In this step, bacteria break down the complex carbohydrates, protein, lipid into their monomer components (amino acids, sugar, and long chain fatty acid). Polymers are break down into soluble monomers through hydrolysis step as shown in Eq 1:

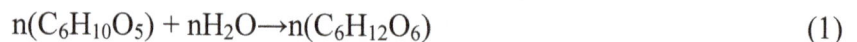

$$n(C_6H_{10}O_5) + nH_2O \rightarrow n(C_6H_{12}O_6) \tag{1}$$

1.2. Acidogenesis

In the second step, acidogenic bacteria decompose the amino acids and sugars into volatile fatty acids (VFAs) and alcohols. A simple substrate such as glucose can be fermented; different products are produced by the diverse bacterial community. Reaction: Eqs 2, 3 and 4 show the conversion process of glucose to acetate, ethanol and propionate (Angelidaki et al., 2007), respectively.

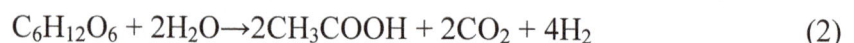

$$C_6H_{12}O_6 + 2H_2O \rightarrow 2CH_3COOH + 2CO_2 + 4H_2 \tag{2}$$

$$C_6H_{12}O_6 \rightarrow 2CH_3CH_2OH + 2CO_2 \tag{3}$$

$$C_6H_{12}O_6 + 2H_2 \rightarrow 2CH_3CH_2COOH + 2H_2O \tag{4}$$

1.3. Acetogenesis

In this step acetogenic bacteria convert the VFAs into acetic acid, carbon dioxide, hydrogen. The conversion process is shown in equation:

$$C_6H_{12}O_6 \rightarrow 2C_2H_5OH + 2C \tag{5}$$

1.4. Methanogenesis

In which the methanogenic bacteria convert acetic acids to methane and CO_2. According to the type of substrate utilized methanogenesis divided into two groups:

(1) Hydrogenotrophic methanogenesis: Hydrogen and carbon dioxide break down into methane according to the following reaction:

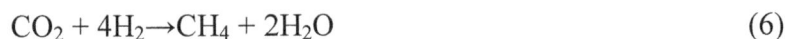

$$CO_2 + 4H_2 \rightarrow CH_4 + 2H_2O \qquad (6)$$

(2) Acetotrophic or aceticlastic methanogenesis: Methane is formed from the conversion of acetate through the following reaction:

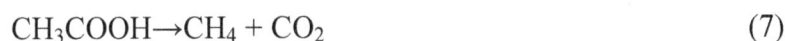

$$CH_3COOH \rightarrow CH_4 + CO_2 \qquad (7)$$

There are many studies available on biogas production from cow dung, solid waste. A study focused on waste reduction that comes from slaughterhouse and also investigated the possibility of biogas production from rumen digesta [9]. The ability of biogas production from slaughterhouse waste i.e., rumen fluid, rumen digsta, cow dung [10]. The initial phase of decomposition compare with methane formation phase can be reduced by using rumen fluid [11]. By using cow manure with rumen fluid and water for increasing production of biogas. This study showed the influence of manure, rumen and water in biogas production [12]. Pretreatment of waste paper with rumen fluid increase methane yield [13]. The effect of rumen fluid of animal ruminant to increase biogas production from cattle manure at mesophilic condition [14]. Therefore, there are few remarkable studies on biogas production from rumen digesta.

The present study focus on production of biogas as a renewable energy source by recycling rumen waste to control environmental pollution.

2. Materials and methods

Careful consideration is needed to select appropriate methods, procedures and design for collecting valid data. This chapter briefly describes the methods and techniques for assess the availability of rumen digesta and production of biogas. The study was based on field survey where the primary data were collected systematically by interview from people work in different slaughterhouse and recorded experimental data of biogas production from rumen digesta in the laboratory work.

2.1. Site selection

The primary data was collected from the different slaughterhouses of Sylhet City Corporation. Twenty slaughter house were selected for data collection from Shibgonj, Mirabazar, Bondar, Ambarkhana, Tilagor, ShahiEidghah.

2.2. Preparation of questionnaire

An interview schedule was prepared according to the objective of the project. All questions written in interview sheet were related to experiment. The interview was taken to know the amount of slaughterhouse, amount of slaughter animal per day/week/month, the amount of rumen digesta in cow/chicken/goat.

2.3. Interview schedule

Primary data were collected through personal interview with people who work in slaughterhouse. The reason of taking interview was clearly explained to respondents. During the interview time each question was explained clearly and tried to know the right answer from respondents. The collected data from the interview are shown in following Tables:

Table 1. Number of slaughterhouse established per year.

Year	Shibgonj	Ambarkhana	Kazitula	Shahi Eidgah	Tilagor	Bondr	Mirabazar
2013	17	2	6	1	2	16	10
2014	17	3	6	3	2	17	10
2015	17	3	7	3	2	19	10
2016	18	3	8	4	2	19	10
2017	18	3	8	4	3	19	11

Table 2. Number of slaughtered cow/goat/chicken.

Name of the place	Cow	Goat	Chicken
Shibgonj	18	25	130
Ambarkhana	3	5	180
Kazitula	7	10	200
Shahi Eidgah	4	8	250
Tilagor	3	6	200
Bondar	19	30	220
Mirabazar	10	20	180
Total	64	104	1360

Table 3. Quantity of rumen digesta from different slaughterhouse.

Name of the place	Cow (kg)	Goat (kg)	Chicken (kg)
Shibgonj	350	25	2
Ambarkhana	50	5	3
Kazitula	120	10	5
Shahi Eidgah	90	8	2.5
Tilagor	45	6	3
Bondar	340	35	5
Mirabazar	180	27	3

2.4. Collection of raw material

Cow rumen digesta was collected from slaughterhouse of Shibganj, Sylhet. Chicken rumen waste was collected from the slaughter house of Eidghah and baluchor bazar, Sylhet. Goat rumen waste was collected from slaughter house of Mirabazar, Sylhet. The collected raw materials were carried by air tight polythene bag. The collected raw materials were stored in the refrigerator at 0 °C to prevent the decomposition before the experiment. During the experiment the materials were kept in ambient temperature to regain its previous condition. The collected rumen digesta contains some undigested food such as grass, straw, grain.

2.5. Digestion sample preparation

Two sets of experiment were carried out in batch anaerobic digestion system. The first set of experiments were done to study the effect of rumen digesta for biogas production from single substrate and second set of experiment was done to study the effect of mixing ratios of cow, chicken and goat rumen digesta on biogas production via anaerobic co digestion. To complete the experimental set up rumen digesta of cow, chicken, goat was weighted. Then three individual sample were made from rumen digesta of cow, chicken, goat mixed with water in the ratio of 1:1. A co-digestion sample of rumen digesta was prepared in the ratios of cow: chicken: goat (1:1:1) and were diluted with water at 1:1 ratio. Due to presence of organic matter, protein and nitrogen in blood were taken from slaughterhouse. The all ratios of materials were mixed out properly by stirring to add the contact of each waste for obtaining homogenous condition. pH of sample was measured which was 6.6, 6.8, 6.6 for cow, chicken and goat respectively, 6.9 for co-digestion substrate with rumen fluid. The size of each digester was 3300 ml. 80% of digester was filled with prepared material. After adding the feedstock, the anaerobic digester was a tightly closed with rubber stopper. The quantity of rumen digesta used in reactors are shown in Table 4:

Table 4. Quantity of rumen digesta used in reactor.

Feedstock	Rumen waste (kg)	Water (kg)	Total weight of material (kg)
Chicken	1.36	1.36	2.72
Cow	1.36	1.36	2.72
Goat	1.36	1.36	2.72
Mixture	0.454×3	1.36	2.72

2.6. Experimental set up

The lab scale experimental set up was installed by using four 3300 ml digester, 1500 ml water-cum-gas chamber and 1500 ml water collector for every observation. The digester was connected with water-cum-gas chamber through 12 mm diameter hose pipe. Pipe was used to permit produced gas to flow from digester to water-cum-gas chamber. Gas created pressure on the surface of water and displaced same volume of water to flow in the water collector through 10 mm diameter hose pipe. One end of the gas pipe was connected to the top of digester with a glass tubing and other end was connected to the top of the water-cum-gas chamber. On the other hand, one end of water collector pipe was connected to the top of the water-cum-gas chamber and other end of pipe was connected to the water collector. Two gas flow control valve were attached with hose pipe. One was to control the flow of produced gas and other was to control the flow of water. Other instruments were used in that experiment include thermometer, pH meter, glass tubing, gas flow control valve, graduated plastic bucket. Biogas production was monitored for 40 days of study period and data was recorded. The temperature range was 25 °C to 30 °C during the study period. The schematic diagram of experimental set up is shown in Figure 2:

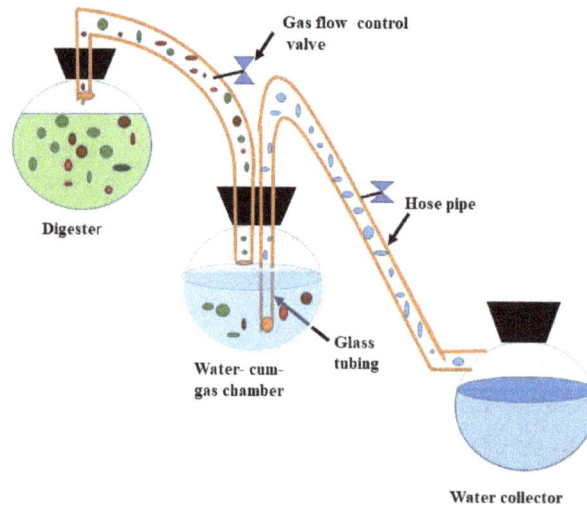

Figure 2. Schematic diagram of experimental set up.

2.7. Data collection of produced biogas

In this experiment the amount of produced gas was assessed by water displacement method. Data was collected every day at 2 pm in the Agricultural and Biosystem Engineering Lab, Department of Farm Power and Machinery, Sylhet Agricultural University.

2.8. Observation

Four observations were started after filling the digester with cow, chicken, goat, and mixed rumen digesta. The digesters were left for anaerobic digestion and gas was started to produce from first operating day and it was almost finished at 39/40th digestion day. Produced gas was measured directly through the measurement of the same volume of expelled water. Observation was continued till the flow of expelled water was stopped.

2.9. Analytical method

Samples were taken and prepared to measure pH of the sample by pH tester (HI98107, HANNA Instruments, Inc. Romania). Thermometer (TP300, China) was used to obtain daily temperature of study period as well as the daily ambient temperature of the environment. Temperature range of TP300 thermometer is −50 °C – +300 °C. Gas composition was analyzed off line by gas chromatography (GC- 8AIT/C159 R8A SHIMADZU Corporation, JAPAN) and Testo-350 portable gas analyzer (Testo AG., Germany). The gas chromatograph was fitted with a

Porapak N 80/100, 274.32 cm, 1/8 mesh 250 × 250 × 145 mm column, a molecular sieve (Mole sieve 5 A 60/80, 182.88 cm, 1/8), maximum temperature 399 °C, temperature stability ±0.1 °C a stainless-steel column and a thermal conductivity detector. Detector type was TCD made by Tungsten rhenium filament. Maximum temperature and sensitivity of the detector was 400 °C and 7000 (mVml/mg) respectively. Argon (Ar) was used as the carrier gas at a flow rate of 30 ml/min. The column temperature was 60 °C and the injector/detector temperatures was 80 °C and current 60 (mA). A 5 ml gas tight syringe was used to take raw biogas samples after releasing the gas. Microsoft Excel 2016 software was used to make the graphical analysis.

3.   Result and discussion

3.1. Present scenario of slaughterhouse

Number of slaughterhouse in Sylhet City Corporation is shown in Figure 3. The bar graph shows that the number of slaughterhouse was increased 2013–2017 in all selected places. It is clearly seen that the maximum number of slaughterhouse was situated in Bondar and lowest number of slaughterhouse in Tilagor. The number of slaughterhouse was almost constant in the years of 2013–2015 might be due to public demand variation in different locations.

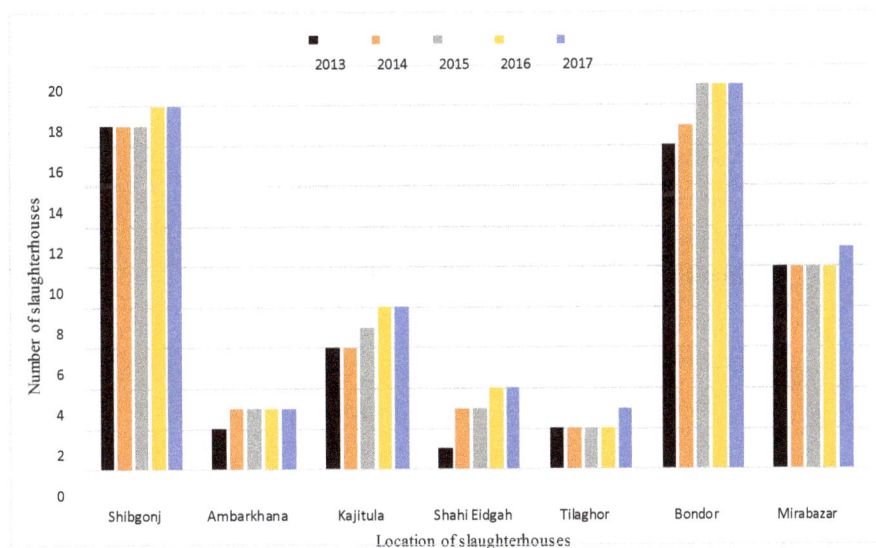

Figure 3. Number of slaughterhouse in different years.

Figure 4 illustrates that number of slaughtered cow,chicken,goat in different locations. The number of slaughtered chicken was high in every slaughterhouse compare to cow and goat. The result indicates that the slaughtered chicken's number was high due to the low cost and readily available. The result also shows that the slaughtered chicken's number was high in Shahi Eidgah slaughterhouse due to the central point of this area.

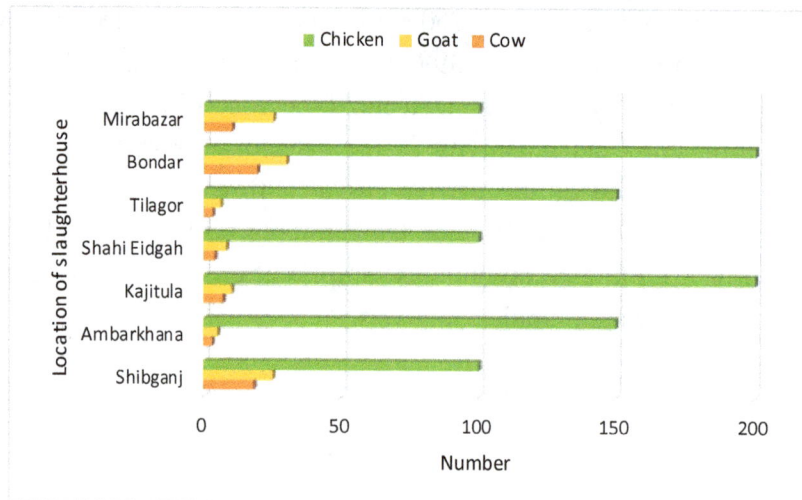

Figure 4. Number of slaughtered animal per day.

Amount of produced rumen digesta is shown in Figure 5. Amount of rumen digesta was high in shibgonj followed by Bondar in case of cow due to transition point from the all access road in Sylhet City Corporation. The Figure shows that the lowest amount of rumen digesta produced in Tilogor. It is clearly shows that chicken rumen digesta was produced in lowest amount among three sources due to size of the rumen.

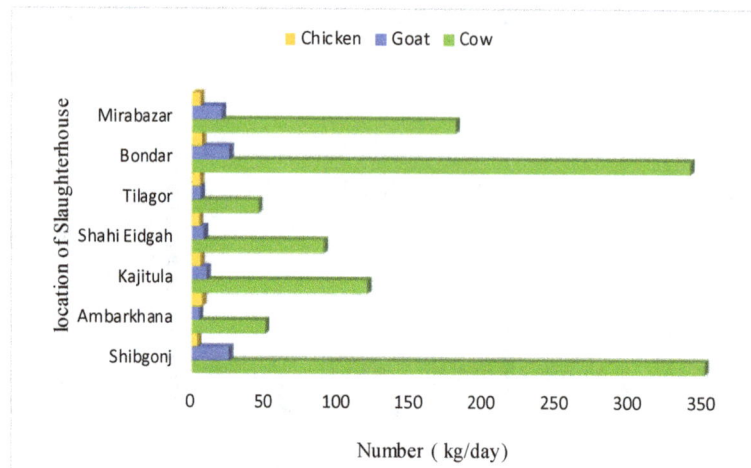

Figure 5. Quantity of rumen digesta.

3.2. Biogas production profile

The biogas production rate under mesophilic condition in batch type digestion system are shown in Figures 6 to 9. It was observed that gas production started from the first digestion day. Figure 6 represents the production of gas from rumen digesta of chicken. It shows that the study period was 40 days and the production starts from the first operating day due to the rapid decomposition of undigested food that means rumen digesta of chicken. Production was gradually increased until 5th day. Then production was decreased due to temperature fall. Maximum gas production was 27.2 ml/day which was recorded in 12th day of study period. The average rate of gas production was 9.865 ml/day.

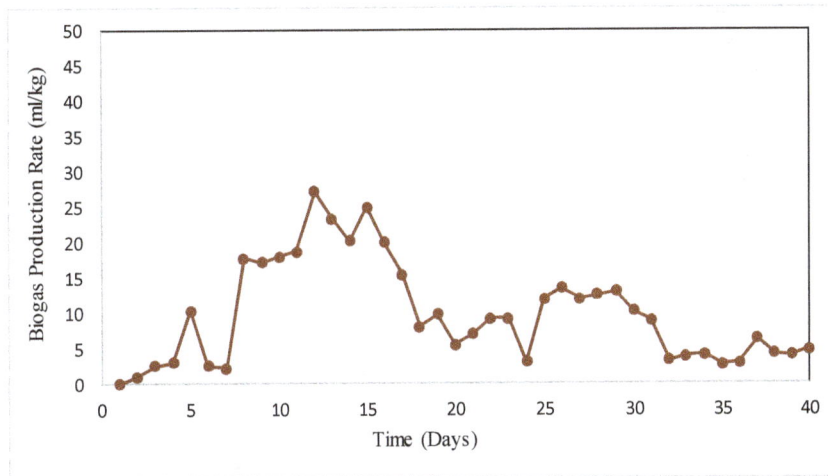

Figure 6. Biogas production from chicken rumen digesta.

The production rate of biogas from cow rumen digesta is shown in Figure 7. Under 40 hydraulic retention time the highest gas production was 3 ml/day at 28th day with high fluctuations due to pH and temperature variation. The average gas production was 1.3125 ml/day. This study shows that the production of biogas from cow's rumen digesta was less than chicken's rumen digesta. Although production starts from the beginning day but production rate was not so high. Because of the high amount of nitrogen in rumen blood. Moreover, microorganism which decompose undigested food need 30 parts of carbon for every part of nitrogen. Rumen digesta of cow which also contain some undigested food such as grass, straw which decomposed slowly which was indicated the upward trend of the graph after 40 days' hydraulic retention time.

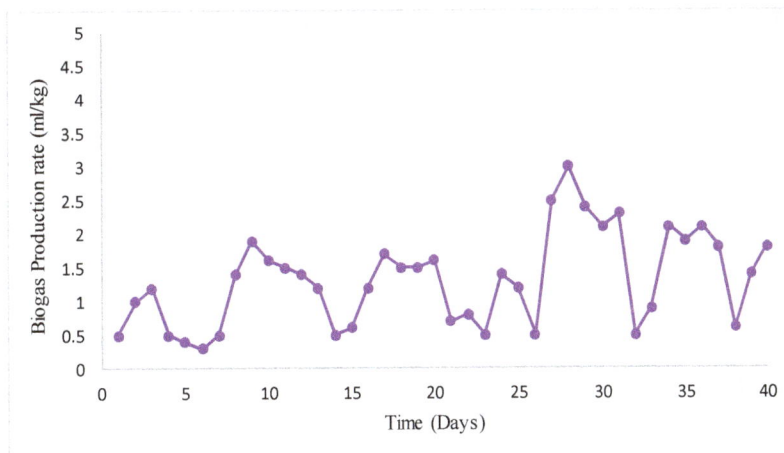

Figure 7. Biogas production from cow rumen digesta.

Figure 8 represents that the production rate of biogas from goat's rumen digesta. The maximum production of biogas was 39 ml/day at 9th day. After 15th day of digestion period production was decreased. The average rate of production was 6.89 ml/day. The undigested waste of goat needed less time to decompose as a result, at the beginning of retention period the biogas production rate was high. It also noticeable that the goat rumen digesta was diluted due to much water content during collection time. The investigation was observed that the biogas production rate was almost stabled from the 16th day to 40th day.

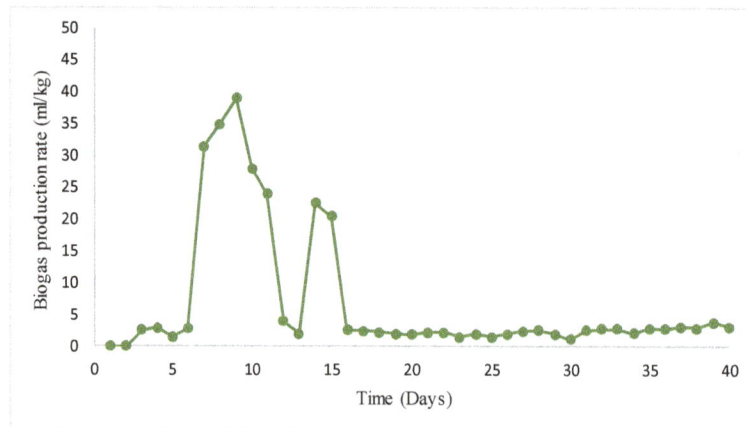

Figure 8. Biogas production from goat rumen digesta.

Production of biogas from mixed rumen digesta of cow, chicken, goat is shown in Figure 9. Mixed sample also contain rumen blood which consists of protein, total organic carbon 31.9%, organic matter 54.8%, nitrogen 12.3%, ammonia 8360 ml results ratio of C/N 2:7. Optimum carbon and nitrogen ratio is one of the key parameter for increasing biogas production rate. Therefore, use of blood increased the production rate at first two days. When all substrate of blood stabled production rate started to decrease. Maximum production was 7.5 ml/day obtained at the 1st day. The average rate of biogas production was 6.35 ml/day. The general prediction is that production rate directly related with the growth of methanogenic bacteria, temperature and pH. The study result reveals that biogas production rate was highly fluctuated over the digestion period.

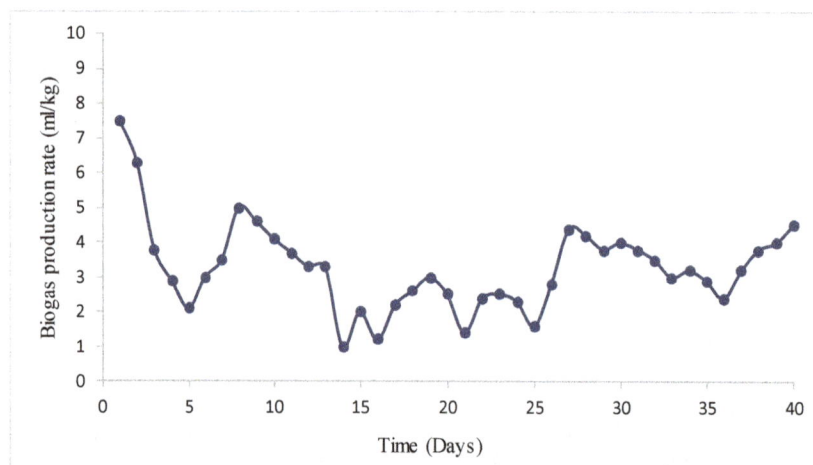

Figure 9. Biogas production from mixed rumen digesta.

3.3. Comparison profile of biogas generation

Figure 10 represents the difference in biogas production after every 10 days. The Figure reveals that after 10 days production was high in every digester except in digester which was filled with cow rumen digesta. The Figure also shows that biogas production from chicken rumen digesta was increased in 2nd 10 days then decreased in 3rd and 4th 10 days because of the highly decomposed rate was occurred after 10 days. Production of gas from cow rumen digesta was increased after every 10 days because of complete decomposition. The Figure shows that Production of gas from goat's rumen digesta was decreased in 2nd and 3rd 10 days and again increased in 4th 10 days means fluctuated over the retention time. The mixed digesta was produced high rate of gas in 1st 10 days because of co digestion of three substrate ratio due to positive synergisms established in the digestion medium and the supply of missing nutrient by the co-substrates. When blood content became stable production rate decreased in 2nd 10 days. Then gradually increased in 3rd and 4th 10 days. Figure also shows that highest gas produced from chicken, cow, goat and mixed rumen digesta was 101.6, 15.4, 143.3, 173.3 ml/day in 3rd, 4th, 1st and 1st 10 days respectively.

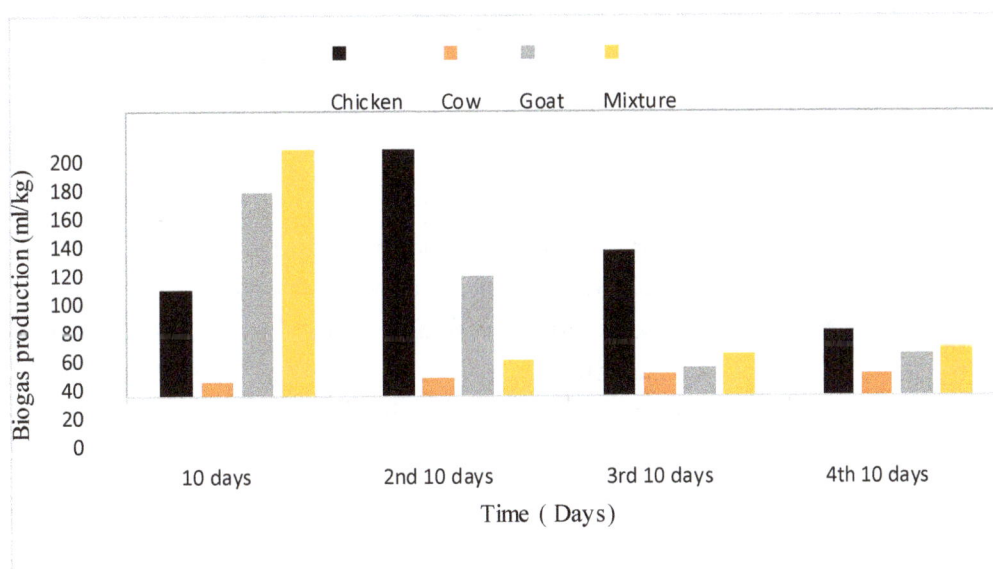

Figure 10. Variation of gas production after every 10 days.

3.4. Cumulative biogas production rate

Figure 11 represents the cumulative biogas production from rumen digesta of cow, chicken, goat and mixed digesta. The Figure shows that cumulative production of 52.5 ml (cow digesta), 394.6 ml (Chicken digesta), 275.6 ml (Goat digesta) and 258.4 ml (mixed digesta) was observed. The result shows that cumulatively highest production obtained at the last 10 days of study period. The average cumulative production of biogas from chicken, cow, goat and mixed digesta were respectively 221.57, 23.53, 186.02, 197.81 ml.

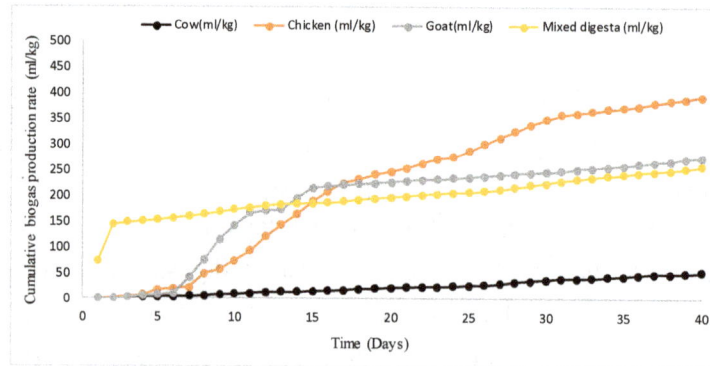

Figure 11. Cumulative biogas production rate.

3.5. Temperature profile

The daily temperature was noted during anaerobic digestion process. The temperature was in mesophilic range during study period. The lowest temperature was 23.8 °C which was obtained on 8th operating day. The highest temperature was 30.2 which was obtained on 20th operating day. The average temperature was recorded 26.8 at the end of 40 day's digestion time.

3.6. Methane content of rumen digesta

Figure 12 represent the percentage of methane present in produced biogas from rumen digesta of chicken, cow, goat and mixed waste. Methane production varied from 55–65%. The Figure also reveals that highest methane produced from chicken, cow, goat and mixed rumen digesta were 65.29%, 62.2%, 63.28%, 62.5% respectively. The Figure also shows that the average methane produced from chicken, cow, goad and mixed rumen digesta were 58.69%, 58.77%, 57.39% and 56.93% respectively. Average methane production was high in cow rumen digesta. Generally, the feedstock chemical characteristics such as the chemical composition (lignin, cellulose, hemicelluloses, starch, total soluble sugars, proteins, and lipids) can determine the gas generation by anaerobic digestion process. Methane yield vary for different chemical constituents of the same feedstock. Fats and proteins produce more methane than carbohydrates and lignin is not biodegradable under anaerobic digestion.

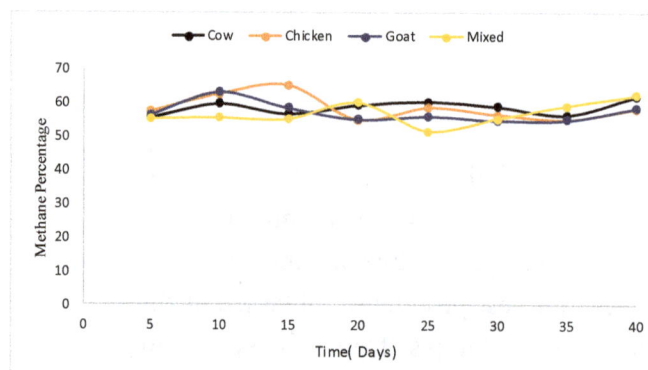

Figure 12. Methane content of produced biogas.

4. Conclusion

Energy plays a vital role in the world. Nothing can be done without the movement or conversion of energy. With the increasing population all over the world, the use of energy also increases. The increasing rate of using energy implies scarcity of energy. In this regard renewable energy is an effective solution. This study was focused on assessment of slaughterhouse number, slaughtered animal number, rumen digesta availability and the production of biogas as a renewable energy source by recycling rumen digesta. The highest production rate of biogas from chicken, cow, goat and mixed rumen digesta were 27.2, 3, 39, 7.5 ml/day respectively. The average production of biogas from rumen digesta of chicken, cow, goat and mixed digesta were 9.865, 1.32, 6.89, 6.35 ml/day respectively. The maximum methane production in biogas from chicken, cow, goat and mixed rumen digesta were 65.29%, 62.2%, 63.28%, 62.5% respectively. The average methane production was 58.69%, 58.77%, 57.39% and 56.93% in produced biogas from chicken, cow, goat and mixed rumen digesta. The results obtained from this study could be used as a basis for designing large scale anaerobic digesters for treatment of rumen digesta. Successful digestion of these substrates is a means of providing renewable energy and environmental friendly waste management system.

Acknowledgement

The authors would like to acknowledge Agricultural and Biosystem Engineering Lab under the Department of Farm Power and Machinery, Sylhet Agricultural University, Sylhet for this experimentation and the extended help to complete the research work.

Conflict of interests

The authors declare no conflict of interest in this paper.

References

1. Dhanalakshmi SV, Ramanujam RA (2012) Biogas generation in a vegetable waste anaerobic digester: An analytical approach. *Res J Recent Sci* 1: 41–47.
2. FAO statistical database (2013) Food and agricultural organization. Available from: http://faostat3.fao.org/home/index.html#download.
3. Wilkie AC (2005) Anaerobic digestion of dairy manure: Design and process consideration. *Dairy Manure Manage Treat* 2005: 301–312.
4. Niesner J, Jecha D, Stehlik P (2013) Biogas upgrading techniques: State of art in European region. *Chem Eng Trans* 35: 517–522.
5. Adesemoye AO, Oper B, Makiwe S (2006) Microbial content of abattoir waste and its contaminated soil in Lagos, Nigeria. *Afr J Biotechnol* 5: 1963–1968.
6. Haryati T (2006) Biogas: Livestock wastes as alternative energy source. Livestock Research Agency, Bogor, Indonesia. *Int J Adv Res* 3: 362–369.
7. Angelidaki I, Ellegaard L, Ahring BK (2003) Applications of the anaerobic digestion process. *Adv Biochem Eng* 82: 1–33.

8.   Lin Y, Wang D, Liang J, et al. (2012) Mesophilic anaerobic co-digestion of pulp and paper sludge and food waste for methane production in a fed-batch basis. *Environ Technol* 33: 2627–2633.

9.   Tefera TT (2010) Potential for biogas production from slaughter houses residues in Bolivia. Royal institute of Technology (KTH), SE-100 44 Stockholm, Sweden.

10. Budiyono B, Widiasa IN, Johari S, et al. (2011) Study on slaughterhouse wastes potency and characteristics for biogas production. *Int J Waste Resour* 1: 4–7.

11. Pertiwiningrum A, Susilowati E, Fitriyanto NA, et al. (2017) Potential test on utilization of cow's rumen fluid to increase biogas production rate and methane concentration in biogas. *Asian J Anim Sci* 11: 82–87.

12. Artanti D, Saputro RR, Budiyono B (2012) Biogas production from cow manure. *Int J Renew Energy Dev* 1: 61–64.

13. Baba Y, Tada C, Fukuda Y, et al. (2013) Improvement of methane production from waste paper by pretreatment with rumen fluid. *Bioresour Technol* 128: 94–99.

14. Sunarso S, Johari S, Widiasa IN (2010) The effect of feed to inoculums ratio on biogas production rate from cattle manure using rumen as inoculums. *Int J Waste Resour* 1: 41–45.

7

Fuzzy PI controller for bidirectional power flow applications with harmonic current mitigation under unbalanced scenario

Nagaraj C[1],* and K Manjunatha Sharma[2]

[1] Research Scholar, Department of Electrical & Electronics Engineering, National Institute of Technology Karnataka (NITK), Surathkal, P.O. 575025, India
[2] Associate Professor, Department of Electrical & Electronics Engineering, National Institute of Technology Karnataka (NITK), Surathkal, P.O. 575025, India

* **Correspondence:** Email: assiniraj@gmail.com

Abstract: The depletion of fossil fuels and environmental concern forces the extraction of power from low carbon fuels causes generation problem due to intermittent solar-wind renewable energy sources and power electronic applications. Furthermore, the significant amount of non-linear loads in the system causes power quality problems. Nowadays, the more and more DC loads like LED lights to save energy consumption are connected to the AC distribution system. These DC loads are connected at DC grid side in order to avoid the extra AC/DC power conversion loss. In this paper, the proposed d-q reference current method applied for shunt active power filter based 3-phase 4-leg bidirectional interfacing converter with fuzzy PI controller to achieve the real power transfer between DC grid side and AC grid side with current harmonics compensation at common connecting point simultaneously under balanced and unbalanced distorted grid and non-linear load conditions. The hysteresis current control comparator without PLL is used to compare actual grid current with reference filter current and generate the switching pulses for driving the bidirectional interfacing converter. The DC grid shunt connected intermittent hybrid solar-wind energy sources are integrating with AC grid utility through bidirectional interfacing converter has been into consideration for simulation studies. The MATLAB/SIMULINK tool is used to yield the improved grid current THD with fuzzy logic controller over PI controller.

Keywords: bidirectional interlinking converter; common connecting point; hysteresis current control; hybrid solar-wind energy sources; i_d-i_q method

1. Introduction

Power generation by renewable energy sources is the most economical one to provide continuous power to the consumer end at the level of distribution system. Due to more and more non-linear loads and power electronic devices, the power quality issues arises in the system. The study of these issues for distinct network configuration is to build appropriate controller to the power converters accordingly in order to supply quality power at the end users in both grid connected and isolated mode [1,2].

The alternative control strategy based shunt active power filter (SAPF) is presented to improve the suppression of harmonic efficiency as well as disturbance caused by unbalanced non-linear loads under unbalanced distorted grid voltage condition [3]. The proposed controller of SAPF based voltage source inverter (VSI) is to address the DG with grid connected issues under balanced undistorted grid with unbalanced non-linear load [4]. The proposed predictive controller for SAPF based 4-leg VSI is analyzed under balanced grid with unbalanced load to compensate unbalanced load current obtained from 1-phase non-linear load, compensation of current harmonics at common connecting point (CCP) and reactive power simultaneously [5].

A direct power control scheme of SAPF based 3-phase 3-leg VSI is proposed with utilization of high selectivity filter instead of classical low pass filter. This allows to control real and reactive power directly injected by SAPF to the AC grid side without using phase locked loop (PLL) and current control loops under any supply voltage conditions [6]. The advanced current controller in d-q frame using combination of PI and repetitive control is presented over traditional current controller with PI control. This method improves the harmonic current compensation irrespective of reduced number of current sensors under source voltage distortion [7]. The optimization algorithm is developed to regulate 4-leg VSI SAPF in order to perform harmonic currents compensation, power factor correction under distorted supply voltage condition. This technique also solved mathematically using Lagrangian formulation and has fast dynamic response [8]. The controller scheme based on Lyapunov control theory is developed for stable operation of SAPF with grid connected mode. This method done the analysis of improvement in harmonic currents compensation, active and reactive power sharing at the same time [9]. The I cosine ANFIS control based dynamic SAPF is presented effectively to enhance the power quality at AC grid source along with real and reactive power sharing with close to unity power factor [10].

In the above papers, the application of SAPF based converter controller are discussed only for DC-AC unidirectional power flow applications with power quality improvement. Also, the PLL used for grid synchronization which creates difficulty in case of unbalanced grid distorted conditions. Nowadays, the more and more DC loads like LED lights to save energy consumption are connected to the AC distribution system. These DC loads are connected at DC grid side in order to eliminate the extra AC/DC power conversion loss. Therefore, in this paper, the DC-AC/AC-DC bidirectional power flow applications with power quality improvement are discussed. The main objective of this study is to eliminate the AC-DC conversion stage and PLL for the cost effective system. The AC-DC conversion stage is eliminated by placing the DC load at DC grid side and making the 3-phase converter as bi-directional. Further, PLL is eliminated in order to avoid the grid synchronization problem during unbalanced distorted grid conditions. The proposed d-q reference current control technique applied to the SAPF based 3-phase 4-leg bidirectional interfacing converter (BIC) with fuzzy logic controller (FLC) resulted in improved power quality performance over PI controller.

Further, the proposed system also resulted in effective real power transfer between DC grid and AC grid side, current harmonics compensation at CCP, power factor compensation and load neutral current compensation. The hysteresis current control (HCC) comparator without PLL is used to generate the switching pulses for driving the gate of BIC. The MATLAB/SIMULINK tool is carried out for balanced and unbalanced distorted grid and non-linear load conditions.

2.　System configuration

Figure 1 shows a block diagram of grid interfacing DC shunted intermittent hybrid solar-wind energy system (HSWES) via BIC in which shunted 48 kW PV array designed at 1000 W/m^2 and 100 kW variable speed PMSG wind turbine designed at rated 12 m/s wind speed which is rectified with unregulated rectifier from AC to DC output power, DC-DC converter and variable DC resistive load are connected to DC grid side whereas unbalanced variable AC non-linear RL load is connected to AC grid side.

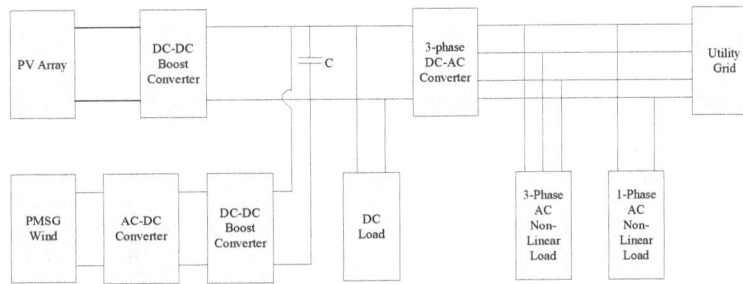

Figure 1. Block diagram of grid connected HSWES.

2.1. Modeling of PV array

The output current of the PV array is modeled by the following mathematical equations and the modeling parameters for PV array are tabulated in Table 1.

$$I_m = I_{pv}N_{pp} - I_oN_{pp}\left[exp\left(\frac{V+R_s\left(\frac{N_{ss}}{N_{pp}}\right)I}{V_taN_{ss}}\right) - 1\right] \qquad (1)$$

where, I_{pv}—photo current which varies continuously depends on the intermittent irradiation level and temperature as follows,

$$I_{pv} = \left(I_{pv,n} + K_i\Delta T\right)\frac{G}{G_n} \qquad (2)$$

V_t—thermal voltage junction is given by,

$$V_t = \frac{KTN_s}{q} \qquad (3)$$

$I_{pv,n}$—photo current generated at actual condition as follows,

$$I_{pv,n} = (R_p + R_s)\frac{I_{sc,n}}{R_p} \tag{4}$$

ΔT—change in actual and reference temperature as follows,

$$\Delta T = (T - T_{ref}) \tag{5}$$

I_0—diode saturation current depends on temperature as follows,

$$I_o = \frac{I_{sc} + K_i \Delta T}{exp\left(\frac{V_{oc,n} + K_v \Delta T}{aV_t}\right) - 1} \tag{6}$$

Table 1. Parameters for PV array model [11,12].

Parameters	Values
Short circuit current	I_{sc} = 3.87 A
Open circuit voltage	V_{oc} = 42.1 V
No. of series cells	N_s = 72
No. of series string modules	N_{ss} = 20
No. of parallel string modules	N_{pp} = 20
Ideality factor	a = 1.3997
Boltzmann constant	K = $1.38e^{-23}$ J/K
Electron charge	q = $1.602e^{-19}$ C
Current temperature coefficient	K_i = $3.2e^{-03}$ A/K
Voltage temperature coefficient	K_v = −0.1230 V/K
Reference temperature	T_{ref} = 298.15 K
Nominal temperature	T = 298.15 K

2.2. Modeling of PMSG variable wind speed

The modeling parameters for variable speed PMSG wind are tabulated in Table 2 and the aerodynamic power obtained from wind turbine depends on wind speed is expressed by,

$$P_m = 0.5\rho V_w^3 A C_p(\lambda, \beta) \tag{7}$$

The power coefficient, C_p (λ, β) decides the total power available from wind and is given by,

$$C_p(\lambda, \beta) = 0.5176\left(\frac{116}{\lambda_i} - 0.4\beta - 5\right)exp^{\frac{-21}{\lambda_i}} \tag{8}$$

Tip speed ratio,

$$\lambda = \frac{\omega_m R}{V_w} \tag{9}$$

$$\lambda_i = \left[\frac{1}{\lambda + 0.08\beta} - \frac{0.035}{\beta^3 + 1}\right]^{-1} \tag{10}$$

Blade pitch angle,

Fuzzy PI controller for bidirectional power flow applications with harmonic current mitigation...

87

$$\beta = \frac{1}{\frac{1}{\lambda + 0.08\theta} - \frac{0.035}{\theta^3 + 1}} \tag{11}$$

The function of β is to control the input mechanical power of wind turbine generator at high wind speed.

The PMSG model in dq frame is given by,

$$\frac{di_{ds}}{dt} = \frac{1}{L_d}\left[-V_{ds} - R_s i_{ds} + \omega L_q i_{qs}\right] \tag{12}$$

$$\frac{di_{qs}}{dt} = \frac{1}{L_q}\left[-V_{qs} - R_s i_{qs} - \omega L_d i_{ds} + \omega \phi_m\right] \tag{13}$$

If the rotor is cylindrical $L_d = L_q = L_s$, the electromagnetic torque is,

$$T_e = 1.5P\phi_m i_{qs} \tag{14}$$

Table 2. Parameters for PMSG wind model [13].

Parameters	Values
Stator phase resistance	2.875 Ω
d & q-axis phase inductance	8.5 m H
Inertia	$0.8e^{-03}$ kg-m^2
Torque constant	12 N-m/A peak
No. of pole pairs	8
Wind power output	100 kW
Rated wind speed	12 m/s

3. Control methods

The overall control diagram of the proposed system is depicted in Figure 2 and discussed in this section.

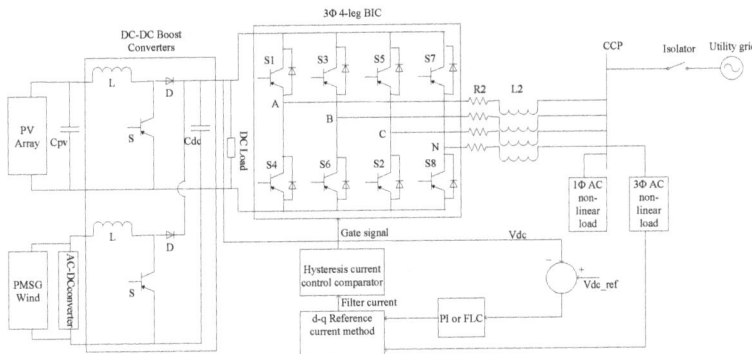

Figure 2. Overall control diagram of the proposed system.

3.1. d-q reference current method

The control diagram of d-q reference current method is shown in detail in Figure 3. The determination of angle θ with respect to α-β frame is obtained by d-q frame. The transformation of d-q reference frame is illustrated as,

$$\begin{bmatrix} i_d \\ i_q \\ i_o \end{bmatrix} = \begin{bmatrix} cos\theta & sin\theta & 0 \\ -sin\theta & cos\theta & 0 \\ 0 & 0 & 1 \end{bmatrix} \begin{bmatrix} i_\alpha \\ i_\beta \\ i_o \end{bmatrix} \tag{15}$$

The above each current has DC average value and AC oscillating value,

$$i_d = \bar{\imath}_d + \tilde{\imath}_d \tag{16}$$

$$i_q = \bar{\imath}_q + \tilde{\imath}_q \tag{17}$$

In the above equations, only the magnitude of currents is transformed and power calculations are performed on d-q-axis components. The zero sequence current components remain unchanged when the d-axis has same direction as like voltage space vector $\bar{v}$. Hence,

$$\begin{bmatrix} i_d \\ i_q \\ i_o \end{bmatrix} = \frac{1}{v_{\alpha\beta}} \begin{bmatrix} v_\alpha & v_\beta & 0 \\ -v_\beta & v_\alpha & 0 \\ 0 & 0 & v_{\alpha\beta} \end{bmatrix} \begin{bmatrix} i_{L\alpha} \\ i_{L\beta} \\ i_{Lo} \end{bmatrix} \tag{18}$$

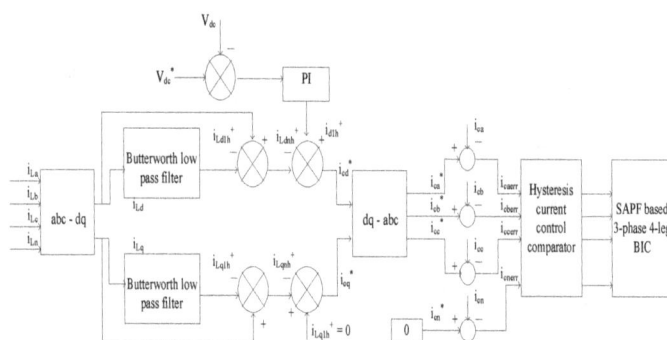

Figure 3. d-q reference current method.

The reference supply current will be determined as follows,

$$i_{sd} = \bar{\imath}_{Ld}; i_{sq} = i_{so} = 0 \tag{19}$$

$$i_{Ld} = \frac{v_\alpha i_{L\alpha} + v_\beta i_{L\beta}}{v_{\alpha\beta}} = \frac{P_{\alpha\beta}}{\sqrt{v_\alpha^2 + v_\beta^2}} \tag{20}$$

The DC average value of Eq 20 is,

$$\bar{\imath}_{Ld} = \left(\frac{P_{L\alpha\beta}}{v_{\alpha\beta}} \right)_{dc} = \left(\frac{P_{L\alpha\beta}}{\sqrt{v_\alpha^2 + v_\beta^2}} \right)_{dc} \tag{21}$$

Multiplying Eq 20 with unit vector because of reference supply current in phase with voltage at CCP,

$$i_{sref} = \bar{i}_{Ld}\frac{1}{v_{\alpha\beta}}\begin{bmatrix} v_\alpha \\ v_\beta \\ 0 \end{bmatrix} \tag{22}$$

$$\begin{bmatrix} i_{s\alpha ref} \\ i_{s\beta ref} \\ i_{soref} \end{bmatrix} = \left(\frac{P_{L\alpha\beta}}{v_{\alpha\beta}}\right)_{dc}\frac{1}{v_{\alpha\beta}}\begin{bmatrix} v_\alpha \\ v_\beta \\ v_o \end{bmatrix} \tag{23}$$

$$\begin{bmatrix} i_{s\alpha ref} \\ i_{s\beta ref} \\ i_{soref} \end{bmatrix} = \left(\frac{P_{L\alpha\beta}}{\sqrt{v_\alpha^2+v_\beta^2}}\right)_{dc}\frac{1}{\sqrt{v_\alpha^2+v_\beta^2}}\begin{bmatrix} v_\alpha \\ v_\beta \\ v_o \end{bmatrix} \tag{24}$$

The main benefits of this d-q scheme are that the angle θ is determined directly from grid source voltage without using PLL to synchronize with utility grid. The PI control is applied to the d-axis in d-q reference frame to control the small real component current and then the current controller controls this current to regulate the DC link voltage constant across capacitor.

Figure 4 represents a process of DC voltage regulation with FLC. It comprises of mainly three blocks, first one is fuzzification using "continuous universe of discourse", second one is inference mechanism based on "fuzzy implication using mamdani's operator" and the third one is defuzzification using "centroid method".

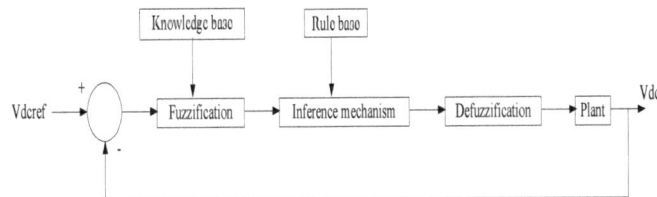

Figure 4. DC voltage regulation with FLC [14].

The FLC was designed with triangular membership function (TMF) due to low computation time. The TMF of 3 × 3 or 5 × 5 is not able to maintain DC link voltage constant and creates an error voltage due to voltage difference. In order to avoid such difficulties, we have considered 7 × 7 TMF to optimize the error efficiently which gives best results in terms of current harmonic filtering under steady and transient state conditions. The development of FLC algorithm is with two inputs and one output each having "seven fuzzy sets". Firstly, the input error "e" and change in error "ce" have been placed of the angular speed to be the input variables of FLC as shown in Figure 5. Then the output variable of FLC is given by the current control "I_{max}" as shown in Figure 6. These numerical variables are converted into linguistic variables and the fuzzy rule mechanism are constructed in Table 3.

(a) Input variable error 'e' triangular membership function

(b) Input variable change in error 'ce' normalized triangular membership function

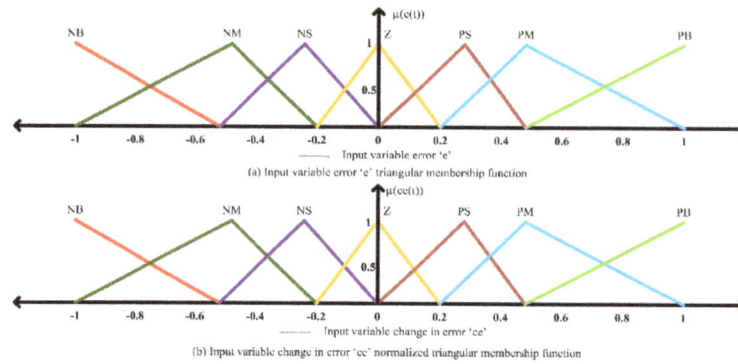

Figure 5. Input variables "e" and "ce" of FLC [14].

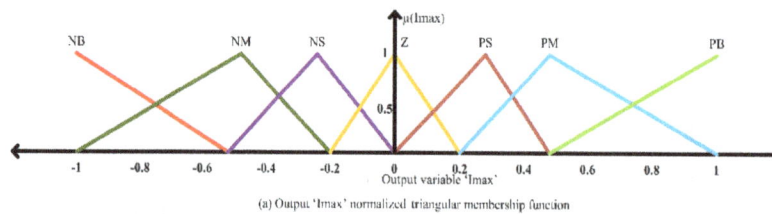

(a) Output 'Imax' normalized triangular membership function

Figure 6. Output variables "I_{max}" of FLC [14].

Table 3. Fuzzy rule mechanism [14].

e ce	NB	NM	NS	Z	PS	PM	PB
NB	NB	NB	NB	NB	NM	NS	Z
NM	NB	NB	NB	NM	NS	Z	PS
NS	NB	NB	NM	NS	Z	PS	PM
Z	NB	NM	NS	Z	PS	PM	PB
PS	NM	NS	Z	PS	PM	PB	PB
PM	NS	Z	PS	PM	PB	PB	PB
PB	Z	PS	PM	PB	PB	PB	PB

3.2. Hysteresis current control comparator

The transfer of real power between AC-DC buses with utility grid depends on DC link voltage control. The working of HCC comparator without PLL as compared to HCC comparator with PLL [3] is discussed as follows.

The reference signals i_{ca}^*, i_{cb}^*, i_{cc}^* and i_{cn}^* obtained from d-q method are compared with actual grid currents i_{ca}, i_{cb}, i_{cc} and i_{cn} as shown below to get the current errors which is given to the HCC comparator to generate the switching pulses (P_1 to P_8) for driving the BIC.

$$i_{caerr} = i_{ca}^* - i_{ca}$$
$$i_{cberr} = i_{cb}^* - i_{cb}$$
$$i_{ccerr} = i_{cc}^* - i_{cc}$$
$$i_{cnerr} = i_{cn}^* - i_{cn}$$

(25)

If the current errors are above the upper bandwidth, then the upper switch of the corresponding phase leg is turned OFF whereas if not, then the upper switch is turned ON and vice-versa. This HCC comparator without PLL has very rapid response and good accuracy due to sudden change of load and its control diagram is depicted in Figure 7.

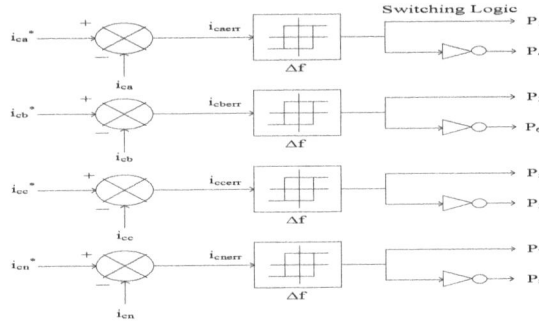

Figure 7. Control diagram of HCC.

4. Case study: Results and discussion for HSWES

The case study results for the given system shown in Figure 2 are simulated using MATLAB/SIMULINK tool under different grid and non-linear load conditions and the parameters used for simulation is tabulated in Table 4.

Table 4. Parameters used for simulation.

Parameters	Values
Grid voltage	V = 400 V
Grid frequency	f = 50 Hz
Grid resistance	$R_2 = 0.1\ \Omega$
Grid inductance	$L_2 = 0.01$ mH
DC link capacitor	$C_{dc} = 7000\ \mu F$
DC link voltage	$V_{dc} = 800$ V

4.1. Performance of d-q method with comparison of PI & FLC under balanced undistorted grid & balanced non-linear load condition at solar irradiation, $I_{rr} = 200\ W/m^2$ & wind speed, $w_s = 8$ m/s in winter

In this case, the performance of the overall system is analyzed with balanced undistorted grid and balanced non-linear load conditions. The overall system main grid voltage with PI control is shown in Figure 8a and fuzzy control is shown in Figure 8b. Initially, before applying the d-q reference current control to the BIC, the grid current is similar to the load current with both PI and FLC (Figure 8i,j). After applying the d-q reference current control to the BIC, the extraction of reference currents is compared with the actual currents by HCC. The DC link voltage regulates constant with the PI controller (Figure 8g) and FLC (Figure 8h), which plays a main role in transferring the power between AC-DC buses. Further, the polluted grid current starts to filter out to become almost sinusoidal is shown in Figure 8c with PI and Figure 8d with FLC and the corresponding grid current THD after filtering is shown in Figure 8k with PI, 1.94% and Figure 8l with FLC, 1.59% respectively.

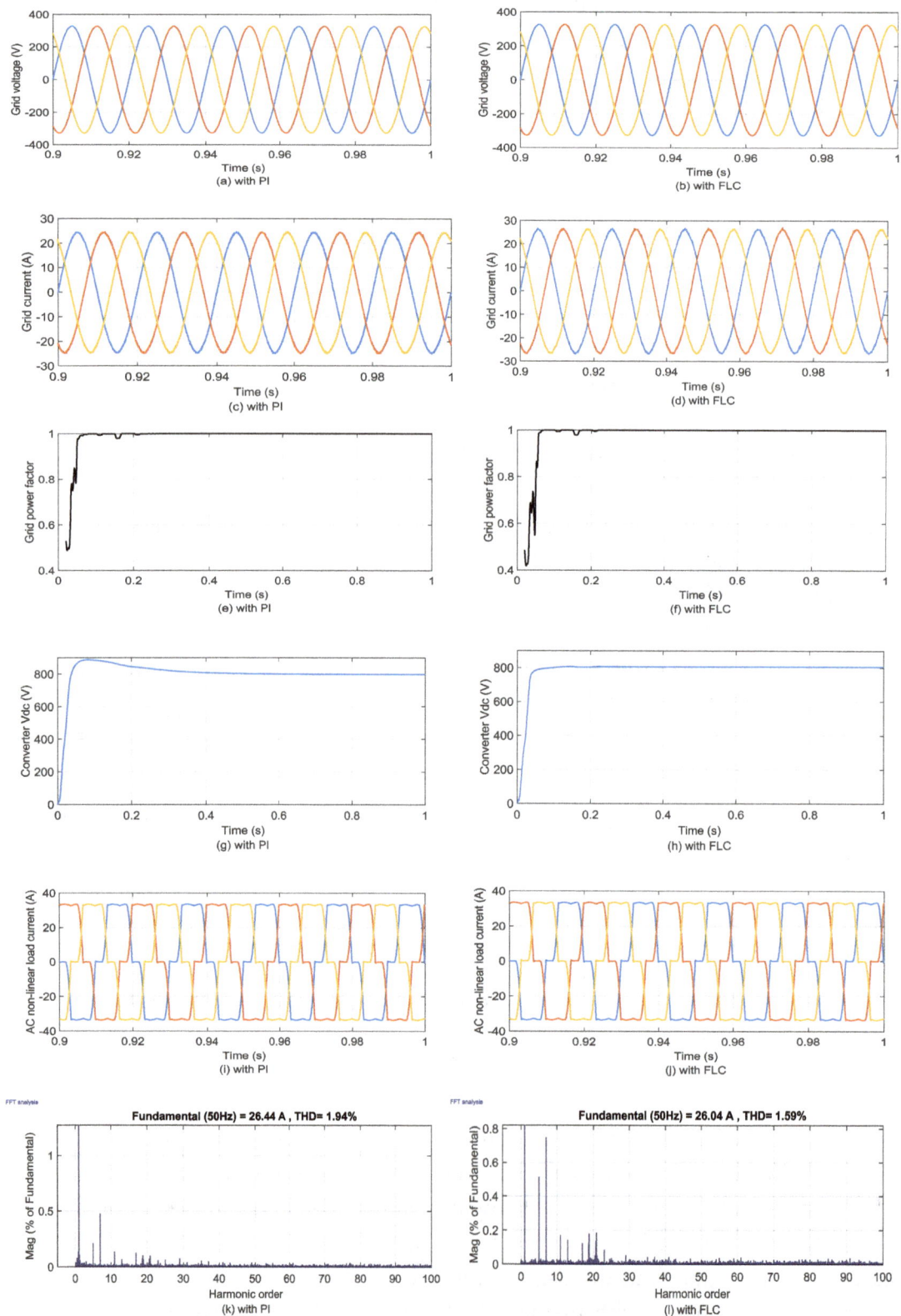

Figure 8. (a) & (b) Grid voltage, (c) & (d) Grid current, (e) & (f) Grid power factor, (g) & (h) Converter V_{dc}, (i) & (j) AC non-linear load current and (k) & (l) Grid THD current.

With both PI and FLC, the power generated by PMSG wind, P_{wind} = 61 kW at 8 m/s and PV array, P_{pv} = 8 kW at 200 W/m^2 in winter. At DC grid side, the HSWES are generating power more than the DC load demand, P_{dcload} = 64 kW. Thus, the remaining surplus power from HSWES is transferring from DC bus to AC bus through BIC, which acts an inverter, P_{conv} = 5 kW (Figure 9b). At AC grid side, the AC load, P_{acload} = 17 kW have been drawn the power from both grid, P_{grid} = 12 kW with Q_{grid} = 0 kW and HSWES through BIC, P_{conv} = 5 kW to meet its demand (Figure 9a). The Q_{grid} indicates that the power factor becomes almost +ve UPF due to grid is supplying the power, which can be observed in Figure 8e, f.

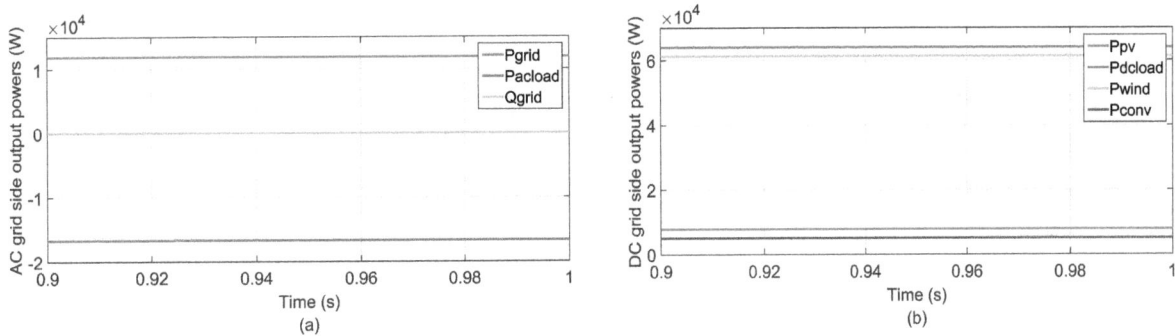

Figure 9. (a) AC grid side output powers and (b) DC grid side output powers.

4.2. *Performance of d-q method with comparison of PI & FLC under unbalanced distorted grid & unbalanced non-linear load condition at solar irradiation, I_{rr} = 800 W/m^2 & wind speed, w_s = 2 m/s in summer*

In this case, the overall system performance is analyzed with unbalanced distorted grid and unbalanced non-linear load conditions. Figure 10a with PI and Figure 10b with FLC presents the main grid voltage of the system. Initially, before connecting BIC to the network, the grid current is similar to the load current with both PI and FLC (Figure 10i,j). After connecting BIC to the network, the reference currents extract from the d-q method and these currents are compared with the actual currents by HCC. The DC link voltage regulates constant with the PI controller (Figure 10g) the FLC (Figure 10h), which plays a main role in transferring the power between AC-DC buses. Further, the polluted grid current starts to filter out to become almost sinusoidal is shown in Figure 10c with PI and Figure 10d with FLC and the corresponding grid current THD after filtering is shown in Figure 10k with PI, 1.62% and Figure 10l with FLC, 1.30% respectively.

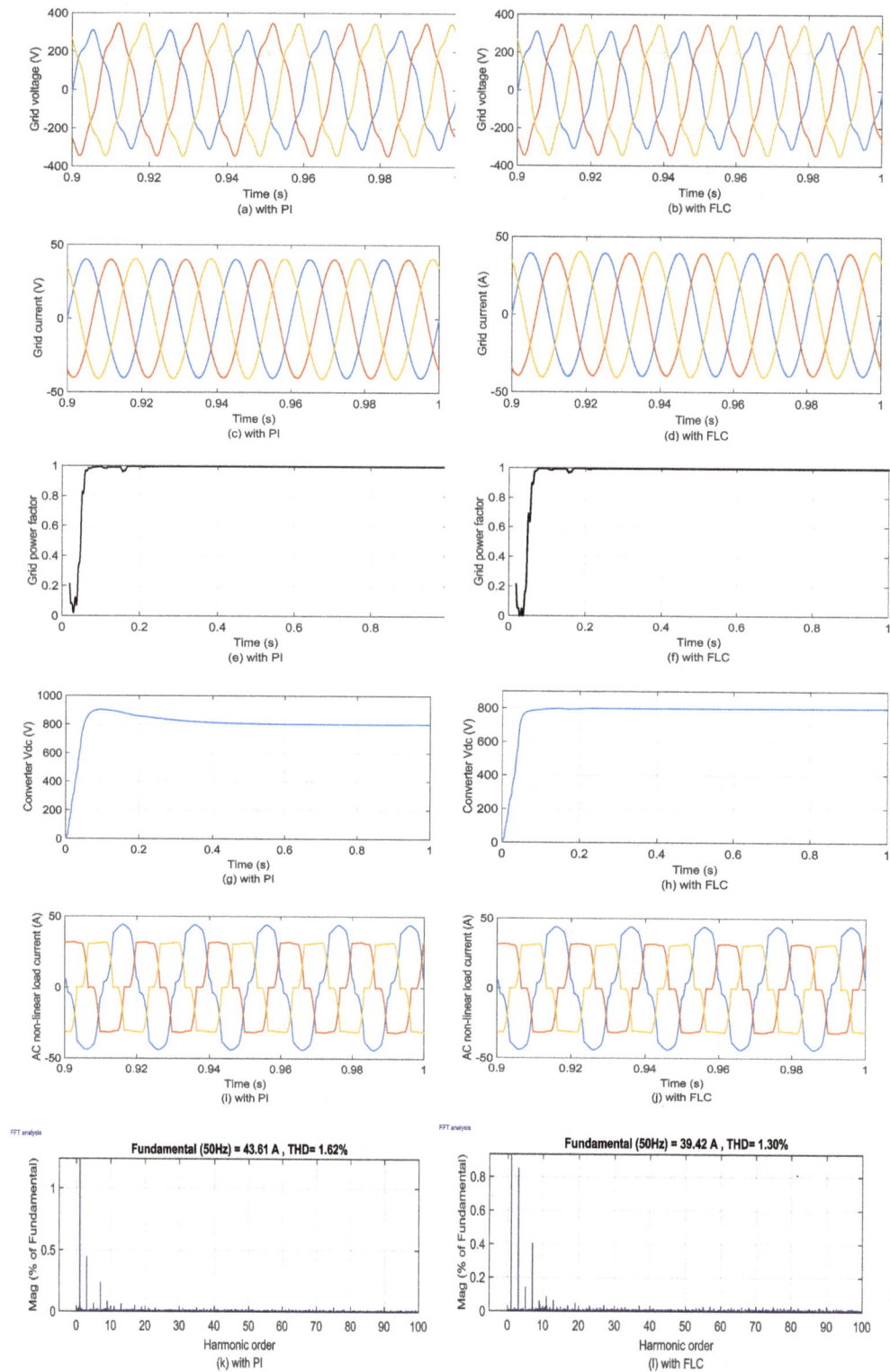

Figure 10. (a) & (b) Grid voltage, (c) & (d) Grid current, (e) & (f) Grid power factor, (g) & (h) Converter V_{dc}, (i) & (j) AC non-linear load current and (k) & (l) Grid THD current.

With both PI and FLC, the power generated by PMSG wind, P_{wind} = 5.5 kW at 2 m/s and PV array, P_{pv} = 38 kW at 800 W/m² in summer. On DC grid side, the generation of HSWES are less than the DC load demand, P_{dcload} = 46 kW. To satisfy the DC load demand, the remaining power drawn from the grid is transferring from AC bus to DC bus through BIC, which acts a rectifier, $P_{conv} = 2.5$ kW (Figure 11b). On AC grid side, the AC load, $P_{acload} = 16$ kW have been drawn the power from only grid, $P_{grid} = 18.5$ kW with $Q_{grid} = 0$ kW through BIC, $P_{conv} = 2.5$ kW to meet its demand (Figure 11a). The Q_{grid} indicates that the power factor becomes almost +ve UPF due to grid is supplying the power, which can be observed in Figures 10e, f.

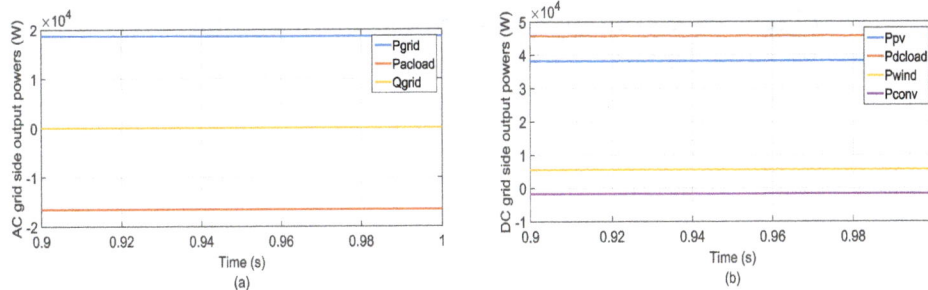

Figure 11. (a) AC grid side output powers and (b) DC grid side output powers.

With both PI and FLC, in case of unbalanced non-linear load on AC grid side, the filter current flows to the CCP through the 4th leg of BIC for compensation of neutral load current as depicted in Figure 12. In both case 4.1 and 4.2, the grid current THD is improved with FLC over PI controller is depicted in Table 5. Thus, the obtained results conclude that the performance of d-q method with FLC holds good over PI controller.

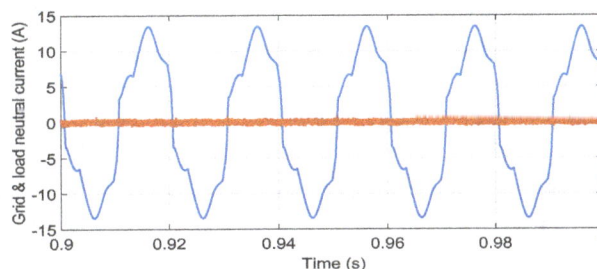

Figure 12. Grid & load neutral current.

Table 5. Power quality improved with FLC over PI controller.

Gird condition	Non-linear load condition	Case	Grid current THD (%)	
			PI Controller	Fuzzy Logic Controller
Balanced Undistorted	Balanced	4.1	1.94	1.59
Unbalanced Distorted	Unbalanced	4.2	1.62	1.30

5. Conclusions

In this paper, the AC-DC conversion stage and PLL has been eliminated in the grid connected HSWES resulted in cost effective system. The BIC has been used to eliminate the extra AC-DC conversion stage. Further, PLL has been eliminated in order to avoid the grid synchronization problem during distorted grid conditions. The proposed d-q reference current control technique using fuzzy PI controller has been applied to SAPF based 3-phase 4-leg BIC. Also, the HCC comparator without PLL has been employed to give the switching pulses to drive the BIC. The extensive simulation analysis has been carried out in MATLAB/SIMULINK tool for balanced and unbalanced distorted grid and non-linear load conditions. For the different case studies, the proposed d-q reference current control technique with FLC resulted in improved power quality performance over PI controller. Further, the proposed system also resulted in effective real power transfer between DC grid and AC grid side, current harmonics compensation at CCP, power factor compensation and load neutral current compensation.

Conflict of interest

All authors declare no conflicts of interest in this paper.

References

1. Arul PG, Ramachandaramurthy VK, Rajkumar RK (2015) Control strategies for a hybrid renewable energy system: A review. *Renew Sust Energ Rev* 42: 597–608.
2. Ullah W, Mekhilef M, Seyedmahmoudian M, et al. (2017) Active power filter (APF) for mitigation of power quality issues in grid integration of wind and photovoltaic energy conversion system. *Renew Sust Energ Rev* 70: 635–655.
3. Özerdem ÖC, Khadem SK, Biricik S, et al. (2014) Real-time control of shunt active power filter under distorted grid voltage and unbalanced load condition using self-tuning filter. *IET Power Electron* 7: 1895–1905.
4. Pouresmaeil E, Akorede MF, Montesinos-Miracle D, et al. (2014) Hysteresis current control technique of VSI for compensation of grid-connected unbalanced loads. *Electr Eng* 96: 27–35.
5. Acuna P, Moran L, Rivera M, et al. (2014) Improved active power filter performance for renewable power generation systems. *IEEE T Power Electron* 29: 687–694.
6. Mesbahi N, Ouari A, Abdeslam DO, et al. (2014) Direct power control of shunt active filter using high selectivity filter (HSF) under distorted or unbalanced conditions. *Electr Power Syst Res* 108: 113–123.
7. Trinh QN, Lee HH (2014) An enhanced grid current compensator for grid-connected distributed generation under nonlinear loads and grid voltage distortions. *IEEE T Ind Electron* 61: 6528–6537.
8. Kanjiya P, Khadkikar V, Zeineldin HH (2015) Optimal control of shunt active power filter to meet IEEE Std. 519 current harmonic constraints under nonideal supply condition. *IEEE T Ind Electron* 62: 724–734.
9. Mehrasa M, Pouresmaeil E, Zabihi S, et al. (2016) A control strategy for the stable operation of shunt active power filters in power grids. *Energy* 96: 325–334.
10. Kasa S, Ramanathan P, Ramasamy S, et al. (2016) Effective grid interfaced renewable sources with power quality improvement using dynamic active power filter. *Int J Electr Power Energy Syst* 82: 150–160.

11. Villalva MG, Gazoli JR, Filho ER, et al. (2009) Comprehensive Approach to Modeling and Simulation of Photovoltaic Arrays. *IEEE T Power Electron* 24: 1198–1208.

12. Rakotomananandro FF (2011) Study of Photovoltaic System. Master of Science. The Ohio State University.

13. Jayalakshmi NS, Gaonkar DN, Kumar KSK (2012) Dynamic modeling and performance analysis of grid connected PMSG based variable speed wind turbines with simple power conditioning system. *IEEE Int Conf Power Electron* 2013: 1–5.

14. Ouchen S, Betka A, Abdeddaim S, et al. (2016) Fuzzy-predictive direct power control implementation of a grid connected photovoltaic system, associated with an active power filter. *Energy Convers Manag* 122: 515–525.

8

Range-extending Zinc-air battery for electric vehicle

Steven B. Sherman, Zachary P. Cano, Michael Fowler* and Zhongwei Chen

Department of Chemical Engineering, University of Waterloo, 200 University Avenue West, Waterloo, ON, N2L 3G1, Canada

* **Correspondence:** Email: mfowler@uwaterloo.ca

Abstract: A vehicle model is used to evaluate a novel powertrain that is comprised of a dual energy storage system (Dual ESS). The system includes two battery packs with different chemistries and the necessary electronic controls to facilitate their coordination and optimization. Here, a lithium-ion battery pack is used as the primary pack and a Zinc-air battery as the secondary or range-extending pack. Zinc-air batteries are usually considered unsuitable for use in vehicles due to their poor cycle life, but the model demonstrates the feasibility of this technology with an appropriate control strategy, with limited cycling of the range extender pack. The battery pack sizes and the battery control strategy are configured to optimize range, cost and longevity. In simulation the vehicle performance compares favourably to a similar vehicle with a single energy storage system (Single ESS) powertrain, travelling up to 75 km further under test conditions. The simulation demonstrates that the Zinc-air battery pack need only cycle 100 times to enjoy a ten-year lifespan. The Zinc-air battery model is based on leading Zinc-air battery research from literature, with some assumptions regarding achievable improvements. Having such a model clarifies the performance requirements of Zinc-air cells and improves the research community's ability to set performance targets for Zinc-air cells.

Keywords: range extender; Zinc-air battery; dual energy storage system; vehicle model; metal-air battery; single energy storage system; driving profile; vehicle use pattern

---

Abbreviations: BEV: battery electric vehicle; DOE: Department of Energy; EA: electrical accessory; ESS: energy storage system; EV: electric vehicle; FD: final drive; GHG: greenhouse gas; GM: general motors; HEV: hybrid electric vehicle; HWFET: highway fuel economy test; ICE: internal combustion engine; Li-Ion: lithium-ion; MSRP: manufacturer's suggested retail price; PC: power converter; PC2: power converter 2; PHEV: plug-in hybrid electric vehicle; SOC: state of charge;

TC: torque coupling; UDDS: urban dynamometer driving schedule; UWAFT: University of Waterloo Alternative Fuels Team; VPA: vehicle propulsion architecture; VPC: vehicle propulsion controller; Zn-Air: Zinc-air

1. Introduction

With global greenhouse gas (GHG) emissions rising and the harmful effects of anthropogenic climate change becoming more apparent, there is a need to reduce the use of CO_2-emitting fuels such as coal, oil and natural gas. GHG emissions from the transportation sector are substantial and growing, accounting for 14% of global emissions [1], 26% of US emissions [2] and 23% of Canada's emissions [3]. The market for electric vehicles (EVs) is developing rapidly and EVs have the potential to substantially reduce emissions from this sector. However, EVs have thus far failed to gain widespread commercial market penetration, accounting for only 1.1% of global car stock and even less of a share in Canada, at 0.59% [4]. The main technological barriers to EV market penetration are their limited driving range, long recharging times and high cost compared to conventional vehicles powered by internal combustion engines (ICE) [5]. These concerns arise due to the limitations of the lithium-ion battery technology used in most EVs. While a significant improvement in nickel-metal hydride batteries has been made, and they are used in a limited number of hybrid vehicles, lithium-ion (Li-Ion) batteries dominate for plug-in hybrid (PHEV) and battery electric vehicle applications (BEV). However, Li-Ion packs are still expensive and provide insufficient energy density to make EVs competitive in the passenger vehicle mass market at this time.

One highly anticipated battery technology for electric vehicles are metal-air batteries, particularly Zinc-air batteries. Zinc-air batteries have a zinc anode and an air cathode; consequentially the battery has a very high theoretical energy density, several times higher than that of commercial lithium-ion batteries. Goldstein and coworkers and Toussaint and coworkers estimate Zinc-air batteries to be significantly cheaper than lithium-ion batteries [6,7], because they are easier to manufacture and are made from more common and less costly materials; the low price of the commercially available rechargeable Zinc-air battery from Eos Energy Storage validates these estimates [8]. They are also safer due to Zinc's lower reactivity compared to lithium, which allows the use of non-flammable electrolytes. However, Zinc-air batteries have not been used in commercial EVs because of their low power density and limited cycle life [9], which are key requirements for EV batteries.

Some have proposed novel vehicle architectures in order to overcome the limitations of current EVs. Bockstette and coworkers [10] modelled a two-battery vehicle architecture with a high energy, low power battery charging a high power, low energy battery, which in turn matches the power demand from the vehicle. They showed how this configuration reduced the combined cost and weight of the battery packs while meeting power and energy performance targets. Tesla Inc. has patented a control strategy for a similar architecture [11]. The patent describes a secondary metal-air battery which works in tandem with a primary battery to power the vehicle while avoiding lifetime limiting discharges on the metal-air battery. Catton and coworkers [12] modelled several vehicles with range extenders, including one with a Zinc-air battery pack range extender. The vehicle outperformed a regular battery electric vehicle (BEV) on range, cost and efficiency.

In this work the efficacy of a dual energy storage system (Dual ESS) is evaluated using a full vehicle model. The Dual ESS vehicle employs a small lithium-ion battery pack as the vehicle's primary power source and a large Zinc-air battery pack as a reserve energy source or range extender. Since rechargeable Zinc-air cells for vehicles are not yet commercially available, the Zinc-air battery pack is modelled based on leading literature on Zinc-air batteries with conservative assumptions regarding future improvements to the technology. The vehicle model provides a means to study both the Dual ESS structure and the performance of Zinc-air batteries in an automotive application.

2. Zinc-air batteries

Metal-air batteries have attracted widespread interest for their use in electric vehicles, mainly because of their high energy density but also because they are expected to be lower cost per unit energy than existing lithium-ion battery technology. However, metal-air batteries suffer from some drawbacks, the foremost of which is their limited cycle life. Metal-air anodes have generally not been reported to cycle more than 50–100 times at deep discharge before failure (though a few have demonstrated longer life under highly favourable circumstances), and some anodes such as aluminum have demonstrated even lower cycle life [13,14]. Metal-air batteries also suffer from low power densities due to their low voltages and low current densities, have high self-discharge rates and suffer from unwanted side reactions [13]. Also, because metal-air batteries are expected to take in oxygen from the air, CO_2 and water contamination are issues in some metal air batteries [13].

Of the available anode materials lithium, aluminum and zinc have been the most thoroughly investigated. Lithium-air batteries have the highest energy density of any metal-air battery [13], and as such have undergone extensive investigation. However, the lithium-air chemistry faces numerous challenges including moisture sensitivity, poor rate capability and irreversible side reactions, resulting in low cycle life, safety hazards and low power density [13]. Due to the current challenges with lithium-air batteries, the commercial focus has shifted to other materials. Aluminum-air batteries have gained widespread interest in academia and in industry. The Israeli battery company Phinergy is marketing an aluminum-air battery for electric vehicles and has attracted interest from global auto manufacturers [15]. However, aluminum-air batteries are not electrically rechargeable [13]. In order to reuse or recharge them, they must be recycled at a processing plant [13]. Consequentially, a battery swapping scheme would be necessary to facilitate the use of these batteries in electric vehicles. Such schemes have been proposed, for example by Nixon [16].

Zinc-air battery technology has been the subject of considerable research. Primary Zinc-air batteries are already used in hearing aids and other mature commercial applications [17], and have also been proposed for metal-air powertrains by Goldstein and coworkers and by Catton and coworkers [8,12]. Zinc-air batteries are not as energy dense as lithium-air or aluminum-air batteries, nor as powerful. However, they are electrically rechargeable to a limited number of cycles and do not suffer from side reactions to the same extent as lithium-air batteries [13]. Water does not harm Zinc-air batteries, which typically have aqueous electrolytes (though the cell can dry or flood if the humidity of the incoming air is not controlled) [14]. Consequentially, Zinc-air batteries are good candidates for electric cars, particularly as range extenders.

Zinc-air batteries have a zinc anode, an inert cathode where oxygen reduction and evolution take place, a separator and an electrolyte, which is usually aqueous. During discharge, the zinc undergoes a two-step reaction to form zinc oxide, while oxygen is reduced at the cathode [9].

Negative electrode:

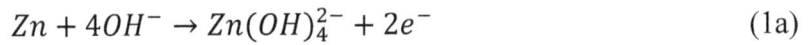

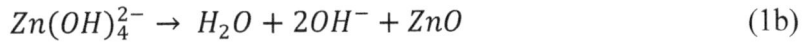

$$Zn + 4OH^- \rightarrow Zn(OH)_4^{2-} + 2e^- \tag{1a}$$

$$Zn(OH)_4^{2-} \rightarrow H_2O + 2OH^- + ZnO \tag{1b}$$

Positive electrode:

$$\frac{1}{2}O_2 + H_2O + 2e^- \rightarrow 2OH^- \tag{2}$$

Overall reaction:

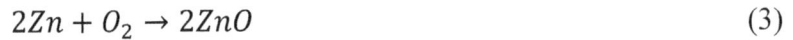

$$2Zn + O_2 \rightarrow 2ZnO \tag{3}$$

Parasitic reaction:

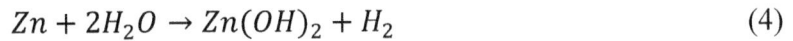

$$Zn + 2H_2O \rightarrow Zn(OH)_2 + H_2 \tag{4}$$

Zinc-air batteries have a number of limitations which have thus far prevented their commercial use in electric vehicles. Their greatest limitation has been their poor cycle life—particularly at the anode, which can suffer from dendrite formation, shape change and passivation [14]. When Zinc-air batteries are electrically recharged, the zinc forms dendrites on the anode surface due to non-uniform local current densities. These high-current areas attract the highly soluble Zinc-oxide reaction products from along the anode surface and cause them to plate on the anode as dendrites [9]. These dendrites frequently break off from the anode, resulting in a loss of battery capacity. The dendrites can also cause cells to short circuit if they form a contact between the two electrodes [9], leading to cell failure and creating safety hazards. Researchers have tried to alleviate dendrite formation using different methods. Vatasalarni and coworkers coated their porous zinc electrode with polyaniline; the electrode exhibited a uniform morphology after 100 cycles and improved capacity over an uncoated electrode [18]. Lee CW and coworkers alloyed zinc with nickel and indium; their anode also showed reversibility after 100 cycles [19]. Parker and coworkers designed a porous anode which would trap the zinc oxide, with the result that during recharge operation the zinc would plate inside the pores [20]. They successfully demonstrated 45 charge-discharge cycles with no morphology change.

Deficiencies at the air electrode have also contributed to short cycle life and have limited specific power and roundtrip energy efficiency. Oxygen reduction requires a triple phase boundary, where the electrode is in contact with both the gas phase and the electrolyte. The inability of air electrodes to maintain a large triple phase boundary is a major impediment to high current density operation [9]. Researchers have investigated a wide range of catalysts, hoping they will prove more effective, more durable and cheaper than precious-metal catalysts. While some researchers have focused on unifunctional catalysts, a Zinc-air battery with unifunctional catalysts generally requires two air electrodes, one for charging and one for discharging. In order to preserve Zinc-air's specific energy and energy density advantages, researchers have increasingly investigated bifunctional catalysts. Jung and coworkers [21] showed that La_2NiO_4 layered perovskites significantly reduced voltage polarization, achieving a discharge voltage of 1.0 V even at 75 mA cm^{-2}. Lee D and coworkers have developed a novel bifunctional catalyst by growing Co_3O_4 nanowires directly onto a steel mesh [22]. A Zinc-air cell using this catalyst maintained a discharge voltage of 0.9 V at

17.6 mA cm^{-2}. The cell successfully demonstrated 1500 pulse cycles (5 minutes for each charge and each discharge) and 100 deep discharge cycles of 3 hours each.

In addition to limited cycle life and low power density, Zinc-air batteries suffer from other, less severe deficiencies. The introduction of air to the cell poses challenges, both because CO_2 can change the pH of the cell electrolyte and affect electrolyte conductivity, and because the cell can dry out or flood if the incoming air is not at the right humidification level [14]. These issues can be managed with an onboard air-management system, including a small CO_2 scrubber. The use of near-neutral electrolytes in Zinc-air cells has attracted interest as these electrolytes do not carbonize. Clark and coworkers developed a model for an aqueous $ZnCl_2$-NH_4Cl electrolyte [23], and Goh and coworkers demonstrated the satisfactory performance of a Zinc-air cell with a near-neutral electrolyte [24]. Eos Energy Storage uses a neutral electrolyte in their commercially available Zinc-air battery as well [25,26]. Gelling the electrolyte has been found to reduce water loss, and Mohamad found that using a 6 M KOH/gel electrolyte improved specific capacity to 657.5 mAh/g compared to a 2.8 M KOH/gel electrolyte [27]. Zinc-air batteries also have the potential for corrosion which results in the production of hydrogen gas. Several strategies have been undertaken in order to reduce zinc corrosion, including alloying with other metals (particularly indium and bismuth), surface coating with aluminum oxide or lithium boron oxide, and the use of chemical additives [9]. In spite of the potential for formation of hydrogen gas, Zinc-air batteries are considered safer than lithium-ion batteries due to the inherent reactivity of lithium, and the potential for lithium-ion batteries to suffer thermal runaway [28].

3. Vehicle powertrains

Commercially available electric vehicles fall into three categories: hybrid, plug-in hybrid and battery electric vehicles. Hybrid electric vehicles (HEV) are primarily powered by traditional internal combustion engines (ICE) but have small batteries able to power the vehicle at low speeds where engines are least efficient, and the battery pack also allows for regenerative braking to increase overall vehicle efficiency. The battery is charged by the engine rather than from an external source. Battery electric vehicles (BEV) have no ICE but rather a single large battery which powers the vehicle, and the batteries are recharged by plugging into a recharging station. At this time commercially available BEVs use lithium-ion batteries, which although better than any other commercially available battery technology are still costly and insufficiently energy dense to meet consumer demand for range and rapid recharging. Consequentially BEVs suffer from short driving ranges and high costs. Plug-in hybrid electric vehicles (PHEV) have powerful, moderately sized batteries as well as small ICEs. They differ from HEVs in that their batteries are powerful enough and large enough to power the vehicle independently for short to medium distances, even at high speeds, and can be charged by plugging the battery into an external outlet. Thus PHEVs have some charge depletion range. Compared to BEVs they are less costly due to the smaller battery and have greater driving range and the ability to quickly refuel due to the presence of the ICE. The ICE effectively acts as a backup power source or 'range extender' energy storage system (ESS) for the vehicle.

Hybrid powertrains employ variations on a few main architecture types, known as series, parallel and series-parallel split. PHEVs typically employ a series architecture (Figure 1a) due to its simplicity and because it allows the primary energy source to operate independently of the backup energy source. BEV architectures are simpler due to their having only one energy source (Figure 1b).

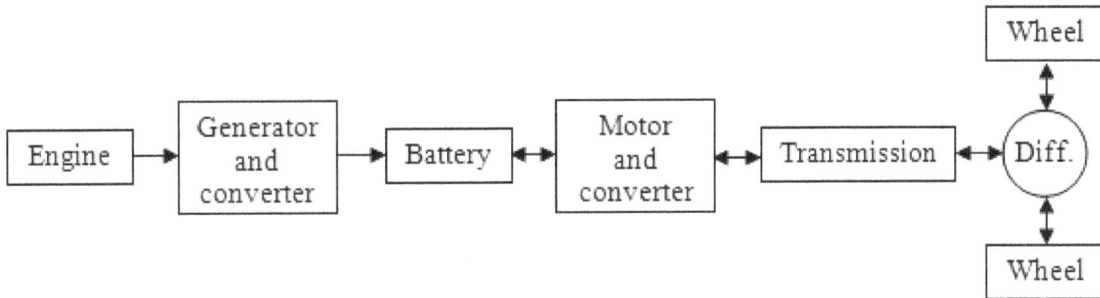

Figure 1a. A series hybrid vehicle architecture commonly employed by PHEVs.

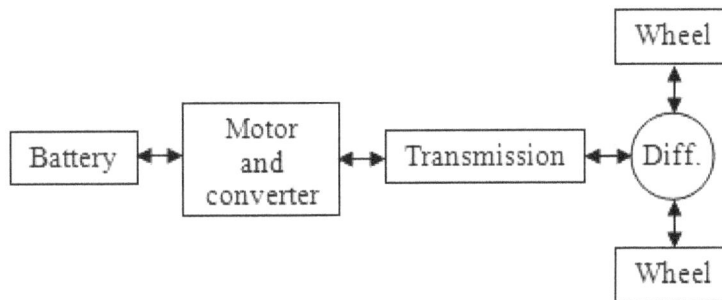

Figure 1b. The architecture of a traditional BEV.

Two vehicle architectures are modelled in this work. The vehicle of interest is the Dual ESS vehicle, which utilizes the PHEV series powertrain configuration except that the engine is replaced with a Zinc-air battery and the generator is replaced with a power converter. The other vehicle is the Single ESS vehicle, which utilizes the BEV powertrain with a large lithium-ion battery. The Single ESS vehicle is used to benchmark the performance of the Dual ESS vehicle. There is a recognition that the Dual ESS can achieve a power boost from a parallel configuration, and this will be explored in future works.

4. Model summary

4.1. Autonomie vehicle models

In order to study the Dual ESS architecture, two vehicle models were created in Autonomie, a vehicle modelling program developed by Argonne National Labs [29]. Autonomie functions by modelling the performance of individual vehicle components in response to the demands placed on the vehicle by the driver. The software feeds the speed and acceleration targets to the model, and the model calculates the performance of the individual components in response to those demands. Figure 2 shows the graphical user interface for the Dual ESS vehicle model.

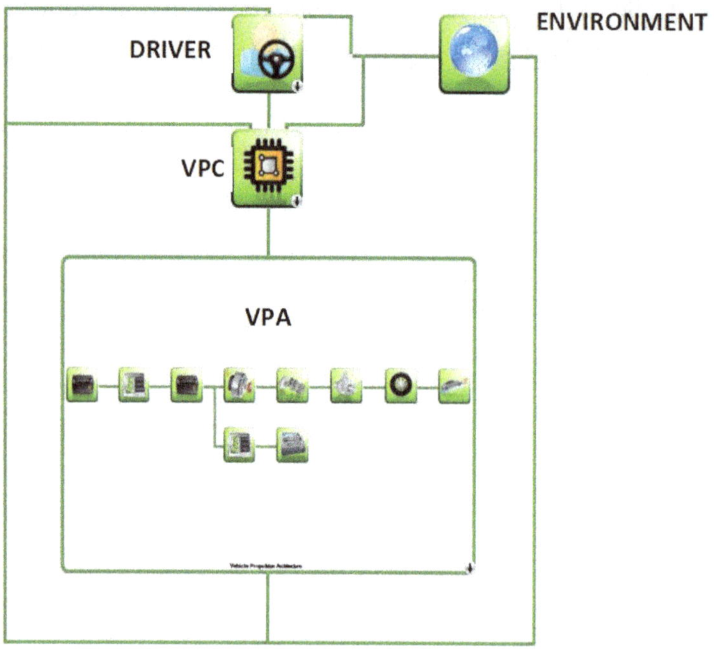

Figure 2. The graphic user interface of autonomie shows how the different components of the Dual ESS vehicle model are connected.

Each model contains four main blocks: the driver, the environment, the vehicle propulsion controller (VPC) and the vehicle propulsion architecture (VPA). The VPA and the VPC receive acceleration-related inputs from the driver block (e.g., target speed, road gradient), while the environment block specifies such inputs as wind speed and temperature. Based on those inputs, the VPC will set the vehicle's mode of operation, which changes the way certain components of the VPA are controlled. Based on all these inputs, Autonomie will calculate the performance required of each component in order to meet target speed and acceleration.

The VPAs of the Dual ESS vehicle and the Single ESS vehicle are shown in Figures 3a and 3b, respectively. Each block models a particular vehicle component. The far-right block contains the vehicle chassis model; the next blocks from right to left represent the wheels, the final drive (FD), the torque coupling (TC) and the motor. To the left of the motor is the block containing the lithium-ion battery model, which provides power to the motor and also to the electrical accessories (EA) through a power converter (PC2). In the Dual ESS VPA, the lithium-ion battery can in turn be charged by the Zinc-air battery (far left block) through a power converter (PC).

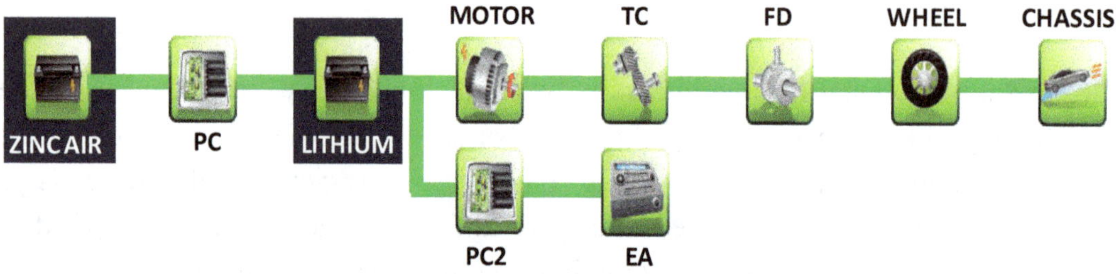

Figure 3a. Dual ESS vehicle propulsion architecture.

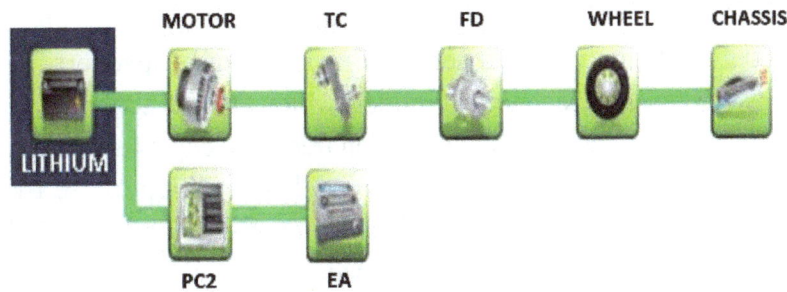

Figure 3b. Single ESS vehicle propulsion architecture.

4.1.1. EcoCAR 3 competition

Both vehicle models are based on the designs of the University of Waterloo Alternative Fuels Team (UWAFT) for the EcoCAR 3 Competition. Sponsored by the U.S. Department of Energy (DOE) and General Motors (GM), the EcoCAR 3 Competition is a green-vehicle design competition with student teams from universities accross North America [30]. For their design UWAFT retrofited a 2016 Chevrolet Camaro to make it a PHEV, installing a battery and a motor but maintaining a backup ICE ethanol engine for extended range. Both powertrains considered in this work use a similar VPA except that the range extender is not an ICE, but a Zn-air battery. Shared components include the chassis which is based on the 2016 Chevrolet Camaro chassis; the 62 kW motor model from the UWAFT design vehicle and the power converters which are 92% efficient as in the UWAFT design vehicle. The lithium-ion battery model also uses the same A123 Li-Ion cells as in the UWAFT design, although the pack configuration is not the same. The only component that is not based on the UWAFT design is the Zinc-air range extender.

4.2. The lithium-ion battery model

The lithium-ion cell model is based on the 20 Ah prismatic cell manufactured by A123. A123 cells have graphite anodes and lithium iron phosphate (LFP) cathodes [31]. This makes them a safer and potentially lower cost chemistry than other lithium-ion cells, but also less energy dense and power dense [28]. This chemistry was selected because the cells were available for the actual vehicle prototyping, and UWAFT was already using them in their vehicle. The cell performance parameters are given in Table 1 below.

Table 1. Lithium-ion cell specifications used in the model [31,32].

Parameter	Unit	Value
Cell Weight	g	496
Cell Capacity	Ah	19.5
Nominal Voltage	V	3.3
Nominal Energy	Wh	65
Specific Energy	Wh/kg	131
Energy Density	Wh/L	247
Cycle Life (1C, 100% DOD)		7000

In order to model the cell's polarization curve a simple equivalent circuit model was used. An equivalent circuit model treats the cell as a series of resistors and capacitors. Such models are less computationally intensive than electrochemical models which attempt to model the internal dynamics of the cell. A modified version of the Rint model, which treats the cell as a simple resistor [33], is used as the basis of the lithium-ion cell model. The Rint model is modified to use two different resistors for charge and discharge. Figure 4 depicts the modified Rint model. The modified Rint model is used because A123 provided UWAFT with detailed resitance values for charge and discharge, indexed by state of charge and temperature. Resistance increases with decreasing SOC and with decreasing temperature for both charging and discharging. A more advanced equivalent circuit model might adequately model the cell as well, but the data for such a model was unavailable. Figure 5 depicts the discharge curves of the A123 cells [34].

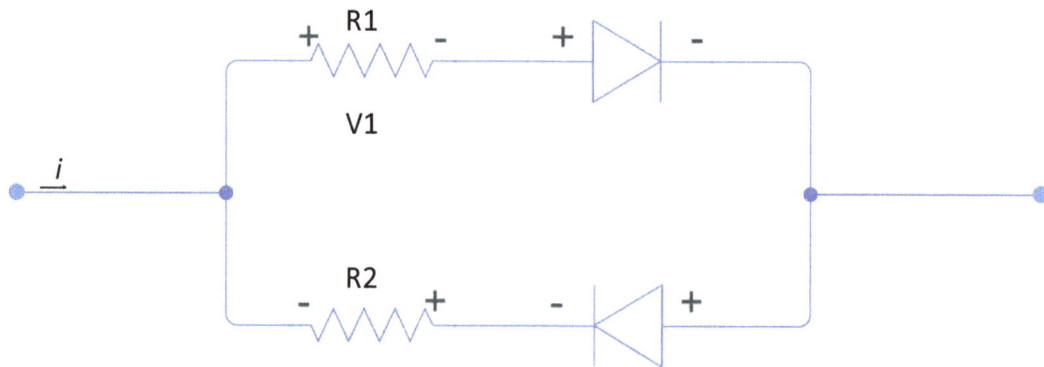

Figure 4. The modified Rint model.

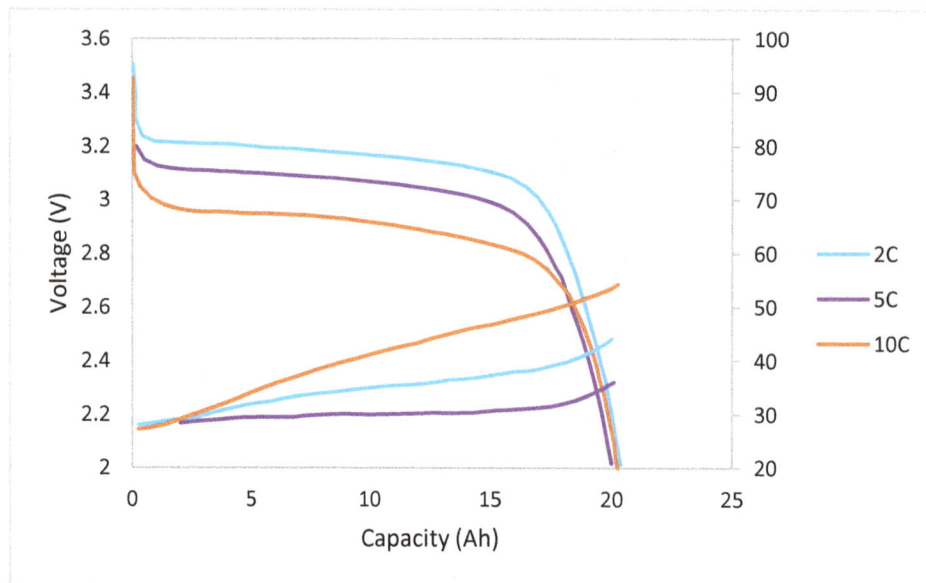

Figure 5a. A123 20 Ah cell discharge curve at various C-rates (data from [34]).

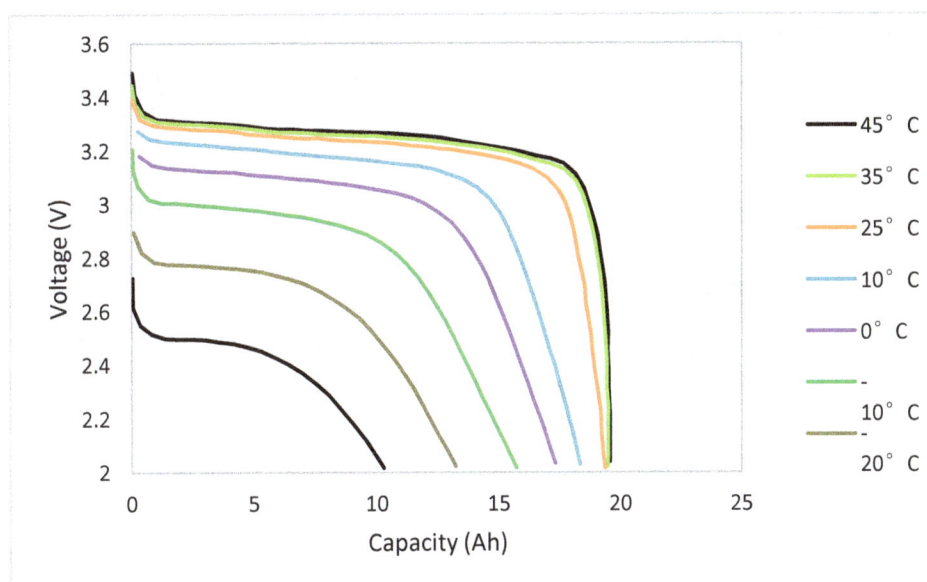

Figure 5b. A123 20 Ah cell discharge curve at 1C at various temperatures (data from [34]).

For both vehicles, the lithium-ion battery pack is comprised of modules each containing 15 individual cells connected in series. These modules are connected in series to form arrays, and the arrays are connected in parallel to form the pack. In the configuration of the current battery pack in the UWAFT EcoCar3 vehicle (of similar size and configuration) the pack is found to be 25% heavier than the combined weight of the cells due to the additional component, packaging and vehicle mount weights. This is factored into the model as a packaging factor. Because A123 cells have extremely good cycle life and because they operate better than other cells at high and low states of charge (SOC) [31] the cells have been set to operate between 100%–5% SOC in the simulation. The battery pack is estimated to cost $230/kWh, based on an average of the market-leading lithium-ion battery pack costs [35]. The effect of capacity fade on vehicle range is ignored for simplicity, but this effect should be incorporated into future work.

4.3. The Zinc-air battery model

Autonomie currently has no model for a Zinc-air battery pack; the potential to incoporate this technology into a vehicle is too recent an innovation. Three different studies were drawn upon in order to create the Zinc-air cell model in Autonomie. Eckl and coworkers [36] describe a mechanically rechargeable Zinc-air cell weighing 349 g, and having a volume of 167 cm^3, with a single zinc electrode sandwiched between two air electrodes. Both air electrodes are used for discharging and charging, rather than one being used for discharging and the other for charging. Because the cell was designed for practical use the cell volume and the mass of the non-active components were taken to be representative of that of a commercial Zinc-air cell. However, because the cell was not demonstrated to be electrically rechargeable other literature data were used to model the cell's electrochemical performance.

Parker and coworkers [20] tested a sponge-like Zinc-air anode, designed to increase cycle life and zinc utilization. Because the anode is specifically designed to be recharged, performs reasonably well under exacting conditions and has a practical thickness and mechanical strength, Parker's anode is used in the Zinc-air battery model. Parker's anode has a specific capacity of 694 mAh/g_{Zn}, a mass density of 1.29 g cm^{-3} and a thickness of 1–4 mm. The model assumes a thickness of 4 mm. Because Eckl's paper breaks down the cell mass by component, the mass of zinc in the cell is changed to reflect the new anode structure. In this work a cycle life of 150–200 cycles to 85% depth of discharge is assumed, which is a modest assumed improvement over the performance demonstrated by Parker. As with the lithium-ion battery, capacity fade is not modelled for the Zinc-air battery. Future work would be well served by modelling this effect.

The air electrode and catalyst described by Lee D and coworkers forms the basis of the air cathode. In the referenced work the electrode is made by growing Co_3O_4 nanowires directly onto a stainless steel mesh, which is placed against a commercial gas diffusion layer [22]. The polarization curve generated in Lee's study forms the basis of the model's polarization curve. However, the polarization curve reported in Lee's study is based on a much larger inter-electrode distance than in the model. Consequentially the polarization curve is adjusted to match the inter-electrode distance modelled in this study. The model's electrolyte is also based on the electrolyte used in Lee's study— 6 M KOH—but is also saturated with ZnO so that ZnO formed during anode discharge would have reduced solubility and improved reversability [37].

The discharge polarization curve reported by Lee D and coworkers [22] was modelled per Equation 5; voltage losses were modelled as a combination of activation polarization (using Butler-Volmer kinetics) and Ohmic (solution and interfacial) resistance loss. The original solution resistance was reported by Lee D and coworkers as 1.76 Ω (obtained via equivalent circuit modelling of impedance measurements), and was multiplied by their electrode area (2.835 cm^2) to obtain the original R_s value of 4.99 Ω cm^2 used in this model. After obtaining values for the V_{OC}, α, i_o and R_{int} parameters through least squares fitting, a new discharge curve was modelled using a modified R_s value. $R_{s,mod}$ was calculated according to Eq 6, which reflected the cell geometry used in the present investigation. Parameter definitions and values are listed in Table 2. The new discharge polarization curve is shown in Figure 6. Due to the relative voltage stability of Zinc-air cells within the prescribed SOC range (15–100%), the effect of SOC on the model has been ignored.

$$V_{operating} = V_{OC} - \frac{RT}{\alpha F} \ln\left(\frac{i}{i_o}\right) - i(R_s + R_{int}) \tag{5}$$

$$R_{s,mod} = \frac{\frac{t_{Zn,avg}}{\varphi_{Zn}} + \frac{t_s}{\varphi_s}}{\sigma} \tag{6}$$

The accuracy and sophistication of the Zinc-air cell model leaves room for improvement. A physics-based model of Zinc-air chemistry or an experiment-based model derrived directly from cell testing would serve equally well or better in some respects as the described model. The described model has the advantage of drawing on leading Zinc-air battery technology, however, whereas an experiment-based model derrives only from the best Zinc-air technology from a particular lab. A physics-based model would be acceptable but not necessarily appropriate as the purpose of analyzing Zinc-air performance within a vehicle model is to understand the potential of the technology rather than its internal mechanisms. A physics-based model might also be too computationally intensive for the vehicle model.

Table 2. Parameters for zinc air polarization curve model.

Parameter	Symbol	Unit	Modelled Value
Open circuit voltage	V_{OC}	V	1.4
Universal gas constant	R	$J\,mol^{-1}\,K^{-1}$	8.314
Temperature	T	K	298.15
Charge transfer coefficient	α	-	0.3209
Faraday constant	F	$C\,mol^{-1}$	96585
Exchange current density	i_o	$A\,cm^{-2}$	0.0002298
Solution resistance (original)	R_s	$\Omega\,cm^2$	4.99
Interfacial resistance	R_{int}	$\Omega\,cm^2$	0.604
Average zinc electrode thickness	$t_{Zn,avg}$	cm	0.2
Zinc electrode porosity	φ_{Zn}	-	78.5%
Separator thickness	t_s	cm	0.02
Separator porosity	φ_s	-	0.55
Electrolyte conductivity	σ	$S\,cm^{-1}$	0.6
Solution resistance (modified)	$R_{s,mod}$	$\Omega\,cm^2$	0.458

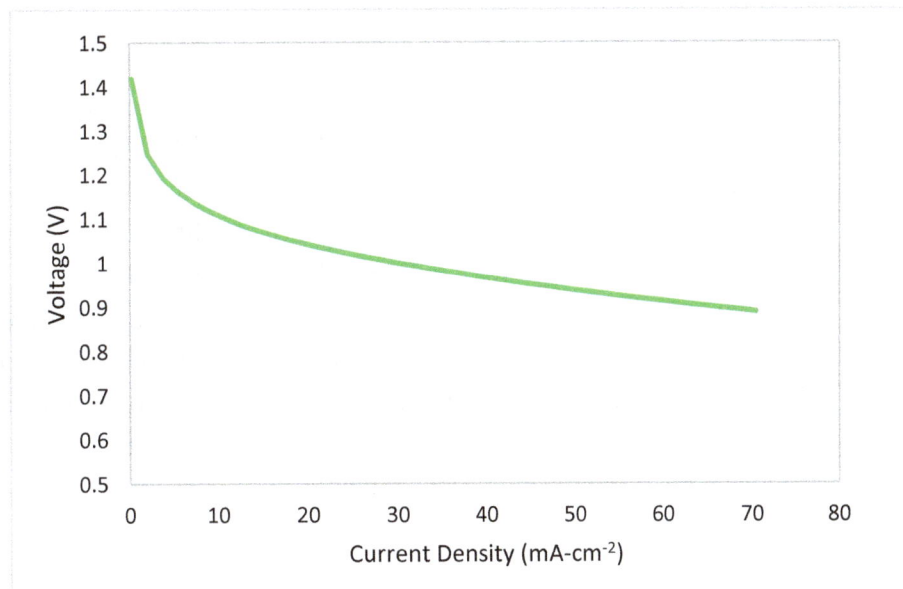

Figure 6. Discharge polarization curve of modelled Zinc-air cell.

By combining the results of the three studies and making reasonable adjustments, the Zinc-air battery model reflects the best anode and cathode performance with a realistic estimate of the total mass and volume of the Zinc-air battery pack. The cell parameters are listed in Table 3. A packaging factor of 25% has been assumed to account for the weight of additional components. There are currently few reliable estimates of what a commercial rechargeable Zinc-air battery might cost. Eos Energy Storage has commercialized a Zinc-air battery which they have been selling for \$160/kWh; they have recently dropped their price to \$95/kWh for orders fulfilled in 2022 [8]. One paper estimated the cost of producing an iron-air battery at \$59/kWh [38] and Electric Fuels Ltd (EFL) published a paper claiming a mechanically rechargeable Zinc-air battery would cost \$60/kWh for a low-power pack (and \$80/kWh for a high-power pack) [6]. Starting from the EFL price of \$80/kWh and adjusting for inflation and different nominal voltage the Zinc-air battery pack cost is estimated to be around $121/kWh. Toussaint and coworkers estimate the cost of a mass-produced, rechargeable Zinc-air battery at €50–100/kWh, corresponding to $61–122/kWh [7]. Thus, the price range of the Zinc-air battery pack is likely between $61/kWh and $160/kWh. In this paper the Zinc-air battery price is taken to be $150/kWh, near the higher cost estimate.

Table 3. The Zinc-air cell specifications.

Parameter	Unit	Value
Cell Weight	g	218
Cell Capacity	Ah	39.4
Nominal Voltage	V	1.0
Nominal Energy	Wh	39
Specific Energy	Wh/kg	181
Energy Density	Wh/L	236

4.4. Dual ESS control logic

Figure 7 illustrates the control logic pertaining to when each battery in the Dual ESS vehicle is activated. When both batteries are fully charged, the Zinc-air battery is disconnected and the lithium-ion battery powers the vehicle independently. When the lithium-ion SOC reaches 20%, the Zinc-air battery is activated and starts charging the lithium-ion battery. If at any point the lithium-ion battery SOC increases to 25% the Zinc-air battery is deactivated until the lithium-ion SOC drops to 20% again. Maintaining the lithium-ion battery SOC between 20–25% ensures that the Zinc-air battery only activates if the lithium-ion battery is in danger of running out of charge, thus minimizing Zinc-air battery use and preserving its longevity. If the Zinc-air battery ever drops below its minimum SOC of 15% it shuts off and is unavailable until the car is plugged in and recharged. The car continues to run on the lithium-ion battery until the lithium-ion battery reaches its minimum SOC of 5%, at which point the car has completely charge depleted.

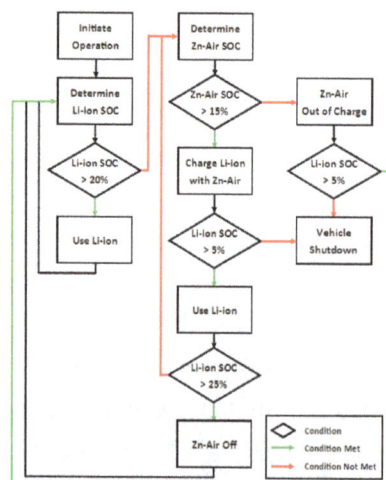

Figure 7. Dual ESS battery pack control logic.

4.5. Drive cycles

In order to test vehicle performance under typical conditions, the US Environmental Protection Agency tests vehicles using specific drive cycles which are representative of typical driving patterns. In this paper, only the two main drive cycles are used. The Urban Dynamometer Driving Schedule (UDDS) is typical of city driving patterns, while the Highway Fuel Economy Test (HWFET) is typical of highway driving patterns. For the purpose of calculating range, fuel economy, cycle life and other vehicle parameters, it is assumed that 55% of driving can be represented with the UDDS and the balance with the HWFET. This is in line with older EPA standard fuel economies, which took a weighted average (55%/45%) of the fuel economies under the two drive cycles [39].

4.6. Zinc-air battery longevity model

A key design consideration is the useful life of the Zinc-air battery pack. Zinc-air cells have a cycle life of only 150–200 cycles, yet must last many years of operation. In order to maximize the Zinc-air battery pack life the pack should be used only sparingly, and thus the basic novelty of this simulation. In this simulation the potential for a Zn-air pack to be used as a range extender over the life of a vehicle is demonstrated, as it can be employed in a limited charge-discharge mode. Data from the 2009 US National Household Transportation Survey (Figure 8) indicates that most US drivers drive only short- to medium-length distances each day. Consequentially a well-designed Dual ESS vehicle will have a lithium-ion battery pack large enough to power the car independently of the Zinc-air battery pack on most days. Taking the US NHTS results to be representative of a typical user's driving patterns, the Zinc-air battery pack's useful life can be estimated based on the frequency of the driver's long-distance trips.

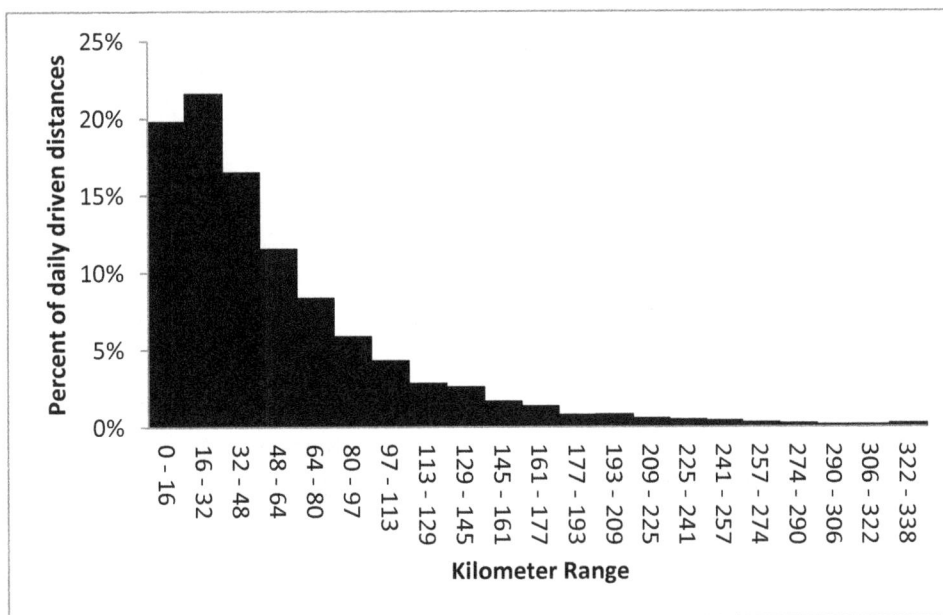

Figure 8. Summary of US driving patterns [40]. In this study, the typical US driver is assumed to have this driving pattern.

5.   Results and discussion

5.1. Battery pack size optimization

Initially the Dual ESS vehicle was tested with 3, 5 and 7 lithium-ion module arrays and 6, 10 and 14 Zinc-air module arrays, for a total of nine battery pack configurations. These configurations were tested with UDDS and HWFET cycles and the results averaged, with a 55% weighting for UDDS results and a 45% weighting for HWFET results. Table 4 shows the key results.

Table 4. Dual ESS battery pack size first optimization.

ID[1]	Li-Ion Energy[2] [kWh]	Zn-air Energy[2] [kWh]	Total Energy[2] [kWh]	Range [km]	Battery Cost [USD]	Zn-air Battery Life[3] [yrs]	Li-Ion Fuel Economy[4] [km/kWh]	Zn-air Fuel Economy[5] [km/kWh]
Li3-Zn6	19	58	77	326	$14,921	7.2	3.83	2.27
Li5-Zn6	32	58	90	374	$18,060	15.9	3.75	2.20
Li7-Zn6	45	58	103	416	$21,198	33.3	3.62	2.11
Li3-Zn10	19	96	116	468	$21,731	10.3	3.46	2.28
Li5-Zn10	32	96	129	506	$24,869	21.4	3.43	2.21
Li7-Zn10	45	96	142	541	$28,007	42.0	3.41	2.13
Li3-Zn14	19	135	154	591	$28,540	12.4	3.22	2.21
Li5-Zn14	32	135	167	622	$31,678	25.1	2.95	2.14
Li7-Zn14	45	135	180	651	$34,817	47.5	2.87	2.08

[1] ID number denotes number of arrays in each pack e.g., Li3-Zn6 has 3 lithium-ion arrays and 6 Zinc-air arrays; [2] Usable energy i.e., based on minimum lithium-ion SOC of 5% and minimum Zinc-air SOC of 15%; [3] Based on 150 total cycles; [4] Vehicle fuel economy when powered by lithium-ion battery, Zinc-air battery off; [5] Vehicle fuel economy when powered by Zinc-air battery.

It should be noted here that the vehicle has lower fuel economy and higher overall battery pack costs as compared to commercial EVs. This is primarily because of the heavy Chevrolet Camaro vehicle base and the conservative battery pack design assumptions that have been made in this simulation for pack weight, as well as the high target range of 500 km which none but the most expensive vehicles are able to attain. The results are more meaningful in comparison to the Single ESS vehicle.

The preliminary results show that for each battery size combination except for Li3-Zn6 the Zinc-air battery lasts at least ten years. Interestingly, regardless of the Zinc-air battery pack size each successively larger lithium-ion battery pack roughly doubles the lifetime of the Zinc-air battery pack. This is the main value of vehicles with larger lithium-ion battery packs, which have shorter ranges compared to similarly-priced vehicles with larger Zinc-air batteries and smaller lithium-ion packs. For example, the Li7-Zn6 combination costs as much as the Li3-Zn10 combination, lasts 23 years longer but travels 50 km less on a single charge. Of course, such a long battery life adds relatively little value since most vehicles are not used for more than 15 years. The large lithium-ion battery simply increases the vehicle's cost while failing to meet the vehicle's range target of 500 km. Similarly, vehicles with 14 Zinc-air arrays have ranges greatly in excess of the target range but also extremely high battery pack costs. The Li3-Zn6 and Li5-Zn6 combinations are inexpensive but with comparatively low range, and in the Li3-Zn6 combination the Zinc-air battery pack lasts only 7.2 years. The most appropriate configurations are therefore the Li3-Zn10 and Li5-Zn10 combinations.

To determine more precisely the best battery pack combination, the Dual ESS vehicle was retested with 3, 4 and 5 lithium-ion arrays and 9, 10 and 11 Zinc-air arrays. Table 5 shows the key results.

Table 5. Dual ESS battery pack size second optimization.

ID[1]	Li-Ion Energy[2] [kWh]	Zn-air Energy[2] [kWh]	Total Energy[2] [kWh]	Range [km]	Battery Cost [USD]	Zn-air Battery Life[3] [yrs]	Li-Ion Fuel Economy[4] [km/kWh]	Zn-air Fuel Economy[5] [km/kWh]
Li3-Zn9	19	87	106	434	$20,028	9.6	3.59	2.29
Li4-Zn9	26	87	113	456	$21,598	14.1	3.53	2.26
Li5-Zn9	32	87	119	475	$23,167	20.5	3.51	2.22
Li3-Zn10	19	96	116	468	$21,731	10.3	3.46	2.28
Li4-Zn10	26	96	122	487	$23,300	15.0	3.47	2.25
Li5-Zn10	32	96	129	506	$24,869	21.4	3.43	2.21
Li3-Zn11	19	106	126	502	$23,433	10.8	3.41	2.28
Li4-Zn11	26	106	132	519	$25,002	15.8	3.42	2.24
Li5-Zn11	32	106	139	537	$26,571	22.5	3.38	2.20

[1] ID number denotes number of arrays in each pack e.g., Li3-Zn6 has 3 lithium-ion arrays and 6 Zinc-air arrays; [2] Usable energy i.e., based on minimum lithium-ion SOC of 5% and minimum Zinc-air SOC of 15%; [3] Based on 150 total cycles; [4] Vehicle fuel economy when powered by lithium-ion battery, Zinc-air battery off; [5] Vehicle fuel economy when powered by Zinc-air battery.

Several of these battery pack combinations offer good performance. In particular, the Li4-Zn9, Li3-Zn10, Li4-Zn10 and Li3-Zn11 configurations are similarly priced, offer good Zinc-air battery life and long range. Although all these configurations could be considered the most suitable, the Li4-Zn10 combination is selected because it essentially meets the target range of 500 km and because the excess Zinc-air battery life would make the vehicle more appealing to drivers worried about having to carefully manage the Zinc-air battery so as not to overuse it.

The CO_2 scrubber for the Zinc-air battery is sized to cover a year of use by the battery. The scrubber was sized per Eq 7, and the amount of water necessary for humidification per Eq 8. Table 6 details the parameter values. Humidification is necessary because the adsorbent, $LiOH$-$Ca(OH)_2$, performs vastly better at high humidification levels [41]. The adsorption coefficient of $LiOH$-$Ca(OH)_2$ is taken to be 313.5 mg CO_2/$g_{adsorbent}$ [41], based on an average of repeated tests. Based on ten cycles a year and a 30% buffer, the CO_2 scrubber should contain 3.7 kg of adsorbent and 49.6 kg of water. This weight is negligible compared to the weight of the vehicle.

$$m_{ads} = \left(\left[\frac{m_{Zn} \times (SOC_{max} - SOC_{min})}{mm_{zn}} \times \left(\frac{n_{O2}}{n_{zn}}\right) \times \left(\frac{n_{air}}{n_{O2}}\right) \times C_{CO2} \times mm_{CO2}\right] \div K_{100}\right) \times N \times (1 + B) \quad (7)$$

$$m_{H2O} = \left[\frac{m_{Zn} \times (SOC_{max} - SOC_{min})}{mm_{zn}} \times \left(\frac{n_{O2}}{n_{zn}}\right) \times \frac{P_{vap}}{0.21 P_{atm}} \times mm_{H2O}\right] \times N \times (1 + B) \quad (8)$$

Table 6. Parameters and numerical data for CO_2 scrubber sizing.

Parameter	Symbol	Unit	Modelled Value
Mass of adsorbent	m_{ads}	kg	3.7
Mass of water	mH_2O	kg	49.6
Mass of zinc in battery pack	mz_n	kg	163.4
Maximum SOC of ZnAir pack	SOC_{max}	-	100%
Minimum SOC of ZnAir pack	SOC_{min}	-	15%
Molar mass of zinc	mmz_n	kg/kmol	65.4
Ratio of reacting oxygen to reacting zinc	(nO_2/nz_n)	-	0.5
Ratio of moles of air to moles of oxygen in the air	(n_{air}/nO_2)	-	4.76
Concentration of CO_2 in the air	CCO_2	ppm	400
Vapour pressure of water (at 30 °C)	P_{vap}	kPa	4.25
Atmospheric pressure	P	kPa	101.325
Molar mass of CO_2	$mmCO_2$	kg/kmol	44
Molar mass of water	mmH_2O	kg/kmol	18
Adsorption capacity of adsorbent	K_{100}	$mgCO_2/g_{adsorbent}$	313.5
Number of Zinc-air cycles per year	N	-	10
Buffer	B	-	30%

The zinc air battery pack lifetime is worth analyzing in more detail, since it is highly dependent on the user's driving patterns and the cycle life of the Zinc-air battery. In Figure 9 the Zinc-air battery cycle needed to last ten years (based on the driver profile outlined in Figure 8) is plotted as a function of the number of lithium-ion arrays.

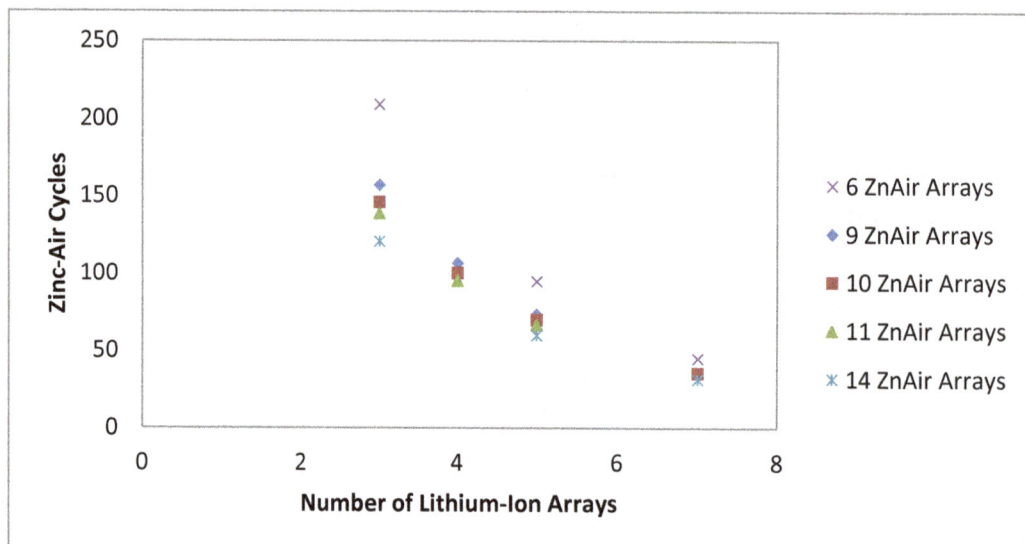

Figure 9. Number of Zinc-air cycles needed for ten years of operation.

The Zinc-air cycle life required to reach ten years of use reduces by roughly 30% for each additional lithium-ion array (6.5 kWh) included in the powertrain, and this remains true regardless of Zinc-air battery pack size. This reduction occurs because larger lithium-ion battery packs reduce the percentage of trips where the Zinc-air battery pack is required and because it reduces the use of the Zinc-air battery even on longer trips.

In this investigation a cycle life of 150 cycles has been assumed, but based on the results in Figure 9 clearly even 150 cycles are excessive with a large enough Li-Ion battery pack. Parker and coworkers have demonstrated 45 charge-discharge cycles [20] which would be enough to sustain the Zinc-air battery for ten years if the vehicle's lithium-ion battery pack comprises 7 arrays (48 kWh). Having a 4-array lithium-ion battery pack (27 kWh) means the Zinc-air battery can be sustained for ten years with a cycle life of only 100 cycles. This clearly demonstrates the viability of low-cycle life battery range extenders in vehicles when used as a secondary energy source (although only when the vehicle owner mainly drives short-to-moderate distances most of the year). Conversely, a greater Zinc-air cycle life reduces the size of the lithium-ion battery required. A Zinc-air battery with a cycle life of 300 cycles might last ten years even if the lithium-ion battery had only two arrays (14 kWh). A vehicle with such a configuration would have greater safety and lower cost due to the reduced lithium-ion battery size, which demonstrates the value of modest Zinc-air battery cycle life improvements. However, it is important to note that Zinc-air cells have not yet achieved the cycling performance assumed in this investigation. The Dual ESS battery pack was cycled at up to 40 mA cm^{-2} and the cycle life was taken to be 150 cycles, whereas Parker's anode achieved only 45 cycles at 24 mA cm^{-2} [20] and D. Lee's cathode was cycled at 18 mA cm^{-2}, though higher current densities were achieved [22].

5.2. Performance comparison between Dual ESS and Single ESS

The Single ESS vehicle was designed with no Zinc-air battery but instead with a large lithium-ion battery pack comprising 15 arrays and having a nominal energy of 102 kWh. Table 7 compares the battery packs of each vehicle. Table 8 compares the Single ESS and Dual ESS vehicles on performance-related measures.

Table 7. Battery pack specifications for the Dual ESS and Single ESS vehicles.

Parameter	Unit	Dual ESS-Lithium-Ion pack	Dual ESS-Zinc-air pack	Single ESS-Lithium-Ion pack
Cells per module	-	15	4	15
Modules in series	-	7	72	7
Arrays in parallel	-	4	10	15
Total cell weight	kg	208	627	781
Packaging factor	-	1.25	1.25	1.25
Pack weight	kg	260	784	977
Nominal pack voltage	V	347	288	347
Pack capacity	Ah	78	394	294
Pack energy	kWh	27	113	102
Maximum SOC	-	100%	100%	100%
Minimum SOC	-	5%	15%	5%
Pack cost	USD	$6,277	$17,023	$23,538

Table 8. Performance summary of the Dual ESS and Single ESS vehicles.

Properties	Unit	Dual ESS	Single ESS
Vehicle Weight	kg	2721	2651
Range-UDDS	km	459	383
Range-HWFET	km	521	449
Fuel Economy-UDDS	km/kWh	$3.20^1/2.13^2$	3.35
Fuel Economy-HWFET	km/kWh	$3.80^1/2.40^2$	3.93
Battery Pack Cost	USD	$23,300	$23,538

[1] Refers to the fuel economy when vehicle is powered by the Li-Ion battery, [2] Referes to the fuel economy when vehicle is powered by the Zn-Air battery.

There are several points of interest in the results. First, as previously indicated both vehicles achieve lower fuel economy and have higher battery pack costs as compared to commercial EVs due to the heavy vehicle base, conservative battery pack design and large battery packs. However, the vehicles also travel hundreds of kilometers further on a single charge than all but the most expensive EVs. The relevance of the results is in how the Dual ESS architecture compares to a similar vehicle with a Single ESS architecture. In that regard, the Dual ESS vehicle performs well, travelling 75 km further on a single charge compared to the Single ESS vehicle, which costs about the same. Although the vehicles have equal weight, the Dual ESS vehicle has more energy storage due to the Zinc-air battery's higher energy density. This higher energy content allows the Dual ESS vehicle to travel further than the Single ESS vehicle in spite of its low fuel economy. The Zinc-air fuel economy is lower than the lithium-ion fuel economies of both vehicles for a number of reasons. First, Zinc-air cells are much less efficient than lithium-ion cells due to their high levels of voltage polarization. The Zinc-air battery pack has a round-trip energy efficiency of 56–59% when charged with a Level 2 charger (6.6 kW), compared to an 85% round-trip energy efficiency for lithium-ion cells [28]. Second, the energy from the Zinc-air battery must pass through a power converter; 8% of the Zinc-air's energy is lost at this stage. Third, some of the energy from the Zinc-air battery is put towards charging the lithium-ion battery rather than directly powering the car. Lithium-ion batteries are highly efficient, but they still have efficiency loses due to their charge/discharge voltage separation. Thus, some energy is lost by having the Zinc-air battery power the vehicle through the lithium-ion battery. The Dual ESS vehicle is also less fuel efficient than the Single ESS vehicle when powered only by its lithium-ion battery. This is because the Dual ESS vehicle's lithium-ion battery is smaller than the Single ESS vehicle's and thus has to operate at a higher C-rate.

Figure 10 shows the energy of each battery of the Dual ESS vehicle during repeated HWFET cycles. The lithium-ion battery powers the vehicle independently until the lithium-ion SOC drops to 20%, at about the 98th kilometer. At this point the Zinc-air battery starts charging the lithium-ion battery at a constant power rate, while the lithium-ion battery continues to power the vehicle. This results in the lithium-ion battery oscillating above and below 20% capacity depending on the demands of the driver. This continues until the Zinc-air battery is depleted (reaches 15% SOC) at about the 505th kilometer. At this point the Zinc-air battery shuts off and the lithium-ion battery powers the vehicle independently again, draining until reaching 5% SOC at the 521st kilometer.

One of the obstacles to electric vehicles is their long recharge time. While regular vehicles can be refilled in a matter of minutes from the gas pump, batteries take much longer to recharge under normal circumstances. And although it is possible to recharge lithium-ion batteries using rapid recharging stations, large power spikes can be difficult for electrical grids to manage. Having even a few vehicles undergo rapid recharging in a 22 MW/9 MVAR distribution grid decreases the grid's voltage stability by 25–45%, according to work by Dharmakeerthi and coworkers [42]. The same work also shows that rapid recharging during an unexpected generator shutdown decreases grid stability by 40–60%. Consequentially designing vehicles so as to not require rapid recharging would be highly beneficial to the electrical grid and the long-term prospects of electric vehicles. In this, the Dual ESS vehicle does not improve upon the Single ESS vehicle. Because of their high voltage polarization Zinc-air batteries take more energy to recharge than lithium-ion batteries, resulting in longer recharge times and wasting more energy. Furthermore, the Dual ESS vehicle's combined battery energy is larger than that of the Single ESS vehicle, further lengthening recharge times. And unlike lithium-ion batteries which can be rapid-charged provided the grid does not destabilize Zinc-air batteries are naturally low-current batteries, limiting their charge rate.

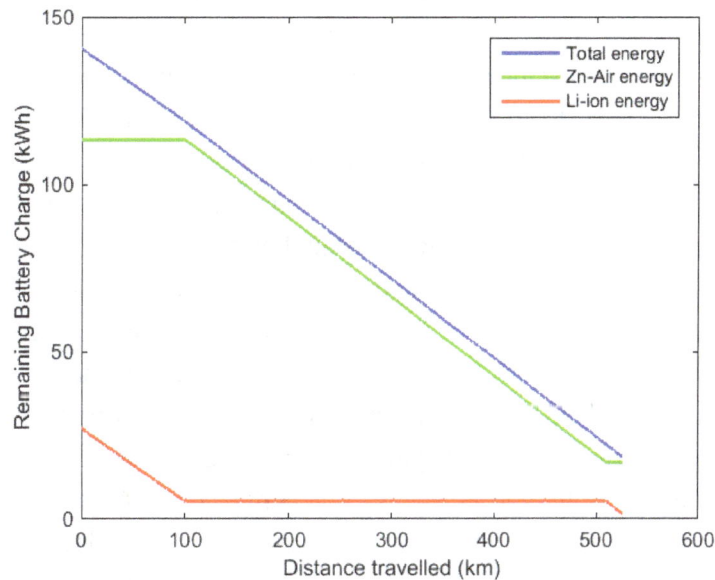

Figure 10. Dual ESS battery energy during continuous highway driving.

5.3. Economic and environmental analysis

In order to analyze the environmental impact and the economics of the Single ESS and the Dual ESS vehicles, these vehicles are compared to a 2016 Chevrolet Camaro and to a traditional PHEV, which has a small lithium-ion battery pack and an ICE as a secondary power source. The results are shown in Table 9. The official Camaro numbers are used in estimating the vehicle's fuel economy [43]. The PHEV Camaro has a lithium-ion battery pack identical to that of the Dual ESS vehicle as described in Table 7, and its fuel economy under charge-sustaining mode is taken to be the same as that of a 2014 Chevrolet Volt [44] (because PHEVs do not need powerful engines, the PHEV Camaro's engine will be more fuel efficient than the 2016 Camaro's engine).

Table 9. Economic and environmental factors.

Measure	Units	Dual ESS	Single ESS	PHEV Camaro	2016 Camaro
Estimated MSRP[1]	USD	$43,660	$43,900	$36,290	$26,700
Fuel Costs[2]	USD/yr	$680	$570	$770	$2,300
Maintenance Costs[3]	USD/yr	$1,540	$1,540	$1,610	$1,920
US CO_2 emissions[4]	kg/yr	2926	2560	2845	5370
CAN CO_2 emissions[4]	kg/yr	1120	980	1504	5370

[1] Based on 2016 Camaro's base price [45], adjusted for battery pack prices, engine costs [46] and cost of battery- and engine-related components, [2] Based on $1.2/L and $0.10/kWh, [3] Based on per-kilometer costs estimated by Propfe and coworkers [47], [4] Based on emissions factors for gasoline and power plants [48] and on the emissions intensity of the overall grid in the US [49] or Canada [50].

The manufacturer's suggested retail price (MSRP) of the Dual ESS and Single ESS vehicles are high, but not so much higher than some of the commercial vehicles on the market. The Chevrolet Bolt, for example, costs $37,500 USD before subsidies but travels over 100 km less on a single charge than the Dual ESS vehicle [51]. The Bolt also serves as an illustrative contrast with the Single ESS vehicle, travelling a similar distance but costing much less and having a much smaller battery pack (60 kWh to 100 kWh). This demonstrates the degree to which the model is constrained by the heavy vehicle base and conservative battery pack design with heavy battery chemistry.

It is particularly notable that the Dual ESS vehicle has higher fuel costs and emissions than the Single ESS vehicle. The reason for this is the Zinc-air battery's poor roundtrip energy efficiency. The individual Zinc-air cells of the Dual ESS battery pack charge at roughly 1.8V and discharge at 1.0 V in a level 1 or level 2 charger, for a 56% roundtrip efficiency. By comparison lithium-ion batteries are typically 85–95% efficient [28]. The Dual ESS vehicle has 20% more power-plant related emissions than the Single ESS vehicle, and in fact emits more GHG emissions than the PHEV if a high percentage of the electricity is generated from fossil fuels.

Maintenance costs are lower for the Dual ESS and Single ESS vehicles due to their not having engines, but not drastically lower as sometimes anticipated for fully electric vehicles. Propfe and coworkers [47] estimate that batteries and their electrical components have sufficiently high maintenance costs that electric vehicles have only a moderate advantage.

6. Conclusions

A Dual ESS vehicle powertrain with two complementary battery packs was modelled and optimized in order to demonstrate the potential of this type of vehicle architecture and of Zinc-air battery technology as a limited cycle range extending pack. The Dual ESS vehicle showed superior range relative to the Single ESS vehicle, going 75 km further on a single charge while costing the same as the Single ESS vehicle. In particular, the model demonstrated that a low cycle life, even as few as 150 cycles, does not preclude Zinc-air batteries from being used in electric vehicles. The Dual ESS vehicle's Zinc-air battery is projected to last 15 years even with only 150 cycles on the Zn-air pack. However, the results do show that improvements in Zinc-air battery cycle life would be highly desirable, and that Zinc-air cells need to be able to cycle well at higher current densities in order to be successful. Improvements in the depth of discharge of anodes and reductions in voltage polarization would also greatly enhance the technology's attractiveness. Further analysis showed that the Dual ESS vehicle produces substantially less emissions than a conventional Camaro and that vehicle owners will pay much less in fuel costs and modestly less in maintenance costs.

Acknowledgements

The authors would like to thank the National Scientific and Engineering Research Council of Canada (NSERC) for funding this work, and the University of Waterloo Alternative Fuels Team for providing a model of their competition vehicle. Thanks are also due to Caixia Wang, Nathan Buckley, Betty Liu and Julia Helter, whose early-stage contributions to the investigation were invaluable. Finally, the authors would like to thank Zi Qi Chen and Murtaza Aziz Fatakdawala for their assistance in generating and recording data from simulations for analysis.

Conflicts of interest

All authors declare no conflicts of interest in this paper.

References

1. Edenhofer O, Pichs-Madruga R, Sokona Y, et al. (2014) IPCC, 2014: Summary for Policymakers. In: Climate Change 2014: Mitigation of Climate Change. Contribution of Working Group III to the Fifth Assessment Report of the Intergovernmental Panel on Climate Change. Available from: http://www.ipcc.ch/pdf/assessment-report/ar5/wg3/ipcc_wg3_ar5_summary-forpolicymakers.pdf.

2   United States Environmental Protection Agency (2016) Inventory of U.S. Greenhouse Gas Emissions and Sinks: 1990-2014. Available from: https://www.epa.gov/sites/production/files/2016-04/documents/us-ghg-inventory-2016-main-text.pdf.

3. Environment and Climate Change Canada (2016) Canadian Environmental Sustainability Indicators: Greenhouse Gas Emissions. Available from: https://www.ec.gc.ca/indicateurs-indicators/F60DB708-6243-4A71-896B-6C7FB5CC7D01/GHGEmissions_EN.pdf.

4. International Energy Agency (2017) Global EV Outlook 2017: Two million and counting.

5. Li W, Long R, Chen H, et al. (2017) A review of factors influencing consumer intentions to adopt battery electric vehicles. *Renew Sust Energ Rev* 78: 318–328.

6. Goldstein J, Brown I, Koretz B (1999) New developments in the electric fuel Ltd. zinc/air system. *J Power Sources* 80: 171–179.

7. Toussaint G, Stevens P, Rouget R, et al. (2012) A high energy density rechargeable Zinc-air battery for automotive application. Meet. Abstr. MA2012-02, 1172–1172. Available from: http://ma.ecsdl.org/content/MA2012-02/11/1172.full.pdf.

8. Buisness Wire (2017) Eos Energy Storage Now Taking Orders at $95/kWh for the Eos Aurora® DC Battery System. Available from: https://www.businesswire.com/news/home/20170418005284/en/Eos-Energy-Storage-Orders-95kWh-Eos-Aurora%C2%AE.

9. Li Y, Dai H (2014) Recent advances in Zinc-air batteries. *Chem Soc Rev* 43: 5257.

10. Bockstette J, Habermann K, Ogrzewalla J, et al. (2013) Performance plus range: Combined battery concept for plug-in hybrid vehicles. *Sae Int J Altern Powertrains* 2: 156–171.

11. Stewart SG, Kohn SI, Kelty KR, et al. (2014) Electric vehicle extended range hybrid battery pack system: US, US 8803470 B2.

12. Catton J, Wang C, Sherman S, et al. (2017) Extended range electric vehicle powertrain simulation and comparison with consideration of fuel cell and metal-air battery. WCX™ 17: SAE World Congress Experience.

13. Rahman MA, Wang X, Wen C (2013) High energy density metal-air batteries: A review. *J Electrochem Soc* 160: A1759–A1771.

14. Fu J, Cano ZP, Park MG, et al. (2016) Electrically rechargeable Zinc-air batteries: Progress, challenges, and perspectives. *Adv Mater* 29: 1604685.

15. CBC News (2014) Electric car with massive range in demo by Phinergy, Alcoa. Available from: http://www.cbc.ca/news/technology/electric-car-with-massive-range-in-demo-by-phinergy-alcoa-1.2664653.

16. Nixon DB (1994) Electric vehicle having multiple replacement batteries: US, US 5542488 A.

17. Passaniti J, Carpenter D, McKenzie R (2011) Button cell batteries: Silver oxide-Zinc and Zinc-air systems. In: Reddy TB (Ed.), *Linden's Handbook of Batteries* (13). Available from: http://www.accessengineeringlibrary.com/browse/lindens-handbook-of-batteries-fourth-edition.

18. Vatsalarani J, Trivedi DC, Ragavendran K, et al. (2005) Effect of polyaniline coating on "Shape Change" phenomenon of porous Zinc electrode. *J Electrochem Soc* 152: A1974–A1978.

19. Chang WL, Sathiyanarayanan K, Eom SW, et al. (2006) Novel alloys to improve the electrochemical behavior of Zinc anodes for Zinc/air battery. *J Power Sources* 160: 1436–1441.

20. Parker JF, Chervin CN, Nelson ES, et al. (2014) Wiring zinc in three dimensions re-writes battery performance-dendrite-free cycling. *Energ Environ Sci* 7: 1117–1124.

21. Jung KN, Jung JH, Im WB, et al. (2013) Doped lanthanum nickelates with a layered perovskite structure as bifunctional cathode catalysts for rechargeable metal-air batteries. *ACS Appl Matter Inter* 5: 9902–9907.

22. Lee DU, Choi J, Feng K, et al. (2013) Advanced extremely durable 3D bifunctional air electrodes for rechargeable Zinc-air batteries. *Adv Energy Mater* 4: 1301389.

23. Clark S, Latz A, Horstmann B (2017) Rational development of neutral aqueous electrolytes for Zin-Air batteries. *Chemsuschem* 2017: 4735.

24. Goh FWT, Liu Z, Hor TSA, et al. (2014) A near-neutral chloride electrolyte for electrically rechargeable Zinc-air batteries. *J Electrochem Soc* 161: A2080–A2086.

25. Eos Energy Storage, Llc. Electrochemical cell with divalent cation electrolyte and at least one intercalation electrode. US 20150244031 A1.

26. Eos Storage. Products and Technology. Available from: https://eosenergystorage.com/products-technology/.

27. Mohamad AA (2006) Zn/gelled 6M KOH/O_2 Zinc-air battery. *J Power Sources* 159: 752–757.

28. Dahn J, Ehrlich GM (2011) Lithium-Ion Batteries. In: Reddy TB (Ed.), *Linden's Handbook of Batteries* (26). Available from: http://www.accessengineeringlibrary.com/browse/lindens-handbook-of-batteries-fourth-edition.

29. Argonne National Labs (2017) Available from: http://www.autonomie.net/expertise/Autonomie.html.

30. Argonne National Laboratory (2016) EcoCar3 Advanced Vehicle Technology Competition. Available from: http://ecocar3.org/.

31. A123 Systems, Inc. Nanophosphate Basics: An Overview of the Structure, Properties and Benefits of A123 System' Proprietary Lithium Ion Battery Technology. Available from: https://www.neces.com/assets/A123-Systems_Nanophosphate-overview-whitepaper_FINAL1.pdf.

32. A123 Systems, Inc. 20Ah Prismatic Pouch Cell. Available from: http://accessengineeringlibrary.com/browse/lindens-handbook-of-batteries-fourth-edition/p2001c2f299713_1001#p2001c2f299713_16002.

33. Nejad S, Gladwin DT, Stone DA (2016) A systematic review of lumped-paramter equivalent circuit models for real-time estimation of lithium-ion battery states. *J Power Sources* 316: 183–196.

34. A123 Systems, Inc. Battery Pack Design, Validation and Assembly Guide using A123 Systems AMP20mlHD-A Nanophosphate Cells. Available from: http://www.formula-hybrid.org/wp-content/uploads/A123_AMP20_battery_Design_guide.pdf.

35 Nykvist B, Nilsson M (2015) Rapidly falling costs of battery packs for electric vehicles. *Nat Clim Change* 5: 329–332.

36. Eckl R, Ehrl B, Lienkamp M (2013) Range extender for seldom use in the electric car mute-Zinc Air battery. Conference on Future Automotive Technology: Focus Electromobility, Munich.

37. Adler TC, McLarnon FR, Cairns EJ (1993) Low-Zinc-solubility electrolytes for use in Zinc/Nickel oxide cell. *J Electrochem Soc* 140: 289–294.

38. Narayanan SR, Prakash GKS, Manohar A, et al. (2011) Materials challenges and technical approaches for realizing inexpensive and robust iron-air batteries for large-scale energy storage. *Solid State Ionics* 216: 105–109.

39. U.S. Department of Energy (2017) Available from: https://www.fueleconomy.gov/feg/fe_test_schedules.shtml.

40. U.S. Department of Transportation, Federal Highway Administration, 2009 National Household Travel Survey. Available from: http://nhts.ornl.gov.

41. Drillet JF, Holzer F, Kallis T, et al. (2001) Influence of CO_2 on the stability of bifunctional oxygen electrodes for rechargeable zinc/air batteries and study of different CO_2 filter materials. *Phys Chem Chem Phys* 3: 368–371.

42. Dharmakeerthi CH, Mithulananthan N, Saha TK (2014) Impact of electric vehicle fast charging on power system voltage stability. *Int J Electr Power* 57: 241–249.

43. U.S. Department of Energy (DOE) (2017) Model Year 2016 Fuel Economy Guide. Available from: https://www.fueleconomy.gov/feg/pdfs/guides/FEG2016.pdf.

44. U.S. Department of Energy (DOE) (2016) Model Year 2014 Fuel Economy Guide. Available from: https://www.fueleconomy.gov/feg/pdfs/guides/FEG2014.pdf.

45. Vijayenthiran V (2015) 2016 Chevrolet Camaro Full Pricing Released. *MotorAuthority*. Available from: http://www.motorauthority.com/news/1099687_2016-chevrolet-camaro-full-pricing-released.

46. Kochhan R, Fuchs S, Reuter B, et al. (2014) An overview of costs for vehicle components, fuels and greenhouse gas emissions.

47. Propfe B, Redelbach M, Santini DJ, et al. (2012) Cost analysis of Plug-in Hybrid Electric Vehicles including Maintenance & Repair Costs and Resale Values, EVS26 International Battery, Hybrid and Fuel Cell Electric Vehicle Symposium, Los Angeles, California.

48. Schlömer S, Bruckner T, Fulton L, et al. (2014) Annex III Technology-specific cost and performance parameters. In: Climate Change 2014: Mitigation of Climate Change. Contribution of Working Group III to the Fifth Assessment Report of the Intergovernmental Panel on Climate Change. Available from: https://www.ipcc.ch/pdf/assessment-report/ar5/wg3/ipcc_wg3_ar5_annex-iii.pdf.

49. U.S. Energy Information Administration. (2016, April). Frequently Asked Questions: What is U.S. electricity generation by energy source? Available from: https://www.eia.gov/tools/faqs/faq.cfm?id=427&t=3.

50. Canadian Electricity Association (2014) *Key Canadian Electricity Statistics*. Available from: http://www.electricity.ca/media/Electricity101/KeyCanadianElectricityStatistics10June2014.pdf.

51. U.S. Department of Energy (2017) Available from: https://www.fueleconomy.gov/feg/Find.do?action=sbs&id=38187.

9

The probabilistic load flow analysis by considering uncertainty with correlated loads and photovoltaic generation using Copula theory

Li Bin[1], Muhammad Shahzad[1,*], Qi Bing[1], Muhammad Ahsan[2], Muhammad U Shoukat[3], Hafiz MA Khan[1] and Nabeel AM Fahal[1]

[1] School of Electrical and Electronic Engineering, North China Electric Power University, Beijing 102206, P.R. China
[2] Department of Electrical Engineering, The Superior University Lahore, Lahore 54000, Pakistan
[3] Department of Electrical Engineering, Government College University Faisalabad Sahiwal Campus, Sahiwal 57000, Pakistan

* Correspondence: Email: shahzadpansota@hotmail.com

Abstract: In this paper, a probabilistic load flow analysis is proposed in order to deal with probabilistic problems related to the power system. Due to increasing trend of penetration of renewable energy sources in power system brought two factors: One is uncertainty, and another one is dependence. Uncertainty and dependence factor increase risk associated with power system operation and planning. In this proposed model these two factors is considered. Gaussian Copula theory is proposed to establish the probability distribution of correlated input random variables. Three sampling methods are used with Monte Carlo simulation as simple random sampling, Box-Muller sampling, and Latin hypercube sampling in order to evaluate the accuracy of the proposed method. The main advantages of this model are as: It can establish any type of correlation between input random variable with the help of Copula theory, it is free from the restrictions of Pearson coefficient of correlation, it is unconstrained by the marginal distribution of input random variables, and uncertainty is established with photovoltaic generation this is the main source of uncertainty. Additional, in order to evaluate the accuracy and efficiency of the proposed model a real load and photovoltaic generation data is adopted. For accuracy evaluation purpose two comparative test system is adopted as modified IEEE 14 and IEEE 118-bus test system.

Keywords: Copula theory; correlation; Monte Carlo simulation; photovoltaic generation; probabilistic load flow analysis; sampling; uncertainty

---

1. Introduction

Currently, the electric power system is confronting many uncertainties. The main source of these uncertainties is penetration of renewable energy sources in power system network. Among these renewable energy sources, photovoltaic (PV) generation and wind energy generation are the dominant sources of uncertainty. Due to randomness nature of uncertainty, it increases the risk associated with electric power system operation modes and further has an influence on load flow analysis. Most of these uncertainties are dependent. This dependence relationship can be linear or non-linear. For example, the loads in the same area can be increased or decrease in a same time due to environmental and social ones. Similarly, the output power of wind energy sources is strongly correlated with the wind speed of neighboring measuring stations [1,2]. On the other hand, the power system highly uncertain due to changing pattern of power demand like the transport (electric mobility) and heat sectors (heat pumps in buildings) [1]. Safeguarding the reliability of the power system is a key enabler for this massive system transformation. For the operation of future power systems, new methods are needed to incorporate the different sources of uncertainty (contingencies, load and renewable energy source forecast errors) during the operational planning stage and to help system operators to steer the system into risk-averse modes of operation [3–5].

The probabilistic load flow (PLF) analysis is a tool that can handle uncertainties in the efficient and effective way. The PLF is a steady-state load flow analysis to determine the network parameters by considering uncertainty with input random variables flexibly. The information obtained by PLF analysis is used for power system security assessment, operation and control purpose. In literature, many PLF methods have been used in order to deal with correlated input random variables. In these methods included: Gaussian mixture model method [6–8], point estimation method [9–11], convolution method [12,13], cumulant method [8,14], unscented transformation [15], and Monte Carlo simulation (MCS) [16–19]. MCS methods are the most popular methods among all of above methods. It is due to its simplicity and ability to handle complex non-linear problems. Most of the above methods only deal with a linear dependence. Additional, the Persons linear correlation coefficient has been used for measuring the degree of dependence. Most of the nature uncertainties are not linear. This proposed model can handle this non-linear dependence more flexible.

All methods used for PLF can be divided into four categories. These categories are as numerical methods, analytical methods, approximate methods, and hybrid methods [20]. A numerical MCS method with simple random sampling (SRS) has been used frequently for PLF power system problems. MCS with SRS can be express as SMCS. SMCS method has been used as a reference for accuracy comparison in PLF studies [5,8,10,15,21–24]. From the literature, it is clear that no one method is more accurate than SMCS in term of accuracy. However, it suffers heavy computational burden to achieve the desired accuracy. To overcome computational burden, in literature many methods have been used such as convolution techniques [12,13]. The convolution base methods are also time-consuming due to discrete point problems. Point estimation based methods have lesser computational burden but less accurate. In fact, there are no holistic criteria for PLF evaluation. Always there is a compromise between computational burden and accuracy. But currently, a third moment has been started to overcome computational burden and accuracy problems. This moment is to hybrid methods like combined cumulants and Laplace transform method [25], combine cumulant and Gaussian mixture method [8], cumulant and multiple integral method [26], second order design method [17]. Actually, the accuracy and computational burden of PLF methods highly depend upon the modelling of input uncertainty and sampling techniques.

The contribution in this research work has many aspects included: Almost all of the methods in literature can only deal with a linear correlation between input random variables and have been considered Pearson linear correlation coefficient. This Pearson linear correlation coefficient some time cannot measure the degree of correlation. For example: 1) it cannot measure non-linear dependence between input random variable; 2) it can't measure dependence for those distributions that have not define standard deviation; 3) it is invariant under strictly increasing transformation. To overcome this problem in this work Gaussian Copula theory is proposed that can handle linear and non-linear problems flexibly. By using Copula theory, a desirable correlation can be achieve between input random variables. The other aspect of this work is modelling of input uncertainty. The PV generation and correlated load are used for modelling the input uncertainty. The special in this work real data about PV generation and load are used for accuracy comparison purpose. Additional, the Box-Muller sampling (BMS) and Latin hypercube sampling (LHS) are used for accuracy and computational burden comparison purpose. The SMCS is used for a benchmark for other methods in term of accuracy and computational burden. MCS combine with BMS can be express as BMCS and MCS combine with LHS can be express as LMCS. For three methods, Copula theory is used for capturing the complex stochasticity of correlated loads and PV generation. The modified IEEE 14-bus and IEEE 118-bus test system is used for obtaining results.

The paper is organized as follows. Section 2 describes the modelling of the probability distribution of input random variable included correlated loads and correlated PV generation. Section 3 describes the methodology, and sampling techniques included SRS, BMS, and LHS. Section 4 describes the PLF analysis procedure. Section 5, evaluate the performance analysis of proposed methods. Finally, conclusions are drawn in Section 6.

2. Modelling of input random variables for uncertainty

The accuracy and computational efficiency of PLF analysis usually depending upon three major factors included: a) Uncertainty handing method; b) Accurate power system model; c) Accurate modelling input random variable uncertainty.

2.1. Modelling of correlated loads for uncertainty

There are normally two methods for measurement of correlation between random variable with respect to degree and structure. First one is Pearson's linear correlation coefficient that can only measure the degree of correlation Eq 1. The second one is rank correlation coefficient and Copula theory that can measure the degree as well as the structure of correlation. In this work, Copula theory is proposed to measure the structure as well as the degree of correlation between random variables flexibly. This theory can measure the linear and non-linear correlation between random variables.

More, it is invariant under strictly increasing transformation. The correlation matrix for n input random variable is shown in Eq 1 as:

$$Cor\_matrix = \begin{bmatrix} 1 & \rho_{21} & \cdots & \rho_{1n} \\ \rho_{21} & 1 & \cdots & \rho_{2n} \\ \vdots & \vdots & \ddots & \vdots \\ \rho_{n1} & \rho_{n2} & \cdots & 1 \end{bmatrix} \tag{1}$$

Where ρ_{ij} is the correlation coefficient and can be calculated as $\rho_{ij} = \text{cov}(W_i, W_j)/\sigma_i \sigma_j$ where σ_i and σ_j is the standard deviation of a random variable W_i and W_j respectively and $\text{cov}(W_i, W_j)$ is the covariance. According to Copula theory, it can be express as in Eq 2 for bivariate:

$$F_{12}(w_1, w_2) = C(F_1(w_1), F_2(w_2)) \tag{2}$$

Where C is a copula function, according to Sklar theorem [27]. It couples multivariate joint distribution CDF to one dimensional marginal CDFs. $F_1(w_1)$ and $F_2(w_2)$ are the marginal CDF of two random variables W_1 and W_2 respectively. $F_{12}(w_1, w_2)$ is the joint CDF of these two random variables. $C(F_1(w_1), F_2(w_2))$ is the Copula density function, and it must follow the Sklar theorem property as mention in Eq 3:

$$U \in [0,1] : P(U \leq u) = P(F(w) \leq u)$$
$$= P\left(w \leq F^{-1}(u)\right) = F\left[F^{-1}(u)\right] = u \tag{3}$$

It can be extended to multivariate as in Eq 4, m is the number of random variables:

$$f(w_1, \ldots, w_m) = c\left(F_1(w_1), \ldots, F_m(w_m)\right) \cdot \prod_{i=1}^{m} f_i(w_i) \tag{4}$$

Where $c\left(F_1(w_1), \ldots, F_m(w_m)\right) = \dfrac{\partial^m C\left(F_1(w_1), \ldots, F_m(w_m)\right)}{\partial F_1(w_1) \ldots \ldots \partial F_m(w_m)}$ is the Copula density function for multivariate random variables.

There are many Copula families, but famous one is included elliptical Copula and Archimedean Copula. Here, Gaussian Copula is proposed to measure the dependence between input random variables that belong to Archimedean Copula family. All loads PDFs are modelled as a Gaussian distribution follows the Eq 5:

$$f(x; \mu; \sigma) = \frac{1}{\sqrt{2\pi\sigma}} e^{-\left(\frac{(x-\mu)^2}{2\sigma^2}\right)} \tag{5}$$

Where x is the required value for which probability distribution function (PDF) needed, μ is the mean value of the load, and σ is the arbitrary value of standard deviation. To show the accuracy of modelled load PDF is compare with actual one day load at MAISY utility customer energy use with per hourly interval available in [28]. Figure 1 shows the PDFs and CDFs of real, generated, and ideal data of loads. Gaussian Copula function is used to establish the desired correlation between different random variables (loads) in the proposed network. The PDFs of multivariate Gaussian distribution with desirable correlation can be found by Eq 6.

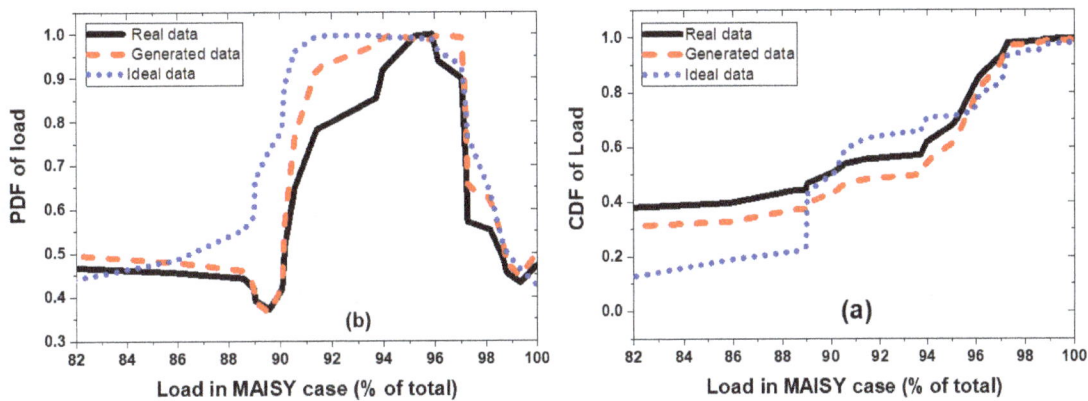

Figure 1. Probability density function (PDF) and Cumulative density function (CDF) of real, generated, and ideal data for loads are shown in (a) and (b) respectively.

$$C_R^{Gauss}(u) = \frac{1}{\sqrt{\det . R}} \exp\left(-\frac{1}{2} \begin{pmatrix} \phi^{-1}(u_1) \\ \cdots \\ \cdots \\ \phi^{-1}(u_d) \end{pmatrix}^T . (R^{-1} - I) . \begin{pmatrix} \phi^{-1}(u_1) \\ \cdots \\ \cdots \\ \phi^{-1}(u_d) \end{pmatrix}\right) \tag{6}$$

Where R is the correlation matrix belong to $[-1,1]^{d \times d}$, C show the correlation density function, ϕ^{-1} is the inverse CDF of standard normal distribution, $u_1, \ldots, u_d$ is the random variables, d is the number of variables, and I is the identity matrix. The correlation of bivariate function over unit domain from weak to strongly correlated loads with the help of Gaussian Copula is shown in Figure 2.

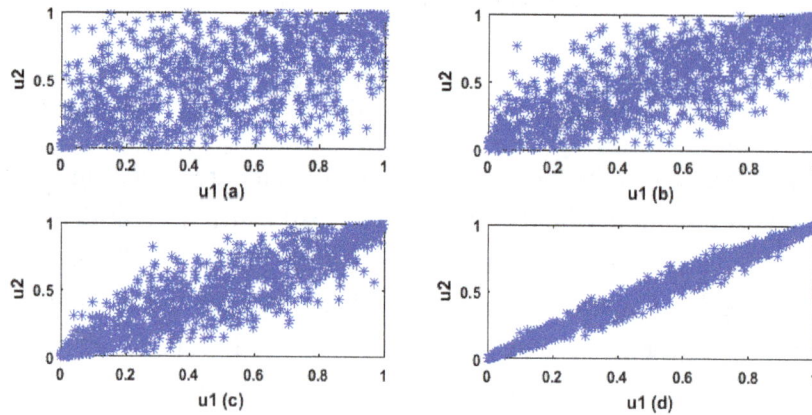

Figure 2. Scatter plots of correlated loads in unit domain (a) ρ = 0.6; (b) ρ = 0.8; (c) ρ = 0.9; (d) ρ = 0.99.

2.2. PV generation modelling for uncertainty

In this work, PV generations are model as a beta distribution with the help of Eq 7. The historical solar power data for integration studies located at California western state is taken to check the accuracy of modelled PV generation available at National Renewable Energy Laboratory (NREL) [29]. One day data (Jun 1, 2006) with a 5-minute interval is taken to check the accuracy of the proposed model.

$$f(x;,\alpha,\beta) = \frac{\Gamma(\alpha+\beta)}{\Gamma(\alpha)\Gamma(\beta)} x^{\alpha-1}(1-x)^{\beta-1} \tag{7}$$

Where x is the observed values or realization, α and β are the shape parameters and always α, $\beta > 0$, Γ is the gamma function. The PDFs and CDFs of real, generated, and ideal values model with the help of Eq 7 are shown in Figure 3.

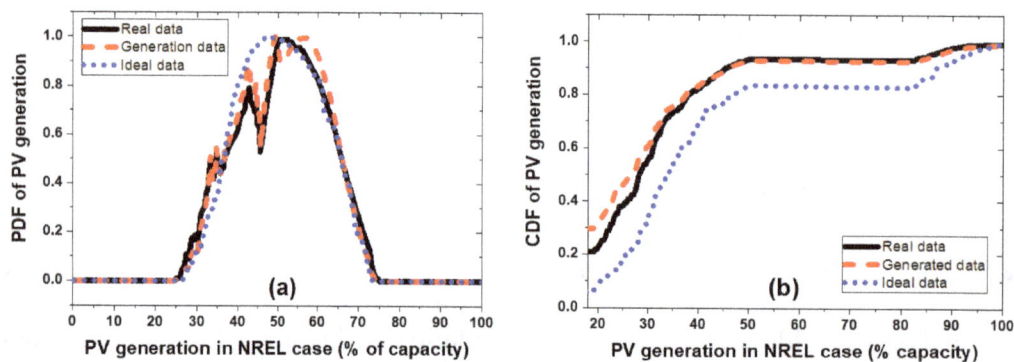

Figure 3. Probability density function (PDF) and Commutative density function (CDF) of PV generation with real, generated, and ideal data are shown in (a) and (b) respectively.

3. Methodology

Monte Carlo Simulation is an iterative approach, mostly used for probabilistic load flow analysis. In this approach cumulative distribution function is prepared of a random variable to obtain the uncertainty. To model, the CDF of a random variable is an essential step in this approach. Accurate modelling of CDF give the accurate final results. In this numerical approach, in order to model the uncertainty as a random variable, three basic steps are needed as shown in Figure 4: (1) to model CDF of random variable; (2) to solve the deterministic load flow model; (3) statistical analysis of results.

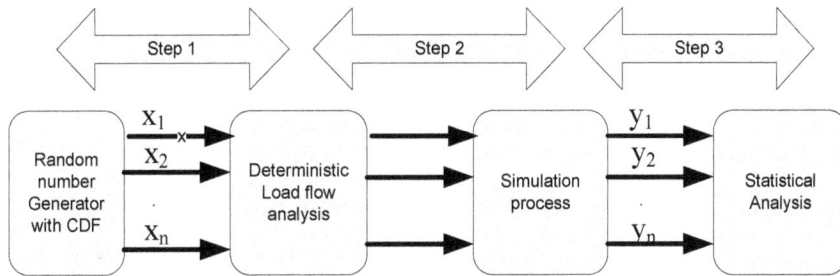

Figure 4. Monte Carlo Simulation steps.

Monte Carlo Simulation process can be described by Eq 8:

$$Y = h(X) \tag{8}$$

$$X = [X_1, ..., X_n]^T \tag{9}$$

$$Y = [Y_1, ..., Y_n]^T \tag{10}$$

Where in Eq 8–10, X is the vector of input random variable, Y is a vector of output random variables vector, and h is the function of the model under study. The object of study is to obtain PDF of Y at given known PDF of X. T shows the transpose.

3.1. Simple random sampling

Simple random sampling techniques is a theoretical basis technique to design other methods. The variance of the mean of output variable is calculated to find the robustness of sampling techniques [30]. The mean of the sample is calculated as in Eq 11:

$$\bar{y} = \frac{1}{N} \sum_{i=1}^{N} y_i \tag{11}$$

Where N is the number of samples, $\bar{y}$ is the expected value and y_i is the output value of *ith* iteration. The variance of the sample can be obtained as in Eq 12:

$$\text{var}(y) = \frac{1}{N}\sum_{i=1}^{N}(y_i - \bar{y})^2 \tag{12}$$

The variance of the estimating mean can be determined by Eq 13:

$$\text{var}(\bar{y}) = \frac{1}{N}\text{var}(y) \tag{13}$$

3.2. Box-Muller sampling

Box-Muller sampling is also as a sampling technique. In order to determine the sample of given iteration, firstly two random number from uniform distribution in the range of (0, 1) is generated. The sample should be determined by following Eq 14:

$$S_i = \mu_i + \cos(2\pi u_1) \times \sigma_i\sqrt{-2ln(u_2)} \tag{14}$$

Where, S_i = is the ith sample; μ = mean value of random variable; σ = standard deviation of a random variable; u_1 & u_2 = uniformly distributed random number in range of (0, 1). For the proposed PLF model, μ_i and σ_i are the base load and standard deviation for input variable.

3.3. Latin hypercube sampling

Latin hypercube sampling (LHS) is the type of stratified sampling. It uses to generate a sample of a random variable from entire distribution. In this technique, random variable distribution is divided into equal probability distribution intervals. Each interval is sample exactly once in its entire range of the random variable. In this way, sample value represents the higher sampling efficiency. The sampling procedure is as follows:

$$Y_n = F_n(G_n) \tag{15}$$

Where in Eq 15, Y_n is the CDF of N input random variables such that $G_n = G_1,........,G_N$ in probability theory, in range $Y_n = [0, 1]$. If the sample size is denoted by K. If the entire range of Y_N is divided into k non-overlapping intervals than each interval length is equal to $1/k$ as shown in Figure 5. The sample value is generated from each interval without replacement randomly or midpoint value. In this work, midpoint value is chosen, and this method is call lattice sampling. The sampling value of G_n can be computed by taking the inverse function of Eq 16. The *nth* sample can be computed as in Eq 16:

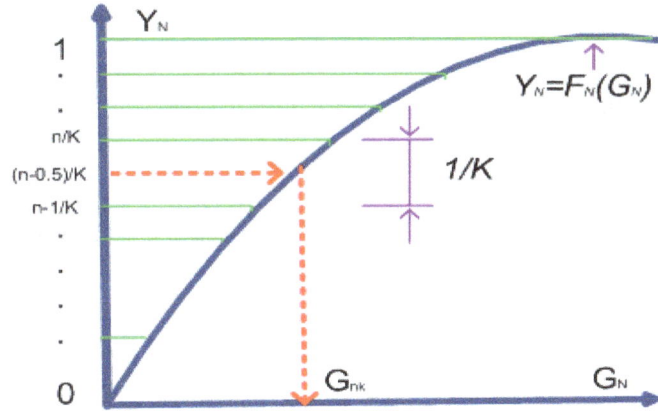

Figure 5. Illustration of latin hypercube sampling procedure.

$$G_{nk} = F_n^{-1}\left(\frac{n-0.5}{k}\right) \tag{16}$$

The sample values of G_n can be assembled as in Eq 17 in the row of sampling matrix:

$$[G_{n1},........,G_{nN}] \tag{17}$$

From this procedure, the $n \times k$ matrix can be obtained.

As described in Section 3.1 the robustness of sampling techniques can be measure on the mean output statistic [30]. Therefore, the mean and variance can be computed by using Eq 11 and Eq 12 respectively. The variance of the estimated mean can be computed with the help of Eq 18:

$$Var(\bar{y}) = \frac{1}{N}\text{var}(y) - \frac{N-1}{N}con(G_1,G_2) \tag{18}$$

Where $con(G_1,G_2)$ is the covariance between random variables.

4. Probabilistic load flow evaluations

The deterministic load flow analysis can be described by Eq 19:

$$Y = h(\mathbf{X}) \tag{19}$$

Where X is the input vector of nodal active and reactive power injections. Y is the output vector of voltage (V), voltage phase angle (θ), active load flow (P_{ij}), reactive load flow (Q_{ij}), $h(X)$ is the load flow function [31]. In case of PLF analysis, correlated nodal loads, PV generation, and conventional generator are the input random variable with their probability distribution functions. The statistical distribution of output variables V, θ, P, and Q are calculated. There are following steps used to perform PLF analysis:

- Prepare the CDF of model correlated loads and correlated PV generations, to set all basic requirement for deterministic load flow analysis, e.g. sample size N.
- To determine the input random variables G.
- To select the desire sampling procedure.
- To generate the sample matrix N × k.
- To set the initial starting point n = 1.
- To perform the deterministic load flow analysis and obtain the values of V, θ, P, Q.
- Set $n = n+1$; if $n < N$ go to step 6. Otherwise, go to next step.
- To do a statistical analysis of output random variable V, θ, P, Q.

5. Performance evaluation of proposed methods

A series of PLF analysis is carried out to determine the performance of SMCS, BMCS, and LMCS. The PLF analysis is carried out on modified IEEE 14-bus test system and IEEE 118-bus test system. The program is developed with "DIgSILENT PowerFactory (15.1) platform. The simulation was performed on PC with AMD A12-9700P, RADEON R7, 10 COPMPUTE CORES UC+6G, 2.5 GHZ processing speed, and 8 GB RAM". The results obtained by the proposed method are compared with correlated SMCS.

To determine the accuracy of proposed model two error indices are introduced in Eq 20 and Eq 21 [9,10] are adopted.

$$\varepsilon_\mu^* = \left| \frac{\mu_{acc} - \mu_{pro}}{\mu_{acc}} \right| \times 100\% \tag{20}$$

$$\varepsilon_\sigma^* = \left| \frac{\sigma_{acc} - \sigma_{pro}}{\sigma_{acc}} \right| \times 100\% \tag{21}$$

Where ε_μ^* is an error of the output random variables with * category (* represents V, θ, P, and Q). Second ε_σ^* is an error of standard deviation with the * category. The results obtained by the SRS method with a sample size of 20,000 are assumed to be accurate and benchmark. The mean and standard deviation of SMCS method are express as μ_{acc} and σ_{acc} respectively. Similarly, the mean and standard deviation of proposed methods are express as μ_{pro} and σ_{pro} respectively. $\bar{\varepsilon}_\mu^*$ and $\bar{\varepsilon}_\sigma^*$ are the average error index and average standard deviation error index that determined the distribution convergence of entire system.

To evaluate the convergence of two methods, each method is run 100 times with certain sample size. In this way, 100 values of ε_μ^-* and ε_σ^-* are calculated with four variable categories. For 100 times calculation, two indices are introduced. First one is the mean of the standard deviation error. The second one is the mean of a maximum of error. The standard deviation error is the standard deviation error obtain from 100 calculations for each output categories and denoted by $\varepsilon^*_{\sigma 100}$. The mean of the standard deviation error is the mean of $\varepsilon_{\sigma 100}$ each specific type of output random variables and denoted by $\bar{\varepsilon}_{\sigma 100}$. This $\bar{\varepsilon}_{\sigma 100}$ show the stability of two methods. The maximum error of each output random variables are calculated in term of its mean and standard deviation. The mean of maximum of error is calculated is express as $\bar{\varepsilon}_{\mu,\max}$ for each output categories. Similarly, mean of the maximum of error standard deviation is calculated and denoted by $\bar{\varepsilon}_{\sigma,\max}$ for each output random variable is calculated. For example, let $V_1,......,V_n$ is the voltage magnitude of the n node test system and $\varepsilon^1_{\mu-V_i},......,\varepsilon^{100}_{\mu-V_i}$ be the error of V_i after 100 calculations $i=1,......,n$. Then $\varepsilon^{V_i}_{\mu,\max} = \max\{\varepsilon^1_{\mu-V_i},........,\varepsilon^{100}_{\mu-V_i}\}, i=1,......,n$. For voltage variable mean of the maximum of error is calculated as $\varepsilon^{-V}_{\mu-\max} = \frac{1}{n}\sum_{i=1}^{n}\varepsilon^{V_i}_{\mu-\max}$. Similarly, other variable statistic can be obtain.

5.1. The modified IEEE 14-bus test system

All deterministic load flow data about this test system is available in [32]. The modified IEEE 14-bus test system is shown in Figure 6. There are total 23 input random variables in this test system. The load distribution is followed Gaussian distribution and discrete, and generator output follows the binomial distribution. The PLF analysis with 20,000 times MCS with SRS sampling consider to be the benchmark for other two methods and consider to be accurate and use to determine the error for other methods as it is usual practice in PLF study [9,10]. Two kinds of correlation is adopted in this test system. First one is between load and other is between active power outputs of PV generations. The constant power factor load model is adopted. The detail load parameters are presented in Table 1.

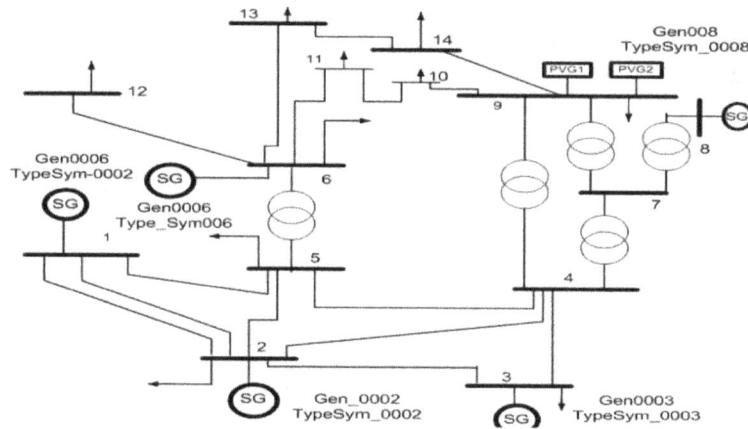

Figure 6. Modified IEEE 14-bus test system.

Table 1. Parameters of loads.

Node No.	Mean (p.u)	StdDev (p.u)	Power Factor
2	0.215	0.065	0.864
3	0.943	0.031	0.991
4	0.481	0.115	0.365
5	0.074	0.036	0.978
6	0.111	0.037	0.833
9	0.295	0.156	0.872
10	0.094	0.036	0.823
11	0.036	0.015	0.889
12	0.061	0.024	0.965
13	0.134	0.048	0.919
14	0.151	0.062	0.949

All loads are divided into three groups as $G_1 = \{2,3,4,5,6\}$, $G_2 = \{10,11,12,13,14\}$ and $\{9\}$. The correlation matrix between G_1 is presented in Eq 22 and correlation matrix between G_2 is presented in Eq 23. The correlation between same group loads is dependent as shown in Eq 22 and Eq 23 and also correlation between different load groups also dependent with 0.5 correlation coefficient.

$$G_1 = \begin{bmatrix} 1 & 0.6 & 0.7 & 0.5 & 0.7 \\ 0.6 & 1 & 0.7 & 0.6 & 0.8 \\ 0.7 & 0.7 & 1 & 0.7 & 0.8 \\ 0.5 & 0.6 & 0.7 & 1 & 0.7 \\ 0.7 & 0.8 & 0.8 & 0.7 & 1 \end{bmatrix} \tag{22}$$

$$
G_2 = \begin{bmatrix} 1 & 0.6 & 0.7 & 0.5 & 0.7 \\ 0.6 & 1 & 0.7 & 0.6 & 0.8 \\ 0.7 & 0.7 & 1 & 0.7 & 0.8 \\ 0.5 & 0.6 & 0.7 & 1 & 0.7 \\ 0.7 & 0.8 & 0.8 & 0.7 & 1 \end{bmatrix}
\tag{23}
$$

Load 9 is correlated with active power output of PV generation available at node 9 with correlation coefficient 0.6. Two PV generation is located at node 9 with installed capacity 0.1 p.u, the active power of PV generation follow the Eq 7 and reactive power assumed to be zero [33,34]. The correlation coefficient matrix between two PV generation and load 9 is presented in Eq 24.

$$
G_{PV-L9} = \begin{bmatrix} 1 & 0.9 & 0.6 \\ 0.9 & 1 & 0.6 \\ 0.6 & 0.6 & 1 \end{bmatrix}
\tag{24}
$$

The simulation is change from 100 to 1000 with step size 100. In order to evaluate the performance of SMCS, BMCS, and LMCS error curve are adopted. All the error indices are calculated for all type of output random variables included voltage magnitude, phase angle, active power, and reactive power. The error indices of only active power are consider for presenting the results for all four categories of variables.

5.2. Simulation results analysis of IEEE 14-bus test system

In the simulation analysis, three types are error indices are calculated as average error, mean of the maximum error, and mean of the standard deviation error indices. The average error index determines the distribution convergence degree of the entire system with corresponding output random variable. The other two error indices as mean of the standard deviation error and mean of the maximum of error are determine the stability of the entire method. For obtaining these error indices simulation is performed 100 times for 1000 samples with 100 step size.

The results of Figure 7 (a–c) illustrate the mean of the maximum of error curves with three methods. To present the results only active power through the line 2–3 was considered for IEEE 14-bus test system. It shows the percentage of the mean of maximum of error for 100 sample size to 1000 sample size with 100 step size of three methods SMCS, BMCS, and LMCS. The mean of the maximum of error of three methods SMCS, BMCS, and LMCS was almost 4.88%, 3.56%, and 0.96% respectively with 100 sample size. Similarly, this error index was 2.15%, 1.16%, 0.15% with 1000 sample size. This error index promptly reduces at 200 sample size, after 200 sample size it remains almost constant. Finally, the results of Figure 7 (a–c) shows that LMCS method is more stable method than other two methods. The results of Figure 7 (d–f) illustrate the average error index for three methods. Form these results, it is clearly shown that the LMCS has 0.10% error, but the other two method SMCS and BMCS have 0.71% and 0.58% respectively with 1000 sample size. This show that the LMCS method is best convergence degree for proposed test system rather than other two methods. The results of Figure 7 (g–i) illustrate the mean of the standard deviation error. This show that the LMCS has 0.11% error and other two methods have 0.42% and 0.32% respectively with 1000 sample size. Form these results, it is clearly shown that the LMCS is more stable method than the other two methods.

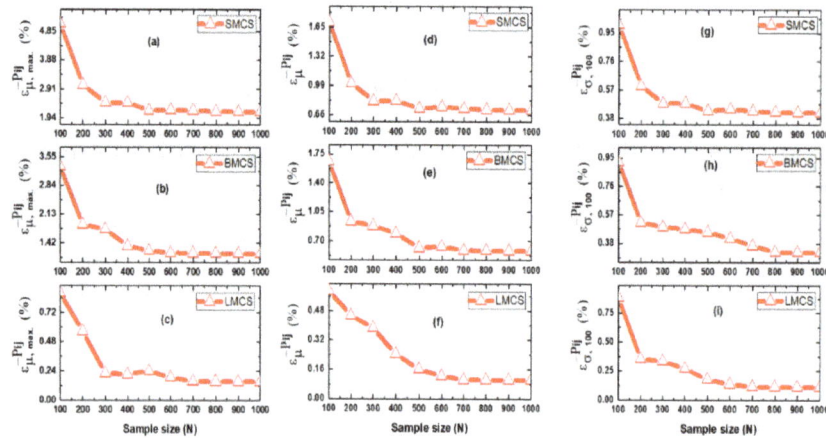

Figure 7. Mean of the maximum error (a–c), average error indices (d–f), and mean of the standard deviation error (g–i) curves for active power through the line 2–3 for corresponding method.

The standard deviation error indices are presented in Figures 8 (a–i). The mean of maximum error of standard deviation is presented in Figure 8 (a–c), average of standard deviation error index is in Figure 8 (d–f), and mean of standard deviation error is in Figure 8 (g–i). All of these indices are calculated with 1000 sample size. The mean of maximum of standard deviation error index of LMCS method was with 1.49%, but the SMCS and BMCS have 3.36% and 2.23%. Similarly, average standard deviation error index of LMCS was 0.40%, but the SMCS and BMCS have 2.23% and 0.44%. Similarly, mean of standard deviation error index of LMCS was 0.16%, but the SMCS and BMCS have 0.28% and 0.20%. All of these results prove that LMCS method is much better than SMCS but almost identical to BMCS method according to distribution convergence and stable point of view.

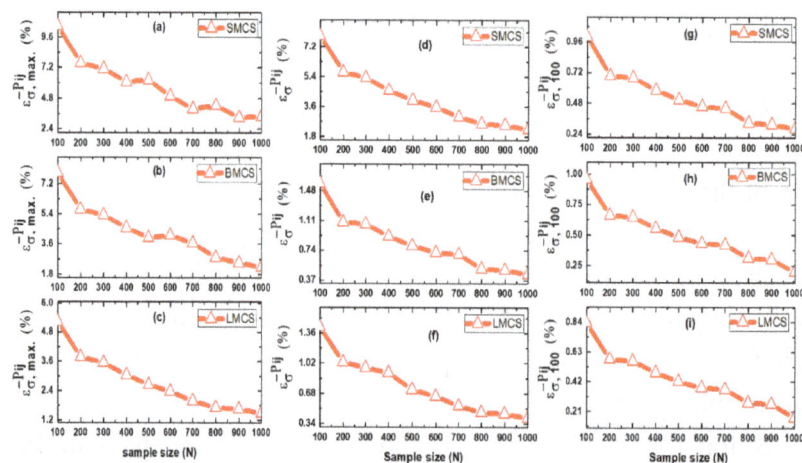

Figure 8. Mean of maximum standard deviation error (a–c), average standard deviation error indices (d–f), and mean of standard deviation error (g–i) curves for active power through the line 2–3 for corresponding method.

All of the above calculated error indices are for active power output random variable. The remaining three output random variables, i.e. voltage magnitude, phase angle, and reactive power error indices are very similar to this. These are not shown here due to space limitations. All of the remaining output random variable's error indices are calculated and presented in Table 2 with sample size 400. The results of Table 2 show that error indices of all output variable are within 10% with SMCS but with BMCS and ILMCS are within 5%. Additional, the PDF and CDF of active power flow through the line 2–3 are illustrated in Figure 9. From the results of Figure 9, shows that the curves are very close to each other for three methods. All of these curves are calculated with a different sample sizes like SMCS with 20,000 samples and BMCS, and LMCS are with 1000 samples. The accuracy of presented results is almost same but BMCS and LMCS with 1000 sample size. The computational time comparison of three methods is shown in Figure 10. The computational time of SMCS method with sample size 20,000 was 981 s seconds. The computational time for both BMCS and LMCS was almost same like 3.17 s but with 1000 sample size. The computational burden of LMCS was very small for desire accuracy. The results of Figure 10 shows the computational superiority of ILHS with respect to same sample size.

Table 2. Error comparison of IEEE 14-bus test system.

Method		SMCS	BMCS	LMCS	Method		SMCS	BMCS	LMCS
ε_μ^v (%)	$\bar\varepsilon_{\mu,max}^v$	1.022	0.018	0.012	$\varepsilon_\mu^{P_{ij}}$ (%)	$\bar\varepsilon_{\mu,max}^P$	2.151	1.164	0.156
	$\bar\varepsilon_\mu^v$	1.042	0.037	0.016		$\bar\varepsilon_\mu^P$	0.717	0.852	0.104
	$\bar\varepsilon_{\sigma100}^v$	0.451	0.021	0.019		$\bar\varepsilon_{\sigma100}^P$	0.421	0.323	0.115
ε_σ^v (%)	$\bar\varepsilon_{\sigma,max}^v$	6.451	1.882	1.667	$\varepsilon_\sigma^{P_{ij}}$ (%)	$\bar\varepsilon_{\sigma,max}^P$	3.365	2.235	1.490
	$\bar\varepsilon_\sigma^v$	2.957	0.924	0.863		$\bar\varepsilon_\sigma^P$	2.235	0.447	0.406
	$\bar\varepsilon_{\sigma100}^v$	1.675	0.643	0.573		$\bar\varepsilon_{\sigma100}^P$	0.285	0.201	0.167
ε_μ^θ (%)	$\bar\varepsilon_{\mu,max}^\theta$	3.056	1.568	1.235	$\varepsilon_\mu^{Q_{ij}}$ (%)	$\bar\varepsilon_{\mu,max}^Q$	5.689	2.235	2.035
	$\bar\varepsilon_\mu^\theta$	1.023	0.756	0.583		$\bar\varepsilon_\mu^Q$	2.865	1.356	1.128
	$\bar\varepsilon_{\sigma100}^\theta$	0.982	0.458	0.394		$\bar\varepsilon_{\sigma100}^Q$	1.225	0.754	0.586
$\varepsilon_\sigma^\theta$ (%)	$\bar\varepsilon_{\sigma,max}^\theta$	9.985	3.895	3.128	$\varepsilon_\sigma^{Q_{ij}}$ (%)	$\bar\varepsilon_{\sigma,max}^Q$	9.972	3.365	2.864
	$\bar\varepsilon_\sigma^\theta$	3.458	1.358	1.023		$\bar\varepsilon_\sigma^Q$	4.258	2.235	2.078
	$\bar\varepsilon_{\sigma100}^\theta$	2.896	0.886	0.748		$\bar\varepsilon_{\sigma100}^Q$	2.458	1.423	1.358

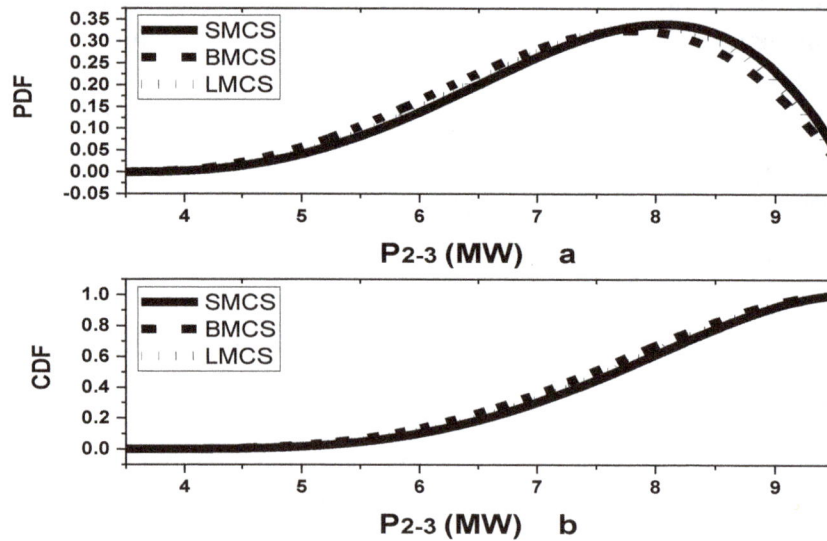

Figure 9. Probability density function (PDF) and cumulative density function (CDF) of active power flow through the line 2–3, (a) and (b) respectively, for three methods as: (a) SMCS, (b) BMCS, and (c) LMCS, for IEEE 14-bus test system.

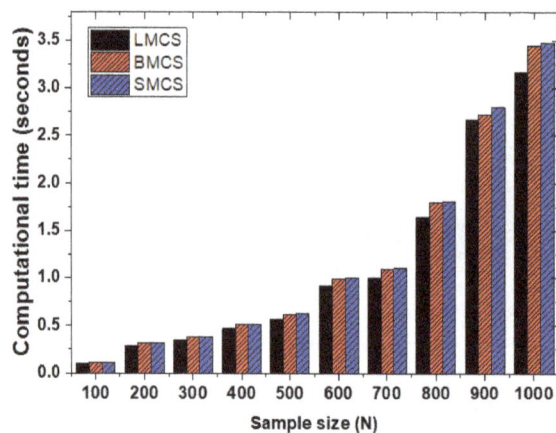

Figure 10. Computational time (seconds) comparison curve for three methods as: (a) SMCS, (b) BMCS, and (c) LMCS, for IEEE 14-bus test system.

5.3. Modified IEEE 118-bus test system

All of the deterministic data about this test system is available in [32] and also shown in Figure 13. All of the assumption and probabilistic data about this test system is same as in [10,31]. The total input random variables in this test system are 170. The reaming assumption are almost same as in previous test system like loads are follow normally distribution and generator output follow binomial distribution. Four PV generator are located at nodes 5, 16, 50, and 53, respectively with 0.2 p.u installed capacity. The correlation matrix of four PV generation is shown in Eq 25.

$$G_{PVG4} = \begin{bmatrix} 1 & 0.7 & 0 & 0 \\ 0.7 & 1 & 0 & 0 \\ 0 & 0 & 1 & 0.7 \\ 0 & 0 & 0.7 & 1 \end{bmatrix}$$

(25)

5.4. The simulation results analysis of IEEE 118-bus test system

By the simulation study, it is shown that the three types of error indices about four output variables categories were almost same pattern as in IEEE 14-bus test system. These indices are not presented here due to space limitations. The error indices of four categories are presented in Table 3 but with 800 sample size. The results of Table 3 show that the error indices of SMCS are within 10%, with BMCS are within 6%, with LMCS are within 4%. These percentage of error change due to increase input random variables. This conclusion shows that LMCS is more stable and distributed convergent method than the other two methods. The PDF and CDF curves are shown in Figure 11 of active power flow through the line 78–79. From these results, it is shown that LMCS is accurate than BMCS and SMCS but in previous test system BMCS was close to LMCS and SMCS methods. The accuracy of BMCS in term of PDF and CDF are bring down may be due to increase number of input random variables. Additionally, the comparison of the computational time of different sample same is shown in Figure 12. The computational time of SMCS method with sample size 20,000 was 5586 seconds. The computational time for both BMCS and LMCS was almost same like 13.92 s with 1000 sample size. The computational burden of BMCS and LMCS was very small as compare to SMCS for achieving desire accuracy. The results of Figure 12 show the computational superiority of ILHS with respect to same sample size.

Table 3. Error comparison of IEEE 118-bus test system.

Method		SMCS	BMCS	LMCS	Method		SMCS	BMCS	LMCS
	$\bar{\varepsilon}^v_{\mu,max}$	0.929	0.016	0.012		$\bar{\varepsilon}^p_{\mu,max}$	1.955	1.012	0.154
$\varepsilon^v_\mu(\%)$	$\bar{\varepsilon}^v_\mu$	0.947	0.032	0.016	$\varepsilon^{P_{ij}}_\mu(\%)$	$\bar{\varepsilon}^p_\mu$	0.652	0.741	0.103
	$\bar{\varepsilon}^v_{\sigma100}$	0.410	0.018	0.019		$\bar{\varepsilon}^p_{\sigma100}$	0.383	0.281	0.114
	$\bar{\varepsilon}^v_{\sigma,max}$	5.865	1.637	1.650		$\bar{\varepsilon}^p_{\sigma,max}$	3.059	1.943	1.475
$\varepsilon^v_\sigma(\%)$	$\bar{\varepsilon}^v_\sigma$	2.688	0.803	0.854	$\varepsilon^{P_{ij}}_\sigma(\%)$	$\bar{\varepsilon}^p_\sigma$	2.032	0.389	0.402
	$\bar{\varepsilon}^v_{\sigma100}$	1.523	0.559	0.567		$\bar{\varepsilon}^p_{\sigma100}$	0.259	0.175	0.165

Continued next page

Method		SMCS	BMCS	LMCS	Method		SMCS	BMCS	LMCS
	$\bar{\varepsilon}_{\mu,\max}^{\theta}$	2.778	1.363	1.223		$\bar{\varepsilon}_{\mu,\max}^{Q}$	5.172	1.943	2.015
$\varepsilon_{\mu}^{\theta}$ (%)	$\bar{\varepsilon}_{\mu}^{\theta}$	0.930	0.657	0.577	$\varepsilon_{\mu}^{Q_{ij}}$ (%)	$\bar{\varepsilon}_{\mu}^{Q}$	2.605	1.179	1.117
	$\bar{\varepsilon}_{\sigma 100}^{\theta}$	0.893	0.398	0.390		$\bar{\varepsilon}_{\sigma 100}^{Q}$	1.114	0.656	0.580
	$\bar{\varepsilon}_{\sigma,\max}^{\theta}$	9.077	5.387	3.097		$\bar{\varepsilon}_{\sigma,\max}^{Q}$	9.065	5.926	2.836
$\varepsilon_{\sigma}^{\theta}$ (%)	$\bar{\varepsilon}_{\sigma}^{\theta}$	3.144	2.181	1.013	$\varepsilon_{\sigma}^{Q_{ij}}$ (%)	$\bar{\varepsilon}_{\sigma}^{Q}$	3.871	2.943	2.057
	$\bar{\varepsilon}_{\sigma 100}^{\theta}$	2.633	0.770	0.741		$\bar{\varepsilon}_{\sigma 100}^{Q}$	2.235	1.237	1.345

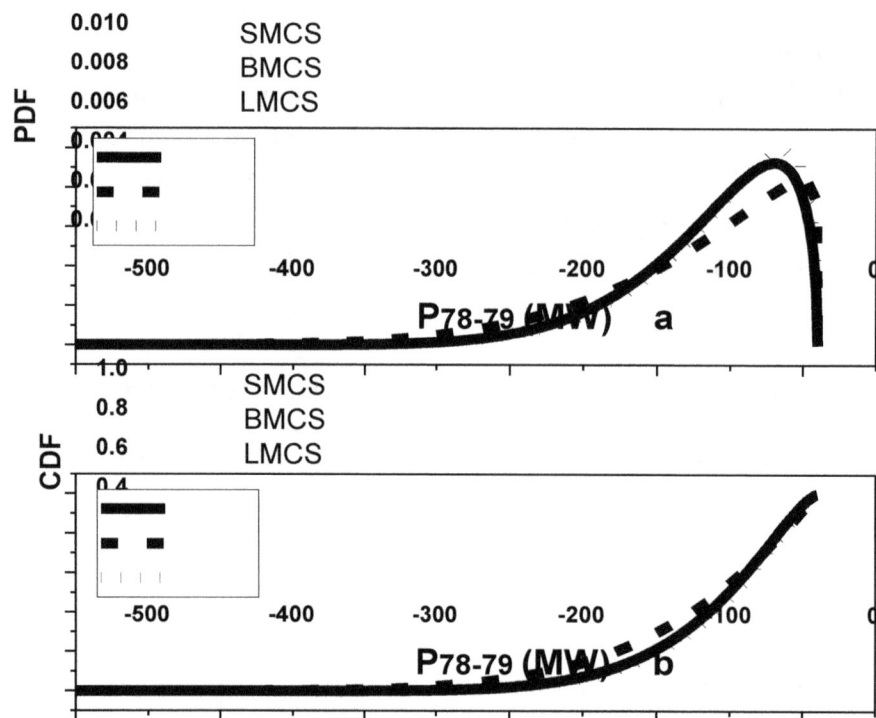

Figure 11. Probability density function (PDF) and cumulative density function (CDF) of active power flow through the line 78–79, (a) and (b) respectively, for three methods as: (a) SMCS, (b) BMCS, and (c) LMCS, for IEEE 118-bus test system.

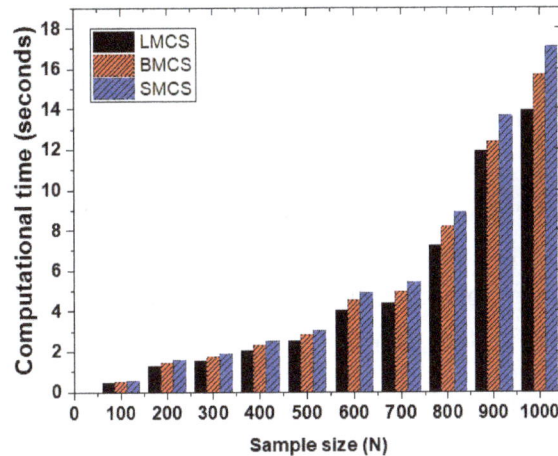

Figure 12. Computational time (seconds) comparison curves for three methods as: (a) SMCS, (b) BMCS, and (c) LMCS, for IEEE 118-bus test system.

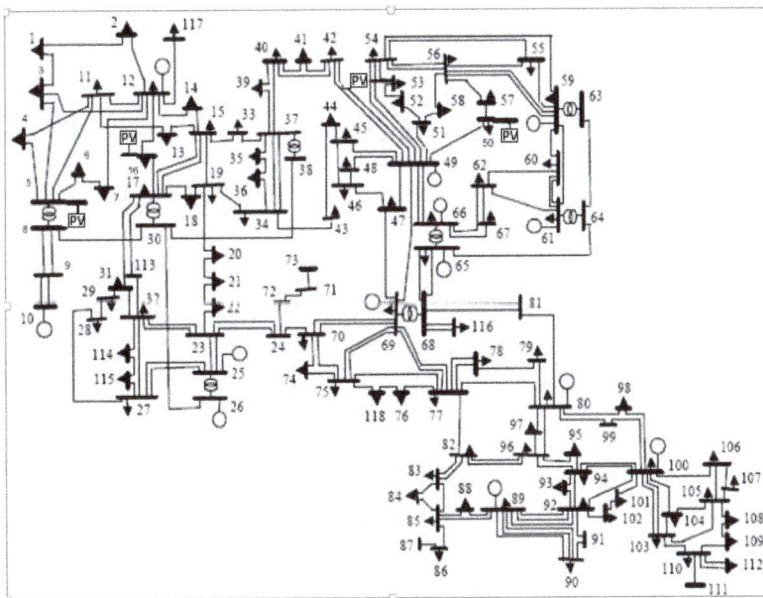

Figure 13. Modified IEEE 118-bus test system.

6. Conclusion

Due to increasing trend of penetration of renewable energy sources into power system has introduced the uncertainties and dependence factors. The modelling of input random variable for the study of uncertainties and dependence effect on power system operation and planning is a challenge for future. In this paper, a probabilistic load flow analysis is presented for correlated loads and photovoltaic generation. Uncertainty is created with correlated loads and photovoltaic generation. Gaussian Copula theory is proposed to establish the correlation between input random variables. Form the results; it has proved that Copula theory can establish correlation between input random variables flexibly and overcome the restriction of Pearson's correlation coefficient. Additional, it is unconstrained by the marginal distribution type of input random variables. Two methods of sampling has been proposed for analysis purpose. The results obtained by LMCS method is more accurate and efficient. When the input random variables were small, the results obtain by BMCS and LMCS almost have same accuracy and robustness. But when the input random variable has increased BMCS become less accurate as compare to LMCS. As an overall result, LMCS is an efficient sampling method for convergence for entire distribution of input random variable, almost independent of input random variable, and flexible too. The LMCS method has a large potential in order to deal with probabilistic problems with renewable energy sources especially penetration of photovoltaic generation in power system network.

Acknowledgements

This work is supported by the "National Key Research and Development Program (2016YFB0901104), the National Natural Science Foundation of China (51307051), the Fundamental Research Funds for the Central Universities (2014ZP03, 2015ZD01) and the Science and technology projects from State Grid Corporation".

Author contributions

Li Bin and Qi Bing conceived the research idea and designed the simulation setup; Muhammad Shahzad did the remaining work.

Conflicts of interest

The authors declare no conflict of interest.

References

1. Bechrakis DA, Sparis PD (2004) Correlation of wind speed between neighboring measuring stations. *IEEE T Energy Conver* 19: 400–406.
2. Villanueva D, Feijóo A, Pazos JL (2010) Correlation between power generated by wind turbines from different locations. Available from: http://proceedings.ewea.org/ewec2010/allfiles2/20_EWEC2010presentation.pdf.
3. Jong MD, Papaefthymiou G, Palensky P (2017) A framework for incorporation of infeed uncertainty in power system risk-based security assessment. *IEEE T Power Syst* 33: 613–621.
4. Kirschen D, Jayaweera D (2007) Comparison of risk-based and deterministic security assessments. *IET Gener Transm Dis* 1: 527–533.
5. Zhang P, Lee ST (2004) Probabilistic load flow computation using the method of combined cumulants and Gram-Charlier expansion. *IEEE T Power Syst* 19: 676–682.
6. Nijhuis M, Gibescu M, Cobben S (2017) Gaussian mixture based probabilistic load flow for LV-network planning. *IEEE T Power Syst* 32: 2878–2886.
7. Valverde G, Saric A, Terzija V (2012) Probabilistic load flow with non-Gaussian correlated random variables using Gaussian mixture models. *IET Gener Transm Dis* 6: 701–709.

8. Prusty BR, Jena D (2016) Combined cumulant and Gaussian mixture approximation for correlated probabilistic load flow studies: A new approach. *Csee J Power Energ Syst* 2: 71–78.

9. Caramia P, Carpinelli G, Varilone P (2010) Point estimate schemes for probabilistic three-phase load flow. *Electr Pow Syst Res* 80: 168–175.

10. Morales JM, Perez-Ruiz J (2007) Point estimate schemes to solve the probabilistic power flow. *IEEE T Power Syst* 22: 1594–1601.

11. Kloubert ML, Rehtanz C (2017) Enhancement to the combination of point estimate method and Gram-Charlier Expansion method for probabilistic load flow computations. *PowerTech, IEEE Manchester* 2017: 1–6.

12. Wang Y, Zhang N, Chen Q, et al. (2016) Dependent discrete convolution based probabilistic load flow for the active distribution system. *IEEE T Sustain Energ* 8: 1000–1009.

13. Zhang N, Kang C, Singh C, et al. (2016) Copula based dependent discrete convolution for power system uncertainty analysis. *IEEE T Power Syst* 31: 5204–5205.

14. Cai D, Chen J, Shi D, et al. (2012) Enhancements to the cumulant method for probabilistic load flow studies. *Pow Energ Soc Gen Meeting* 59: 1–8.

15. Aien M, Fotuhi-Firuzabad M, Aminifar F (2012) Probabilistic load flow in correlated uncertain environment using unscented transformation. *IEEE T Power Syst* 27: 2233–2241.

16. Fang S, Cheng H, Xu G (2016) A modified Nataf transformation-based extended quasi-monte carlo simulation method for solving probabilistic load flow. *Electr Mach Pow Syst* 44: 1735–1744.

17. Ren Z, Koh CS (2013) A second-order design sensitivity-assisted Monte Carlo simulation method for reliability evaluation of the electromagnetic devices. *J Electr Eng Technol* 8: 780–786.

18. Tang L, Wen F, Salam MA, et al. (2015) Transmission system planning considering integration of renewable energy resources. *Pow Energ Eng Conf* 2015: 1–5.

19. Ayodele TR (2016) Analysis of monte carlo simulation sampling techniques on small signal stability of wind generator-connected power system. *J Eng Sci Technol* 11: 563–583.

20. Prusty BR, Jena D (2017) A critical review on probabilistic load flow studies in uncertainty constrained power systems with photovoltaic generation and a new approach. *Renew Sust Energ Rev* 69: 1286–1302.

21. Morales JM, Baringo L, Conejo AJ, et al. (2010) Probabilistic power flow with correlated wind sources. *Let Gener Transm Dis* 4: 641–651.

22. Usaola J (2010) Probabilistic load flow with correlated wind power injections. *Electr Pow Syst Res* 80: 528–536.

23. Gupta N (2016) Probabilistic load flow with detailed wind generator models considering correlated wind generation and correlated loads. *Renew Energ* 94: 96–105.

24. Prusty BR, Jena D (2017) A sensitivity matrix-based temperature-augmented probabilistic load flow study. *IEEE T Ind Appl* 53: 2506–2516.

25. Kenari MT, Sepasian MS, Nazar MS, et al. (2017) The combined cumulants and Laplace transform method for probabilistic load flow analysis. *Let Gener Transm Dis*.

26. Wu W, Wang K, Li G, et al. (2016) Probabilistic load flow calculation using cumulants and multiple integrals. *Let Gener Transm Dis* 10: 1703–1709.

27. Nelsen RB (2007) An introduction to copulas. Springer Science & Business Media.

28. Market Analysis and Information System (MAISY). Utility Customer Energy Use & Hourly Load Databases. Available from: http://www.maisy.com/.

29. National Renewable Energy Laboratory (NREL). Obtain Solar Power data: Western State California. Available from: https://www.nrel.gov/grid/solar-power-data.html.

30. Macdonald IA (2009) Comparison of sampling techniques on the performance of Monte-Carlo based sensitivity analysis. *Eleventh International IBPSA Conference*, 992–999.

31. Yu H, Chung CY, Wong KP, et al. (2009) Probabilistic load flow evaluation with hybrid latin hypercube sampling and cholesky decomposition. *IEEE T Power Syst* 24: 661–667.

32. University of Washington Electrical Engineering: Power System Test Case Archive (January 2018). Available from: http://www.ee.washington.edu/research/pstca.

33. Karaki S, Chedid R, Ramadan R (1999) Probabilistic performance assessment of autonomous solar-wind energy conversion systems. *IEEE T Energy Conver* 14: 766–772.

34. Reddy JB, Reddy DN (2004) Probabilistic performance assessment of a roof top wind, solar photo voltaic hybrid energy system. *Rel Maint S-RAMS* 2004: 654–658.

10

Micro-generation in conflict: The conditions necessary to power economic development in rural Afghanistan

James D. McLellan and Richard E Blanchard*

Centre for Renewable Energy Systems Technology, Loughborough University, UK

* **Correspondence:** Email: r.e.blanchard@lboro.ac.uk

Abstract: Access to reliable electricity eludes many poor rural Afghan communities despite plentiful renewable resources. Micro-generation seems particularly well suited to Afghanistan's mountainous, decentralised society but even with substantial investment since 2001 it has not lived up to expectations. Recognising the causes are likely to dwell in the human (rather than technical) domain, this study takes a qualitative, soft systems approach to deriving and validating the necessary conditions that might improve the success rate of micro-generation projects in enabling sustainable economic development. It acknowledges the governance limitations inherent in fragile states and the significance of the community as the most stable element of society, putting the latter at the centre of its thinking. Those conditions identified as critical are summarised as: a holistic approach that sees micro-generation as a component of broader economic development; an environment safe enough for project build and operation, and for the markets necessary for wealth creation; and external support to build community capacity to fund and maintain schemes through-life. These conditions are likely to have relevance for other fragile states; the next step is to develop them in the field before deployment as part of a comprehensive approach to poverty alleviation in Afghanistan and similar states.

Keywords: micro-generation; Afghanistan; soft systems methodology; critical conditions; rural electrification; failed states

1. Introduction

1.1. Overview

Afghanistan scores poorly against almost every development indicator including access to electricity. One third of the population is recorded as being food insecure and approximately three-quarters of rural Afghan households do not have access to a reliable source of electricity. Grid-level generation and distribution is limited to a few urban centres only. In keeping with the Afghan tradition of decentralisation, energy is typically acquired and used at small community or dwelling level. For most rural Afghans that has meant wood, dung and possibly diesel. Since 2001 Afghanistan has been at, or near, the top of the development agenda for many of the world's most affluent nations [1]. Those states, having committed military forces, followed strategies that viewed rural development and counter insurgency as indivisible [2]. In other words, they paid for development programmes, including rural electrification, to reduce the risk of military and political failure. It is too early to judge the success or otherwise of Western efforts in Afghanistan but rural development, including access to electricity, remain important factors in the viability and development of the Afghan state. Moreover, progress in those sectors has not lived up to expectations; a recent national survey found access to electricity to be the second biggest "local" concern amongst the rural population after unemployment. This despite Afghanistan's hydro potential alone being assessed as up to 25 GW. Installed capacity is well under 1 GW, hydro providing most of Afghanistan's electrical power [3,4]. Despite sustained investment, there remains much to be achieved. Western military intervention has now ended; and thus 2015 marked a new beginning for Afghanistan's civil society. Increasingly it will need to find its own solutions to rural development, and by implication, access to electricity. Through its focus on micro generation, this study is intended to play a part in meeting that challenge. Its unique contribution lies in its methodology and its perspective: it will take a soft systems approach to understanding, from a community viewpoint, the necessary conditions for micro generation to successfully support economic development in rural Afghanistan. Its objectives are:

(1) To understand the Afghan rural community context;
(2) To derive a series of models representing facets of community micro generation. From these, to extract candidate conditions for successful micro generation;
(3) To validate findings by comparing model outputs with real-world studies;
(4) To elicit the critical and desirable conditions for successful micro generation in rural Afghanistan, and to comment on their broader applicability for other fragile states.

1.2. Literature review

1.2.1.   Electrification and poverty alleviation

The link between electrification and poverty alleviation is assumed widely in literature although often with little substantiating evidence. One exception is the IEA's World Energy Outlook 2004, which states that "energy is a prerequisite to economic development" [5]. It then links this statement to electrification by plotting human development indicators against electrical consumption, claiming a strong, non-linear correlation between access to electricity and human development. This means that disproportionate human development gains are associated with modest increases in access to electricity. But the report is careful not to ascribe this to a one-way causal relationship; instead it contends that energy development is a cause and effect of economic growth.[1] The direction of this correlation is the subject of a number of studies, summed up in a review of evidence conducted on behalf of the UK's Department for International Development [6]. It found "a strong causal relationship between consumption of modern energy and GDP growth but is inconclusive as to the direction of the causality and the magnitude of the impact". In an older study, Foley takes a less nuanced view, asserting that "rural electrification does not cause development" [7]. He does, however, acknowledge it is an essential enabler. Supported by some convincing arguments, and in agreement with others (Kirubi et al., World Bank), he goes on to say that timing the introduction of electricity to a community is critical: it must neither be too early (when it may not be economically viable); nor too late (when its absence stifles development) [8,9]. This implies that community electrification must be closely linked with economic development.

1.2.2. Global factors influencing successful electrification

In response to the challenges of poverty and limited infrastructure in the developing world, the cost-effectiveness advantages of standalone, renewable generation are made clear by Mainali et al., Lahimer et al. and others [10,11]. But successful electrification is more than just selecting the optimum technology; it depends also on the aligning of a range of economic, social, political and other human factors. In her study for the IEA, Niez conducted a top-down assessment of government-led electrification programmes in several emerging economies which included a detailed review of conditions required for successful stand-alone systems in a rural setting [12]. She found that a detailed understanding of the population, sustained governmental support, market infrastructure, community support and affordability all underpin success. A similar paper on behalf of the World Bank examined the generic reasons for failure to provide universal access to electricity [13]. Its findings support Niez's assertion that government commitment and financial viability are critical for successful electrification. Indeed Foley, Bailey et al., Peskett, Zalengera et al. and others agree (or at least imply) that organisational, economic and social constraints are likely to be key factors [14–16]. Given this weight of consensus it is perhaps surprising that the human dimension has not always attracted academic interest. As recently as 2012 Schillebeeckx et al. assessed 232 papers on this subject, finding that most focused on technical and institutional perspectives rather than understanding economic viability and user need [17]. This suggests that micro generation in the developing world is still being addressed largely as a top-down policy or engineering problem rather than a more holistic societal issue.

---

[1] Energy development is defined as the transition to modern fuels, and then to mature energy end-use.

1.2.3.   Micro generation in Afghanistan and other failing states

Afghanistan offers an interesting but challenging micro generation case study. Despite plentiful renewable resources [18,19], its rural communities feel the combined pressures of conflict, poverty and weak governance. The limited capacity and reach of government is of particular interest. Much of the literature reviewed here asserts the importance of effective instruments of government to form and implement policy in support of rural electrification. However, states considered to be failed or failing are characterised by the suspension of government functions [20]. Indeed, as a subject, rural electrification in failed states or areas of conflict seems poorly targeted by research despite predictions that intra-state violence will be a re-occurring theme [21]. It is touched upon by Weihe in his examination of the phenomena of local co-operatives where central government cannot deliver economic and social development [22]. He concurs with Neiz's finding that local communities must own their means of power generation, although in the case of failing states it may be aid agencies and other non-state actors that act as facilitators, rather than government. The subject of electrification in rural Afghanistan is less short of commentators. Several wide-ranging reports have been published that address the successes and failures of Afghanistan rural micro generation since 2001 [23–26]. The Afghan government itself has produced several policy documents that show a reasonable grasp of both human and technical challenges [27–29]. In conjunction with more general international literature, these will be used to validate this study.

2.   Methods

2.1. Assumptions

Several assumptions are necessary to deliver meaningful outputs within the larger subject of Afghan development. This is an imperfect approach since development is a multi-faceted challenge; nevertheless, to derive useful conclusions in this study the problem must first be constrained to a sensible degree. These assumptions are described as followed in sections 2.1.1–2.1.4.

2.1.1.   Micro generation contribution to economic development

This study addresses micro generation's contribution to economic development. At its most basic level, it assumes economic development is a positive and desirable goal. This may seem self-evident but in the conflict of ideas in Afghanistan these views are not necessarily shared by the entire population.[2] It focuses on improving micro generation's contribution to near term (<5 years) economic development because the need is severe and urgent. The interaction of micro generation with longer term evolvers of aspiration, such as education, female emancipation and exposure to media, is implied but not considered in detail. These aspects would be appealing as a second stage to this paper (indeed Niez asserts such issues are highly relevant to long term economic development [12]).

---

[2] Some of the more extreme fundamentalist organisations in Afghanistan and the Middle East might be said to be pursuing anti-development agendas.

2.1.2. Community as the primary stable entity

Afghanistan is classified as "highly fragile". Such states are characterised by several dimensions of instability including: factionalisation; public service failure; legal violations; security failures; and economic instability. Other examples include Iraq, Syria, Yemen, Central African Republic, Somalia, and to add complexity, these states are often not in equilibrium; that is to say they may be becoming more or less stable, the direction and speed of which will also affect micro generation and economic development [30]. The literature review points to a gap here in that most papers addressing developing world electrification focus on high-level policy-centric solutions. These imply that governments control their own territory, provide and enact policy, and act in the interests of the population. But in highly fragile states this is often not the case. In Afghanistan, long term instability means villages and tribal groups have frequently been left to devise and implement their own "policies" and to chart their own course through conflict. This is not to belittle the substantial progress made by the Afghan government in recent years, however parts of Afghanistan at least may have an uncertain future in terms of governance, see Appendix 1 [31]. In such volatile circumstances the community becomes the most stable element; the anchor of certainty for individuals. Accordingly, a key element of this study's unique contribution is its focus on the community as the primary organisational and executive entity, situated in a context of political, economic and other instability. There is estimated to be ~24,000 such rural communities in Afghanistan, each unique but having some affinity at least with the following characteristics (further details at Appendix 1):

(1) Traditional leadership structure;
(2) Severe poverty-subsistence economy, food insecurity, very low incomes and levels of capital;
(3) Under-employment;
(4) Very low levels of education and skills.

2.1.3. External organisations

The capital and educational limitations hold particular relevance for micro generation and lead to the next assumption. It is very unlikely a typical rural Afghan community will have the finances and technical wherewithal to develop a micro generation scheme without the support of an external organisation. This may be a governmental body. But in another district or province, for reasons of limited capacity, security or allegiance, it may be a different organisation. In highly fragile states the government may be too weak to fulfil this role or may even deny development opportunities deliberately. This has been claimed by at least one minority group in Afghanistan and seems likely in other fragile states with deep social divides. The trend in Afghanistan seems to be toward increasing government capacity. But given the range of bodies, including Western Governmental Organisations (GOs), International Organisations (IOs), Non-Governmental Organisations (NGOs) and even NATO military forces, that have hitherto implemented micro generation projects, this demands a more open approach to defining which organisations might set policy and assist communities. For this study, the following external organisations are defined. The ambiguity is deliberate and in keeping with the primacy of the community in this study:

Overseeing Organisation (OO)—that which has decided to commission the delivery of micro generation to groups of rural communities, and which sets the policies for implementation. It is the guiding mind and the intellect behind the process. Typically, a central government department or a GO/IO/NGO.

Enabling Partner (EP)—that which is providing practical assistance to individual communities to develop micro generation schemes. It may be one organisation, or many working together in a similar timescale for overall effect. Typically, a provincial or district government office, local NGO or contracted service provider.

2.1.4.   Other assumptions

Micro-generation is a common but ill-defined term. This report uses the meaning: "...***generation of electricity ... on a small scale***, typically for domestic use and by methods that do not contribute to the depletion of natural resources..." [32]. There is no globally agreed scale and, in this context, it seems sensible to define by user community size. The Afghan Government delineation of 5000 individuals will be used here, although most rural communities will be smaller [29]. Furthermore, in pursuing the goal of economic development this study targets community, rather than household level, micro generation as the latter risks marginalising the poorest. Conversely, inter-community schemes are a possibility and could work well where resource ownership is shared (for example a river). They are not addressed specifically here but will have much commonality with the conclusions.

2.2. Soft systems approach

Engineering feasibility studies and technology cost-effectiveness comparisons are well represented in developing world electrification literature however they address only one part of the problem. Reasons for success or failure in such challenging environments are typically traced to less deterministic factors: social; economic; political; and the like. In seeking to address these influences, this study is by nature a qualitative assessment. It steers away from quantitative problem-solving; instead, it seeks to find the combination of qualitative conditions that offers the best likelihood of success in the context of micro generation as an enabler for economic development [33].[3] Establishing these conditions requires an approach suited to the complex, untidy (in a systems sense) environment that is rural Afghanistan. A range of frameworks and methodologies were reviewed, including simpler "PESTLE" and variants, and "hard" and "soft" systems approaches. Checkland's "Soft Systems Methodology" (SSM) was finally selected as it offered a process optimised for the kinds of unstructured human problems toward which the body of literature points. In Checkland's words, these are conundrums that are characterised by a "feeling of unease, which cannot be explicitly stated without oversimplifying the problem" or put another way, a "mismatch between actuality and what is perceived might become actuality" [34]. This is a good description of the nature of this study, which does not seek to define a required outcome. Instead it acknowledges a

---

[3] The term *condition* is defined as *'a stipulation; something upon the fulfilment of which something else depends.'*

feeling of unease, something like: surely micro generation could, and should, make more of a contribution to economic development in rural Afghanistan? It then sets about finding the conditions that could help unlock micro generation as a more effective contributor in narrowing this perceived mismatch. In essence, this study seeks to develop a range of conditions, based on a detailed understanding of the context. These independently-derived conditions are then compared with field reports, to both corroborate the independent conditions, and to spot potential new conditions that have not been noted in other studies. This process is shown in more detail at Figure 1.

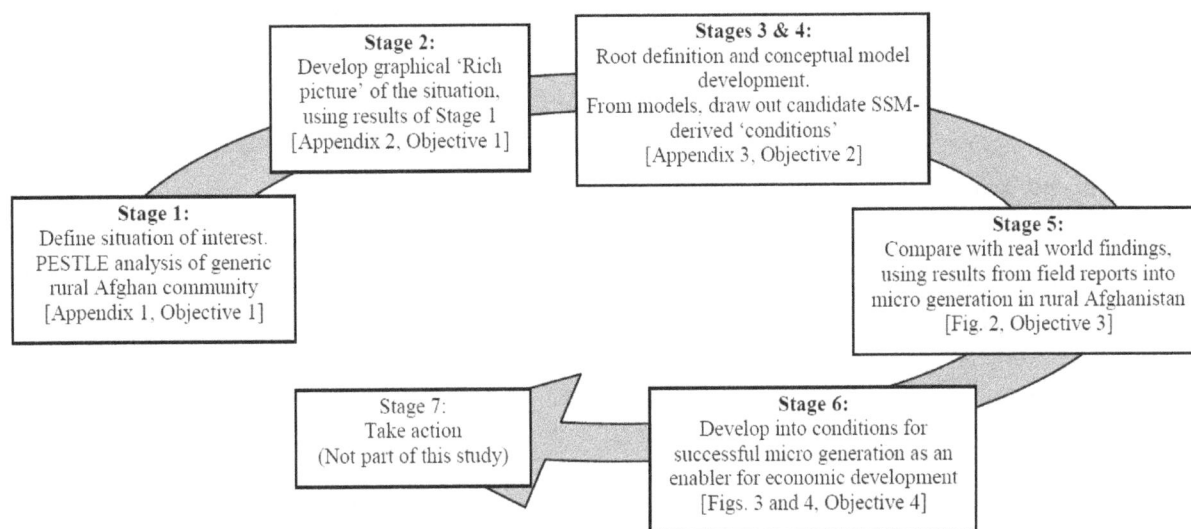

Figure 1. SSM applied to micro generation in rural Afghanistan.

Stage 1 was derived using a range of online sources, in a PESTLE format, to describe in detail the context. Stage 2 took forward the findings from Stage 1, into a graphical representation of a generic rural Afghan community. Combined, Stages 1 and 2 forced an intimate understanding of a generic rural Afghan community (within the constraints of a desktop study). Stages 3 and 4 are at the heart of the SSM approach. Stage 3 saw the development of a series of "holons" (or "plausible perspectives") associated with micro generation. Twenty-two holons were devised, through an iterative process based on the contextual understanding developed at Stages 1 and 2. In keeping with the nature of this study, holons were rooted primarily in the community, however in recognition of the criticality of both OO and EP, several holons addressed perspectives relevant to those organisations. Thereafter, at Stage 4, individual holons were developed into a system description, then a conceptual system model. Each modelling process prompted new insights into other holons and their conceptual models; in this way, the models were refined through several cycles of review. From examination of each system model, a total of 151 candidate conditions for success (or opposing threat conditions) were identified. Predictably, many candidate conditions were duplicated across holons, necessitating the grouping of similar or identical candidate success and threat conditions. Again, this process was qualitative and iterative, care being taken not to lose the meaning of the original candidate conditions. These 46 grouped conditions were checked for feasibility, uniqueness and completeness then scored for their criticality to the success of micro generation as an enabler for economic development, from the perspective of a rural Afghan community. A variety of micro generation studies were then selected against which to validate these independently-derived conditions. These were chosen to include a broad range of viewpoints, purposes and data regarding the success or otherwise of micro generation projects and these are described below.

Specific to Afghanistan:

(1) United States Agency for International Development: audit of 60 micro hydro schemes [26].
(2) German International Development Agency: renewable energy sector capacity assessment [25].
(3) Afghan Government Ministry of Rural Rehabilitation and Development: micro generation policy [27,28].
(4) Journal of Sustainable Development of Energy, Water and Environment Systems: survey of 421 micro hydro installations [24].
(5) Afghan Government: energy sector strategy assessments of sector weaknesses [29].
(6) United States Department of Defense: results from 68 schemes were used to derive a distributed energy model [23].

International:

(1) International Energy Agency: rural electrification policies in emerging economies [12].
(2) World Bank: addressing the electricity access gap in the developing world [13].
(3) UK Department for International Development and World Bank: best practices for sustainable micro hydro in developing countries [35].

From these reports, systematic analysis identified 86 validation conditions. Each SSM condition was scored 0–5 for the level of correlation with one or more validation conditions. The complete validation process is shown at Figure 2. It is fundamentally a qualitative comparison, in keeping with the nature of SSM and its emphasis on perspectives. Scoring allows importance and correlation to be ranked but further numerical manipulation is avoided as it risks losing meaning.

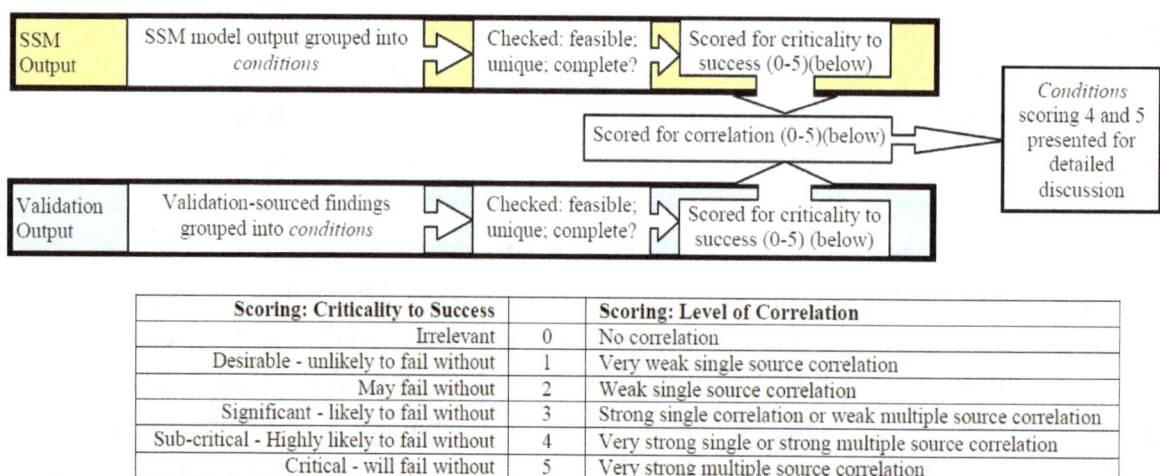

Scoring: Criticality to Success		Scoring: Level of Correlation
Irrelevant	0	No correlation
Desirable - unlikely to fail without	1	Very weak single source correlation
May fail without	2	Weak single source correlation
Significant - likely to fail without	3	Strong single correlation or weak multiple source correlation
Sub-critical - Highly likely to fail without	4	Very strong single or strong multiple source correlation
Critical - will fail without	5	Very strong multiple source correlation

Figure 2. Scoring and validation process.

3. Results

The findings from Stage 6 of the SSM process are shown below. Figure 3 shows conditions scoring in the highest tier; these are considered critical for success in rural Afghanistan. Figure 4 shows conditions scoring "4" (referred to as "sub-critical"), without which success is deemed highly unlikely. Both are shown against an outline project timeline.

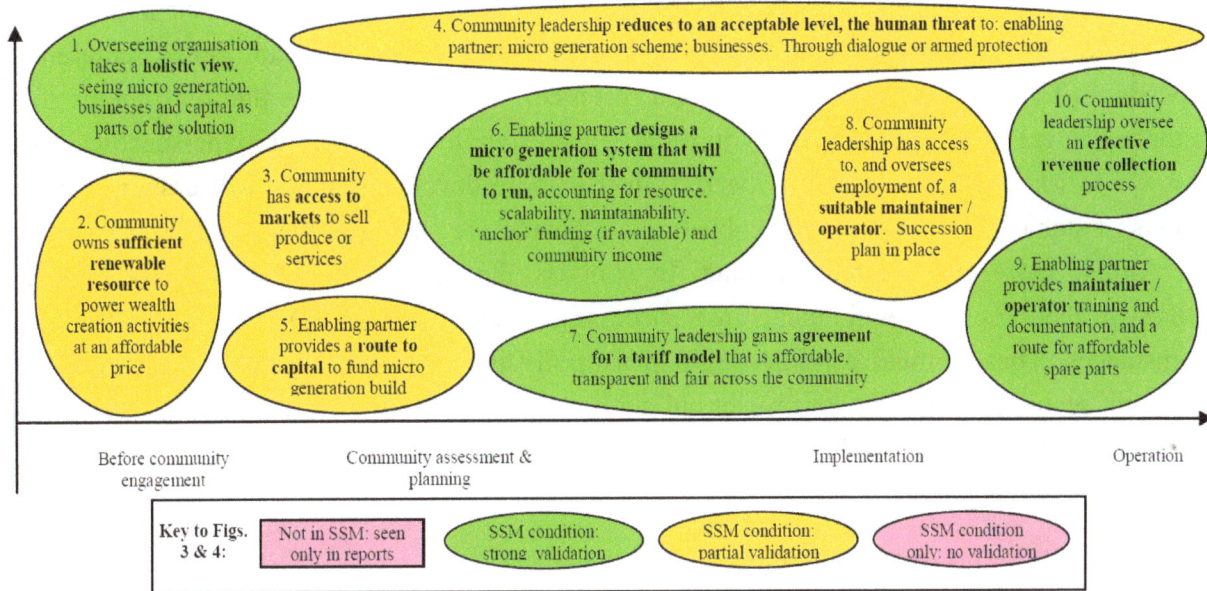

Figure 3. Critical conditions for micro generation to contribute to economic development in rural Afghanistan.

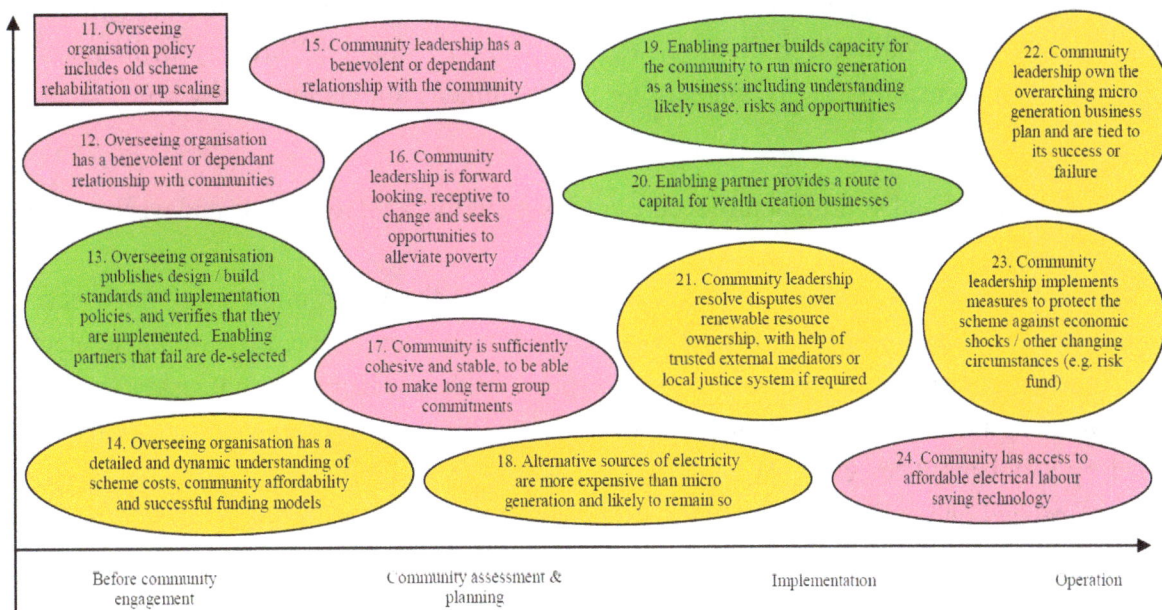

Figure 4. Sub-critical conditions for micro generation to contribute to economic development in rural Afghanistan.

4. Analysis and discussion

4.1. Critical conditions

Ten critical conditions were established. All correlated, entirely or partially, with conditions identified in field or policy reports and all are necessary but none, alone, is sufficient for success. Of course, the qualitative nature of this study and the critical and sub-critical conditions means that few derived conditions are binary in nature. Nevertheless, this does not detract from their importance. It will be apparent that many are universal in nature, in the sense that they are global requirements independent of context (for example sufficient resource). This study adds value by bringing to the fore those that are of the highest priority in rural Afghanistan, from a community perspective. This is analogous to Maslow's "hierarchy of needs", in that more developed states will assume the basic elements of stable government, literacy, business infrastructure etc are in place and instead focus on higher order, more evolved requirements. In contrast, the conditions identified here provide a glimpse of the baseline requirements for successful micro generation in the most tenuous of circumstances. These critical conditions group naturally into 3 themes: a holistic policy view; a sufficiently secure environment for project build, operation and wealth creation; and external support to build community capacity to fund and maintain schemes through-life.

4.1.1.   Condition 1: Holistic policy view

The genesis of success in the context of this study, is a holistic view, originating from the OO, which sees micro generation as a component of a broader approach to economic development. Indeed, because it relies on wealth to fund its operation, electrification as an end in itself is almost certain to fail if not accompanied by activity to develop the community's means to generate that wealth. This condition validated well. It is self-evident and recognised broadly in literature as a critical condition for creating the environment for the sustainable alleviation of poverty.

4.1.2.   Conditions 3 and 4: Sufficiently secure environment for build, operation and wealth creation

These conditions are a mark of fragile states. Firstly, the community must have access to markets through which to sell the goods or services that have been enabled by a reliable electricity supply. This stems from the assumption that the typical Afghan community possesses very little in the way of indigenous wealth. It must rely on the export of textiles or dried fruit, or the service provided by a flour mill for example, to the wider world, to import wealth. Condition 3, access to markets, correlated only partially perhaps because it is assumed. However, the least developed, most fragile societies may experience, through violence or other causes, the regular disruption or even prolonged absence of markets and transport infrastructure that underpin movement of wealth. Even with the help of electricity to produce goods more efficiently, the typical community is still surrounded by similarly poor communities. Without markets to move goods and import wealth from further afield, it is unlikely to make significant economic gains [36].[4] Moving to Condition 4, through the life of a micro generation scheme violence is likely to touch the community itself and the businesses it tries to develop. The Taliban has been more accepting of development in recent years but there remains a significant threat to EPs (particularly to international staff) and the emergence of Islamic State could point to a more rejectionist, anarchic future across the region. Whatever the threat group (and it may be government forces in some states), the community leadership will need to keep the human threat below an acceptable threshold. Condition 4 is hard to judge; who can tell how a conflict will ebb and flow? Geography and allegiance will be factors, as will the leadership's ability to steer a safe course between the warring sides (see section 4.2.2) [31]. Dialogue or armed protection may be sufficient but in the worst circumstances, the community's best efforts may simply be overwhelmed by the tide of fighting.

4.1.3.   Conditions 2, 5–10: External support to build capacity to fund and maintain scheme through life

Affordability begins with community ownership of sufficient renewable resource. Condition 2 correlated only partially, again perhaps because it is assumed. Nevertheless, it must be satisfied and although much of rural Afghanistan is well served by rivers there will be many communities that are limited to more expensive and complex conversion technologies like solar PV. These may be less effective in meeting reliably and affordably; the capacity demands of businesses. Moving to Condition 5, severe poverty is a central and oft-repeated assumption in this study. It therefore follows that capital must be provided from an external source, in part or full, to fund the micro generation scheme build. This correlated well with Afghan policy documents but less so with other Afghanistan reports which tended to address lessons from implementation and operation phases, making the assumption that capital had already been provided. Condition 6 requires affordable operation by design. Implicit in this condition is that the EP designs the micro generation scheme since another characteristic of rural Afghan communities is an absence of education and technical skills. This is improving as the post-2001 education system infiltrates but rural Afghanistan still has one of the lowest literacy rates in the world. Explicit in this condition is that the design effort is focused on through life affordability. This is critical: in fragile states, like Afghanistan, lifetime subsidies are very unlikely to be practical. Through the life of a UK renewable energy tariff agreement (25 years), Afghanistan has had several violent regime changes and long periods of civil war. These conditions make tariff subsidies inappropriate; instead, to give the scheme the best chance of success, investment must be up-front, and the design optimised for low running costs. This condition also makes reference to "Anchor" funding, meaning a source of energy funding from outside the community, usually in support of a service like a school or health centre upon which the community can rely to provide uninterrupted "baseline" funding. This is highly desirable but, like subsidies, it may not be relied upon in the most fragile states. Conditions 8 and 9 focus on aspects of through life operation and maintenance. The former addresses more nebulous but nonetheless important requirements of having access to, and selecting suitable maintainers and operators. The

---

[4] There is a view that highly fragile states develop a form of extreme, unregulated market capitalism such as existed in Afghanistan during 1990-94. But this period was characterised by extreme poverty, concentration of wealth and the marketisation of violence; all highly undesirable outcomes.

phrase "access to" is a reference to the low skills base from which the maintainer must be found. Is the community large enough or educated enough to provide individuals with an aptitude for this work? "Selecting" refers to nepotism or power-base relationships which commonly drive appointments of this kind in Afghanistan. To give the best chance of success the maintainer/operator must be selected on merit rather than family ties. A succession plan is also vital; instability and limited healthcare make it prudent to have at least 2 individuals trained through the life of the scheme. Condition 9 lists more practical needs, specifically training, documentation and a route to spare parts. These are global requirements but made more difficult in Afghanistan by the general lack of education and the vulnerability and affordability of the supply chain. Finally, if the scheme is to be maintained and operated successfully it must be funded through life, to pay the maintainer and to buy parts. This is the subject of Conditions 7 and 10. For reasons discussed at Condition 6, through life subsidies are very unlikely to succeed. Instead, scheme operation must be self-funded. Condition 7 requires the agreement of a tariff that is affordable, transparent and fair. In an environment rich with corruption, transparency will lend credibility, particularly if the rate is linked to an external index. Fairness implies that it will benefit the whole community, which may point to different tariffs depending on ability to pay. Local circumstances will dictate which model best suits but pan-community buy-in will help lead to effective revenue collection, which is the subject of Condition 10. This requires similar transparency and fairness in the act of revenue collection itself.

4.2. Sub-critical conditions

4.2.1.   Condition 11: Rehabilitation of existing micro generation infrastructure

Condition 11 was missed by the SSM approach which did not consider improving existing inadequate or failed schemes. This is an important policy component, particularly in states like Afghanistan, where infrastructure has been damaged by conflict, as it offers a cheaper way to develop latent capacity, both physical and human.

4.2.2.   Conditions 12, 15, 16, 17: Nature of the OO, leadership and community

Conditions 12, 15, 16 and 17 did not correlate directly with any validation source. These are typical "soft" human characteristics and show the uniqueness of an SSM approach. Without the checks and balances inherent in a more stable state, the nature of, and relationships between, the OO, the community leadership and the people it serves, appear highly relevant for micro generation and development. Research focused on economic outcomes in small Afghan villages suggests that effective, progressive leadership matters greatly in steering communities through instability and the breakdown of the state [31]. The same research backs up the notion of accountability, particularly where the leadership is economically less secure (or put another way, more equal), which makes it more likely to act in the interests of the whole community. This should not be a surprise: accountability driving better outcomes is a global constant (it underpins democracy). The relationship between OO and community is of similar consequence; the closer their goals are aligned, the greater the likelihood that the OO will act in the interests of the community. But where the

OO (which might be central government), or the community leadership, is isolated, predatory or seeking short-term gains, for example pacification in counter-insurgent operations, a positive outcome is expected to be less likely. Community stability also appears very relevant for sustainable success. A community that cannot unite in supporting and funding a scheme through life is unlikely to reap sustainable economic benefits, and there is plenty of potential for generational dissent in the introduction of technology into such a traditional environment. It is acknowledged that these conditions are both dynamic and very difficult to assess objectively. This perhaps explains why they don't feature explicitly in the micro generation studies used for validation which tended to pick out more tangible conditions such as a lack of training or spare parts. Nevertheless, they could be inferred from higher level policy reports and are brought out in more general societal research. Thus, it is contended here that the nature of the OO, the leadership, and the community itself, and their relationships, is highly relevant to the likelihood of micro generation successfully enabling economic development.

4.2.3. Conditions 13 and 14: OO Understands the context, and sets and enforces policies

Conditions 13 and 14 make requirements on the OO to understand the environment in which it operates and to mandate various policies that increase the likelihood of success for individual communities. Condition 13 also requires that these policies are audited, and non-compliant EPs removed, to act an as incentive to drive up build standards.

4.2.4. Condition 18: Alternative sources of electricity

Condition 18 requires that alternative sources of electricity, including grid connection, are more expensive than micro generation. It can be argued that economic development is agnostic as to the source of its electricity so long as it is affordable (Conditions 5 and 8). But since this study is addressing micro generation specifically, it will have failed if displaced by a different source of cheaper electricity.

4.2.5. Conditions 19–24: "Business" approach, dispute resolution and access to electrical devices

The importance of a holistic "business" approach within the community is recorded at Conditions 19, 20, 22 and 23. Condition 21 (dispute resolution) is typical of less developed states where land and water rights may not be recorded. Suspension of government justice functions in highly fragile states exacerbates such disputes, whose resolution will have a bearing on longer term success. Finally, Condition 24 draws attention to the importance of electrical devices in turning electricity into profitable output. Without access to affordable machinery such as mills or looms, communities are limited to the economic benefits afforded by electrical lighting only.

4.3. Lower order conditions

Summarising those conditions lower down the scale of importance, strong correlation was noted for those relating to community mobilisation, to encourage support for the scheme and to understand how to use electricity productively. These were deemed significant but did not attract a higher importance because the relatively high penetration of electrical goods like mobile phones and televisions indicates Afghans are already familiar with electricity and its uses [4]. Rushed implementation was deemed a risk to success, but only where the higher order conditions are not met (for example building capacity to run businesses). The dangers of corruption correlated well with other studies. It is, unfortunately, endemic in many fragile states but communities and businesses survive despite it. Of those conditions unique to validation literature, the creation of autonomous, apolitical organisations at each organisational level, featured often in higher level papers. This is a wise goal in the long term but of greater interest in this context are the natures of the various organisations and their relationships, and the running of the scheme in a "business-like" manner. It is the authors' view that at this stage in Afghanistan's development, in an environment of instability and traditional relationships, this typically-western approach will often be unrealistic at a community level. Similarly, the creation of open markets for micro generation is a sensible aspiration for longer term affordability, but less relevant in highly fragile states, where the investment necessary to sustain and develop such markets is unlikely to materialise. It will grow in importance when states transition from conflict, Rwanda being a good example, now an order further up the stability scale and seeing the benefits of private sector involvement [37].

4.4. Relevance for other highly fragile states

To assess broader relevance for other highly fragile states, it is necessary to consider the assumptions made about Afghanistan that led to the aforementioned conditions. By definition fragile states share macro characteristics such as factionalisation; public service failure; legal violations; security failures; and economic instability. More significant in assessing wider relevance are the specific assumptions made about rural Afghan communities, since they are likely to act as discriminators. They are: traditional leadership structure; severe poverty; under-employment; and very low levels of education and skills. Perpetually fragile states, (for example Somalia, Sudan, Yemen) will share most or all of these characteristics and the conditions identified are likely to be applicable in their entirety. Other states that have descended into fragility from more stable antecedents offer a greater contrast. Iraq and Syria are good examples of such "fallen states", sitting well above Afghanistan in terms of wealth, education and infrastructure. Re-examining the critical conditions identified at Figure 3, Table 1 gives a summary of relevance to these "fallen states". It shows that half the critical conditions (marked in green) are essentially universal in nature, applicable across all fragile states. The other half (marked in yellow) remain important, but the challenge they present is likely to be tempered by pre-existing national characteristics, for example wealth, education and infrastructure, which may mean that fallen states require less intervention in order to meet these conditions.

Table 1. Critical condition applicability to "Fallen States".

Condition	Applicability
1. OO takes a holistic view of micro generation and business	Relevant, although the nature of the challenge will be different. Fallen states are likely to have pre-existing business capacity, as well as more capital and infrastructure, than Afghanistan.
2. Sufficient renewable resource	Highly relevant in all circumstances.
3. Access to markets	Relevant. Physical threat to markets will be similar however more developed transport and business infrastructure in fallen states will ease the challenge of selling goods.
4. Human threat reduced to acceptable level	Highly relevant, unfortunately.
5. Capital for micro generation build	Relevant, although richer states like Iraq are more likely to be able to afford part or all of these costs.
6. EP designs a micro generation system with affordable running costs	Highly relevant. Subsidies are unlikely in fragile states so the community must bear the costs of operating the scheme.
7. Agreed tariff model	Highly relevant. Agreeing funding for running costs is critical.
8. Suitable maintainer selected	Relevant. Nepotism and power relationships are factors across fragile states although with higher levels of education generally, these factors are less likely to influence long term success.
9. EP provides maintainer training & documentation	Relevant, although higher levels of education will reduce the training burden.
10. Effective revenue collection	Highly relevant. Again, funding running costs is critical to success.

4.5. Study limitations

SSM is most often used in an iterative fashion through conversations with stakeholders. Applying it in a desktop study of this kind is valid, but its output is constrained to that derived, judged and validated by the authors. Without the direct voices of stakeholders, the "original thought" element is limited to secondary evidence and so this study can be considered a first step only in the evolution of useful output, a field phase being the next logical step in developing the proposed conditions. However, by way of balance, this shortcoming is common to all remote study methodologies. Furthermore, the authors assert that approaching this problem in a qualitative manner, validated against real-world studies, is worthwhile despite the challenges as it adds a new, softer perspective to a problem that has tended to be dominated by hard-systems analysis. And whilst the majority of understanding is derived from the references, one of the authors has first-hand experience of Afghanistan, which helped inform the soft systems process by lending a supporting validation perspective. Several other points are worthy of mention. The military intervention has generated an enormous quantity of literature however much of it is English language and western-authored with the bias that implies. Some Afghan government sources were used, but these are not necessarily representative of conditions on the ground. And not all the validation documents are from field studies; some are derived from policy or similar documents, removed one step at least from field work. Nevertheless, so far as possible within the constraints of a desktop study, it is contended that a reasonably accurate representation of rural Afghanistan has been derived, and that a meaningful outcome has been achieved.

5.   Conclusions

Bringing electrical power to bear usefully in fragile states is amongst the most demanding human problems in the field of renewable generation. Physical violence is the most obvious challenge, but the lack of stable governance arguably has greater influence. It prevents a conventional government-centric approach, instead forcing the community and other organisations such as GOs and NGOs, to the fore. Such human complexity demands a qualitative "soft systems" approach and it is through this methodology, validated by field reports, that a set of critical conditions for successful micro generation in support of economic development has been developed. By thematic group, they are:

(1) A holistic approach that sees micro generation as a component of broader economic development;

(2) An environment safe enough for project build and operation, and for the markets necessary for wealth creation;

(3) External support to build community capacity to fund and maintain schemes through-life.

A range of sub-critical and lower order conditions were also derived. Most validated well however the SSM approach produced a new set of conditions highlighting the importance of the nature of the OO, the community, its leadership, and their relationships. Conversely, it played down the importance of the more progressive ideas of creating autonomous bodies and opening up markets, finding these more relevant for states that have already emerged from conflict. The critical conditions were also found to have broad relevance for other highly fragile states. Recognising the limitations in the desktop nature of this study, the next step is to develop the higher order conditions, in the field, before deploying them as part of a comprehensive approach to poverty alleviation in Afghanistan and other similar states.

Acknowledgements

The authors are most grateful to those individuals who braved Afghanistan's hinterlands to compile the studies and reports upon which we have relied.

Conflict of interests

The authors declare no conflict of interest in this paper.

References

1. OECD (2014) Compare your country–Aid statistics by donor, recipient and sector, OECD. Available from: http://www.compareyourcountry.org/aid-statistics?cr=302&cr1=oecd&lg=en&page=1.

2. NATO (2011) Allied Joint Doctrine for Counter Insurgency, NATO, Allied Joint Publication 3.4.4, 3-18. Available from: https://info.publicintelligence.net/NATO-Counterinsurgency.pdf.

3. World Food Programme (WFP) Afghanistan, WFP. Available from: https://www.wfp.org/countries/Afghanistan/Overview.

4. Asia Foundation (2012) A Survey of the Afghan People. Asia Foundation, 31 & 171. Available from: https://asiafoundation.org/resources/pdfs/Surveybook2012web1.pdf.

5. IEA (2004) World Energy Outlook 2004. OECD/IEA, 329 & 338-339. Available from: http://www.worldenergyoutlook.org/media/weowebsite/energydevelopment/WEO2004Chapter10.pdf.

6. Energy Consulting Associates (2014) Correlation and causation between energy development and economic growth. Energy Consulting Associates on behalf of DfID (CEIL PEAKS). 15 & 18, Available from: http://dx.doi.org/10.12774/eod_hd.january2014.eca.

7. Foley G (1992) Rural electrification in the developing world. *Energ Policy* 20: 145–152.

8. Kirubi C, Jacobson A, Kammen DM, et al. (2009) Community-Based electric micro-grids can contribute to rural development: Evidence from Kenya. *World Dev* 37: 1208–1221.

9. World Bank (1995) Rural Electrification: A Hard Look at Costs and Benefits. WB Operations Evaluation Department No 90. Available from: http://www-wds.worldbank.org/external/default/WDSContentServer/WDSP/IB/2004/04/19/000011823_20040419101927/Rendered/PDF/28516.pdf.

10. Mainali B, Silveira S (2013) Alternative pathways for providing access electricity in developing countries. *Renew Energ* 57: 299–310.

11. Lahimer AA, Alghoul MA, Yousif F, et al. (2013) Research and development aspects on decentralized electrification options for rural household. *Renew Sust Energ Rev* 24: 314–324.

12. Niez A (2010) Comparative Study on Rural Electrification Policies in Emerging Economies, IEA, 9-12, 99-105. Available from: http://www.iea.org/publications/freepublications/publication/rural_elect.pdf.

13. World Bank (2010) Addressing the Electricity Access Gap, Background Paper for World Bank Group Energy Sector Strategy, 12-18. Available from: http://siteresources.worldbank.org/EXTESC/Resources/Addressing_the_Electricity_Access_Gap.pdf.

14. Bailey M, Henriques J, Holmes J, et al. (2012) Providing village-level energy services in developing countries. Malaysian Commonwealth Studies Centre, p. iii. Available from: http://www.easac.eu/fileadmin/PDF_s/reports_statements/Report_220113_PDF.pdf.

15. Peskett L (2011) The history of mini-grid development in developing countries, Global Village Energy Partnership Policy Briefing, September 2011, p. 4. Available from: http://www.gvepinternational.org/sites/default/files/policy_briefing_-_mini-grid_final.pdf.

16. Zalengera C, Blanchard RE, Eames PC, et al. (2014) Overview of the Malawi energy situation and A PESTLE analysis for sustainable development of renewable energy. *Renew Sust Energ Rev* 38: 335–347.

17. Schillebeeckx SJD, Parikh P, Bansal R, et al. (2012) An integrated framework for rural electrification: Adopting a user-centric approach to business model development. *Energ Policy* 48: 687–697.

18. Ludin GA, Amin MA, Aminzay A, et al. (2016) Theoretical potential and utilization of renewable energy in Afghanistan. *Aims Energy* 5: 1–19.

19. Sediqi MM, Howlader HOR, Ibrahimi AM, et al. (2017) Development of renewable energy resources in Afghanistan for economically optimized cross-border electricity trading. *Aims Energy* 5: 691–717.

20. Thurer D (1999) The failed state and international law. International Review of the Red Cross, no 836, Available from: https://www.icrc.org/eng/resources/documents/misc/57jq6u.htm.

21. UK Ministry of Defence (MoD) (2010) Future Character of Conflict. UK MoD Development, Concepts and Doctrine Centre, February 2010, 30-31, Available from: https://www.gov.uk/government/publications/future-character-of-conflict.

22. Weihe T (2004) Co-operatives in Conflict and Failed States, US Overseas Co-operative Development Council, 2004, 2, 11-12, & 36. Available from: http://www.uwcc.wisc.edu/info/intl/weihe.pdf.

23. Haag K (2012) DoD creates 'distributed renewable energy' model to uplift rural communities in Afghanistan, US Department of Defense, 3-6. Available from: https://ases.conference-services.net/resources/252/2859/pdf/SOLAR2012_0855_full%20paper.pdf.

24. Bhandari R, Richter A, Möller A, et al. (2015) Electrification using decentralized micro hydropower plants in North-eastern Afghanistan. *J Sust Dev Energ Water Environ Syst* 3: 49–65.

25. GTZ (2009) Capacity Assessment in the Subsector of Rural Electricity Supply through Renewable Energy Technologies. GTZ for ESRA, 13-24. Available from: http://www.irena.org/documentdownloads/2012/november/capacityneedsassessments/esra_report_summary.pdf.

26. USAID (2006) Micro-hydropower in Afghanistan: An Audit, Lessons and Conclusions. Nexant Inc on behalf of USAID, pp. 5-1–5-3, pp. 7-1–7-3. Available from: http://pdf.usaid.gov/pdf_docs/PNADJ142.pdf.

27. Ministry of Rural Rehabilitation and Development (MRRD) (2013) Rural Electrical Implementation Guidelines, MRRD, 11-43. Available from: http://mrrd.gov.af/Content/files/Rural%20Electrical%20Implementation%20Guidelines.pdf.

28. Ministry of Rural Rehabilitation and Development (MRRD) (2013) National Solidarity Programme Power Sector Engineering Manual. MRRD. Available from: http://www.nspafghanistan.org/default.aspx?sel=139.

29. Government of the Islamic Republic of Afghanistan (GIRoA) (2008) Afghan National Development Strategy: Energy Sector Strategy 2007/8-2012/13., GIRoA, p. 19, & 46. Available from: http://mew.gov.af/Content/files/Energy_Sector_Strategy-English.pdf.

30. Messner J, Haken N, Taft P, et al. (2015) Fragile State Index 2015, The Fund for Peace, Washington, p. 17. Available from: http://library.fundforpeace.org/library/fragilestatesindex-2015.pdf.

31. Pain A, Kantor P (2012) Village-level Behavior Under Conditions of Chronic Conflict. *MT Res Dev* 32: 345–352.

32. Oxford Dictionaries. Micro-generation. Available from: http://www.oxforddictionaries.com/definition/english/micro generation.

33. Condition definition from Fowler HW, Fowler FG, The Concise Oxford Dictionary of Current English, Eighth Edition, Oxford, 1990.

34. Checkland P (1985) Systems Thinking, Systems Practice, John Wiley & Sons Ltd, 154-155, p. 170.

35. Kennas S, Barnett A (2000) Best Practices for Sustainable Development of Micro Hydro Power in Developing Countries for UK Department of International Development and World Bank, 51-55. Available from: https://practicalaction.org/media/download/6537.

36. Schetter C (2004) The 'Bazaar Economy' of Afghanistan. Sudasien-Informationen No 3, February 2004. Available from: http://crossasia-repository.ub.uni-heidelberg.de/68/1/nr3_bazaar.pdf.

37. Pigaht M, Plas RJVD (2009) Innovative private micro-hydro power development in Rwanda. *Energ Policy* 37: 4753–4760.

11

Solid state transformer technologies and applications: A bibliographical survey

M. Ebrahim Adabi and Juan A. Martinez-Velasco*

Universitat Politecnica de Catalunya, Barcelona, Spain

* **Correspondence:** Email: martinez@ee.upc.edu

Abstract: This paper presents a bibliographical survey of the work carried out to date on the solid state transformer (SST). The paper provides a list of references that cover most work related to this device and a short discussion about several aspects. The sections of the paper are respectively dedicated to summarize configurations and control strategies for each SST stage, the work carried out for optimizing the design of high-frequency transformers that could adequately work in the isolation stage of a SST, the efficiency of this device, the various modelling approaches and simulation tools used to analyze the performance of a SST (working a component of a microgrid, a distribution system or just in a standalone scenario), and the potential applications that this device is offering as a component of a power grid, a smart house, or a traction system.

Keywords: distribution system; high-frequency transformer; microgrid; modular multilevel converter; power quality; semiconductor loss; smart grid; solid state transformer

1. Introduction

The future smart grid is being designed to mitigate or avoid consequences derived from power quality events (e.g., voltage sags), improve reliability indices (e.g., by reducing the number of interruptions and their duration), and increase the system efficiency (e.g., by reducing losses). The increasing penetration of renewable generation and a fast implementation of the electric vehicle are two trends that can stress the current grid by causing voltage variations larger than those the system can withstand. A solution for many of these problems is the Solid State Transformer (SST).

Transformers are widely used to perform functions such as voltage transformation and isolation. Although the conventional transformer has been, and still is, the traditional link between end-users and the distribution network, the high-frequency SST design could cope with many of the challenges of the future smart grid since it can enhance power quality performance and expand the capabilities of the conventional transformer: voltage sag compensation, instantaneous voltage regulation, harmonic compensation, power factor correction, auto-balancing, short-circuit protection, variable-frequency output, bidirectional power flows. Since the size of a conventional copper-and-iron based transformer is inversely proportional to the operating frequency, an increase of this frequency would provide a higher utilization of the magnetic core and a reduction in transformer size. In addition, the SST can be used as a link between standard ac power-frequency systems and systems operating with either dc or ac at any power frequency.

The first patent on a device that could be seen a predecessor of the current SST designs was presented in 1992 [1]. Since then several patents have been presented; see, for instance [2–5].

The SST design can be seen as a universal interface that can provide not only power quality improvements but efficient management of distributed resources. By incorporating the SST, utilities can integrate various power requirements, monitoring, and communications into a universal customer interface such as the SST, which can also provide some operational benefits (e.g., reduced environmental concerns by introducing a design that does not use liquid dielectrics, efficient management of distribution resources by incorporating online monitoring and other automation functionalities). The goal of this paper is to provide a bibliographical review of the work carried out to date on the SST. The main contribution of this work is a list of references ordered by publication year. However, since there are many aspects of the SST that make this device so attractive as a component of the future power systems, several short sections have been included to discuss some important features of the SST. Each section is aimed at summarizing the current status with a selection of relevant works. Readers interested in an introduction to the SST can consult references [6–8].

Although the list of references covers SST designs of various voltage levels and different applications, it might be assumed by default that the primary SST application is to function as a medium voltage/low voltage (MV/LV) distribution transformer. Since standardized voltages used for MV distribution grids are usually equal or higher than 10 kV, multilevel topologies must be considered for the MV side of the SST if conventional Si-based semiconductors are used. In general, it can be assumed that if the highest SST voltage is equal or above the lowest standardized voltage (i.e., 3.3 kV), a SST design must be based on a multilevel converter configuration at the MV side.

Different topologies of multilevel converters have been proposed for SST applications. Irrespective of the selected topology, the operation of a multilevel converter has to face important challenges (e.g., capacitor voltage balancing and complex control strategies). Once the converter topology has been selected, the selection of a proper control strategy for each SST stage becomes crucial for a correct performance of the device. In addition, the SST capabilities (e.g., bidirectional power flow, harmonic compensation, current balance) and performance (e.g., reliability, efficiency) are closely connected with the selected configuration and control strategies; Section 2 provides a summary of the work related to multilevel converter designs for SST implementation and the corresponding control strategies. A fundamental component of the SST is the high-frequency transformer (HFT); Section 3 summarizes the current state of HFT designs for SST implementation.

It is widely accepted that the efficiency of current SST designs is lower than that of their conventional iron-and-copper counterpart; Section 4 discusses this aspect and reviews the main works related to analyze SST efficiency.

The performance and benefits of the SST as a part of large distribution system can be predicted by implementing and testing reliable and accurate computer models; Section 5 summarizes the work carried out on modelling and simulation of the SST, including the experience collected with real-time simulation platforms.

The SST can be seen as a replacement of the conventional transformer. However, the foreseen applications of the SST cover an area wider than that of the conventional transformer. Section 6 summarizes the work dedicated to date for fixing the potential applications of the SST.

Other important aspects, such as the semiconductor technologies that could be adequate for this device are not covered here. Readers are referred to the literature; see for instance [6–12].

Although other designations have been used to name this device (i.e., Intelligent Universal Transformer, Electronic Power Transformer), this paper exclusively uses the acronym SST to name it.

2. SST topologies and control strategies

2.1. Introduction

If it is assumed that the SST can be used to link DC and AC systems running at medium and low voltage levels, a very high number of combinations may result; for instance, the SST can be used to link MV and MV, MV and LV, or LV and LV DC and/or AC systems. Remember that in case of AC systems, they can be single- or multi-phase. In fact, a myriad of configurations have been proposed and even analyzed under the concept SST (or any other designation used to date for this device). To facilitate the study, those configurations will be classified taking into account the number of stages and the configuration of converters to be installed at both MV and LV sides.

Figure 1 shows three examples of single-phase SST configurations that can be used to illustrate the concept of stage when applied to this device. One can observe that a common component to all of these configurations is the HFT. For a discussion about SST configurations taking into account the number of stages, see [13]. SST MV-side converters must be multilevel. Although many configurations have been proposed, multilevel converters can be broadly classified into two main groups, depending on whether the configuration is based on a cascaded connection of converters or not; Figure 2 shows two examples.

On the other hand, note that Figure 2b displays only the MV input stage. As for LV converters, those for three-phase systems can be classified into two groups depending on whether they have three or four wires (i.e., they include the neutral). The control strategies to be used in a SST will depend of the overall configuration, the topology selected for each SST stage, and the desired functionalities. The rest of this section provides a short summary of the work carried out to date on these topics. For a discussion of SST configurations and control strategies, see reference [14]. Table 1 provides a short list of the SST prototypes presented to date.

Other SST prototypes, not listed in the table, were presented in [15–50].

a) Single-stage SST

b) Two-stage SST

c) Three-stage SST

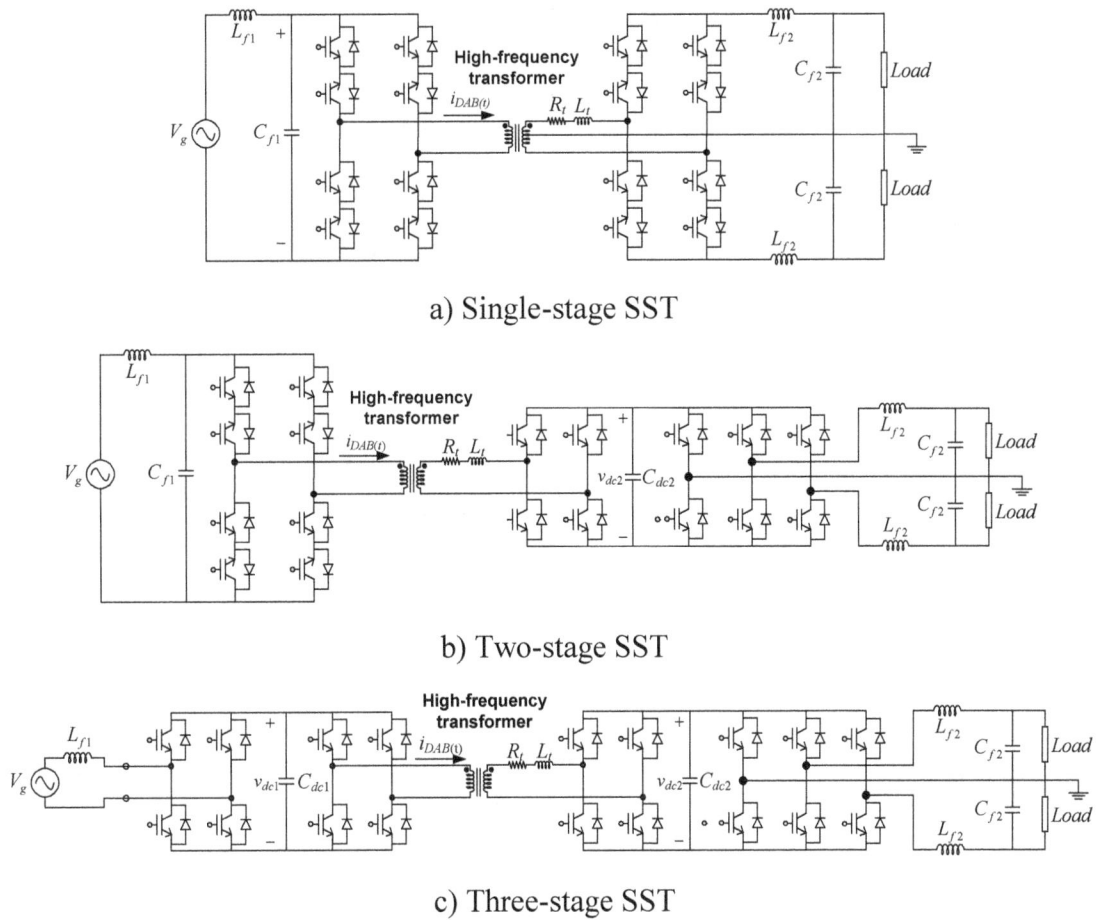

Figure 1. Different single-phase SST topologies.

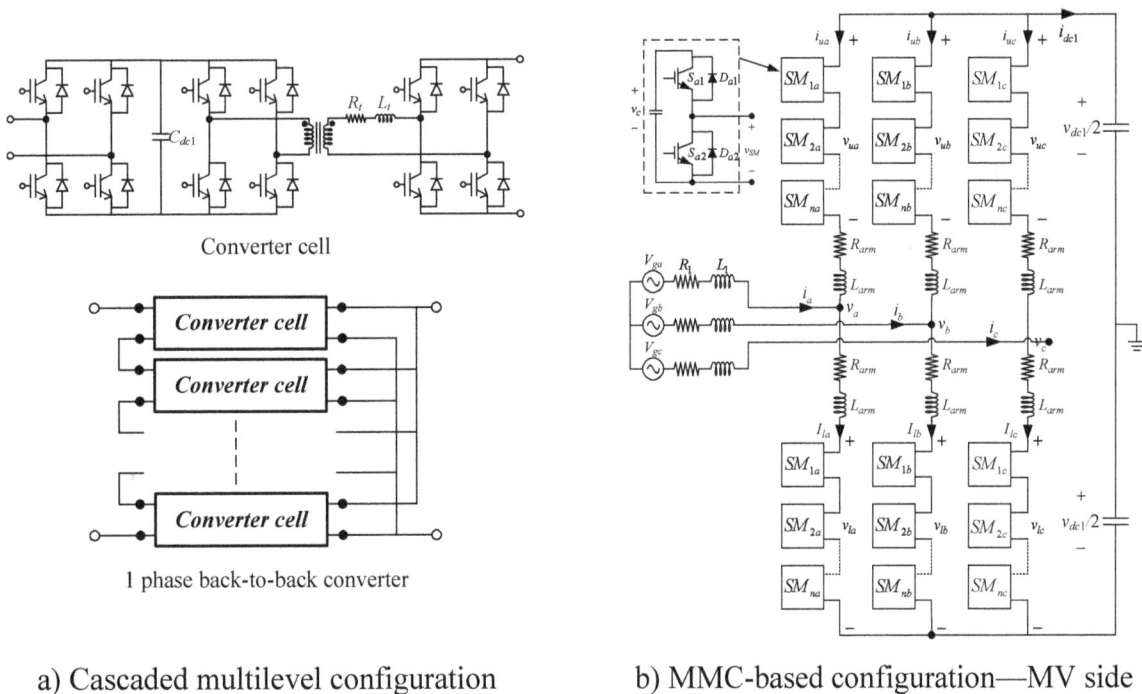

a) Cascaded multilevel configuration   b) MMC-based configuration—MV side

Figure 2. Different configurations for multilevel converters to be used at the SST MV side.

Table 1. SST Prototypes.

Ratings	Configuration	Control	Capabilities	Laboratory tests	Refs.
10 kVA, 7.2 kV/240V	Three stage, cascaded H-bridge	PWM	Unidirectional power flow, power-factor correction	Steady state, load unbalanced	[51,52]
20 kVA, 2.4kV/120-240 VAC or 48V DC	Single phase, three stage, NPC multilevel, multiport-output DC/AC inverter	PWM	Unidirectional power flow	Steady state, load change, load unbalance, voltage sag, nonlinear load,	[53–55]
50 kVA, 2.4 kV/240V/ 120V	Three stage, Three level NPC in MV side	PWM	Voltage sag compensation, fault isolation	voltage sag, load variation, load unbalance	[56]
5 kVA, 220 V/380 V	Two stage, direct AC/AC high-frequency link, dual bridge matrix converter topology	PWM	bidirectional power flow, low harmonic distortion	Load unbalance, unbalanced input voltage	[57]
100 kVA, 13.8 kV/120V or 240 V	Three phase, three stage, cascaded blocks	PWM	Unidirectional power flow	Steady state, load unbalance, voltage sag, non-linear load, load variation, capacitor switching transient,	[58]
1.5 kW, 230 V/39V	Three stage, cascaded H-bridge	PWM	Bidirectional power flow, harmonic voltage compensation, reactive power compensation	Voltage sag, nonlinear load	[59]
2 kW, 110V/20V	Single-stage, AC/AC, two level	PWM	Bidirectional power flow, maximum power-point tracking	Steady state	[60]
20 kVA, 7.2 kV/240 V	Three stage, cascaded H-bridge	PWM	Bidirectional power flow	Steady-state	[61–71]
54 kW, 1.5 kVAC/60VDC	Two stage, cascaded H-bridge	PWM	Bidirectional power flow	Steady state, load variation, power flow reversal	[72,73]
1 kW, 208V/120V	Three-stage, two level	PWM	Bidirectional power flow	Start-up transient	[74]
100 kW, 10kVAC/750VDC	Two stage, AC/DC/DC, MMC	PWM	Bidirectional power flow	Steady state	[75]
2 kVA, 1.9 kV/127 V	Three stage, multilevel converter	PWM, ZVS	Bidirectional power flow, voltage sag compensation	Steady state, voltage sag, power flow reversal	[76,77]
10 kW, 3.6 kV/120 V	Three stage, two level	PWM	Bidirectional power flow, harmonic voltage compensation	Steady state, nonlinear load, load variation	[78–82]
1 kW, 353.55/220	Two stage AC/AC, MMC	PWM	Unidirectional power flow	Steady state	[83]
600 kVA, 3.3 kV DC/3.3 kV DC	Three stage, cascaded H-bridge in MV side	PWM	Bidirectional power flow	Steady state	[84]
5 kW, 3300V AC/ 380VDC	Two stage, cascaded H-bridge	Phase shift modulation	Bidirectional power flow	Steady state	[85]
2 kVA, 380V/120V	Three stage, cascaded	PWM	Bidirectional power flow	Steady state, power flow reversal, startup transient	[86]
5.8 kVA, 5KVDC/ 800VDC	Three stage, NPC with SiC	PWM	Bidirectional power flow	Steady state	[87]

Continued on next page

Ratings	Configuration	Control	Capabilities	Laboratory tests	Refs.
2 kW, 300 V/60 V	Three stage, two level	PWM	Bidirectional power flow	Steady state, load variation	[88]
50 kVA, 480V/480V	Single stage, Dyna-C AC/AC topology	PWM	Bidirectional power flow	Steady state	[89,90]
150 kVA, port1:750VDC port2:375VDC port3:750VDC	Triple active-bridge with energy storage	Phase shift modulation	Bidirectional power flow	Steady state	[91,92]
2 kW, 400V/208V	Single stage, AC/AC, matrix based	Predictive Control	Bidirectional power flow	Steady state, load variation, unbalanced voltage and current	[93,94]
3-kVA, 2.4kV/127V	Three stage, two level	PWM	Unidirectional	Steady state, nonlinear load	[95]
10 kVA, 208 V	Single stage AC/AC, two level	ZVS	Bidirectional power flow	Steady state	[96]
2 kW, 600VDC/200VDC	Three phase modular multilevel dc/dc converter	PWM, ZVS, dual-phase-shift method	Bidirectional power flow	Steady state	[97]
10 kVA, 3.8 kVDC/ 200VDC	Three stage, single phase single converter cell based SST for wind energy conversion system	PWM	Bidirectional power flow	Steady state, load variation	[98]

2.2. Converter topologies for SST application

The possible configuration of a SST has been analysed in many works. References [13,99–109] are some of the works in which the SST configurations were analysed and/or compared. With respect to the number of stages, the main conclusions from the present literature can be summarized as follows: single- and two-stage topologies provide limited functionalities as compared to three-stage topologies, which can provide all the desired SST functionalities while simplifying the control design. A list of selected references in which some of the most popular SST configurations were analyzed is presented below.

Three possible topologies for the higher voltage side of a SST were identified in [101]: the diode clamped multilevel converter, the flying capacitor multilevel converter, and the series stacked converter. Reference [110] analyzed different topologies that can provide a reliable energy management with the SST. References [111,112] proposed a two-level three-stage bidirectional SST.

For the MV side converter configuration, references [113–115] proposed a neutral point clamped (NPC) topology; references [10,116] presented a cascaded H-bridge multilevel inverter configuration, while references [106,107,109,117] suggested a modular multilevel converter (MMC) configuration.

References [89,90] proposed a bidirectional SST configuration, named as dynamic-current (or Dyna-C), with a minimal device count: the topology has two current-source inverter stages with a high-frequency galvanic isolation, and 12 switches for four-quadrant three-phase ac/ac power conversion. The input and output stages can work with arbitrary power factors and frequencies. Dyna-C can be configured as isolated power converters for single- or multi-terminal dc, and single-

or multiphase ac systems. Its modular nature allows Dyna-C to be connected in series and/or parallel for high-voltage high-power applications.

References [118,119] presented a 270 kVA SST based on 10 kV SiC MOSFET: five levels were needed in order to support a 24 kV input voltage; each device has a 10 kV blocking capability. Three flying capacitors enable the operation of zero voltage switching (ZVS) with phase shift control. Reference [120] introduced a new SST topology that included a reduced number of SiC MOSFETs and smaller switching losses.

The configuration of the SST has been the subject of many other works; see [107,121–158].

2.3. Control strategies

A three-stage SST includes up to four different converters (see Figure 1c). Dozens of strategies have been proposed for controlling the various converters of a SST; they primarily depend on each converter configuration and the SST functionalities (e.g., bidirectionality). A summary of control strategies used with lab prototypes is provided in this subsection. For a discussion on strategies to be used with a three-stage SST, see [14].

Reference [159] presented a linear-quadratic-regulator with integral action to improve dynamic performance; the integral action is added to cancel the steady-state errors.

A source-based commutation method for a HFT controlled through a matrix converter was proposed in [160–162].

Reference [163] presented a simple predictive control technique for multilevel configurations either on the line-side (high voltage) or on the load-side (low voltage); the control is performed in two steps: (i) generation of the reference value of the primary current; (ii) evaluation of the optimized delay-angle between primary and secondary voltages using a predictive algorithm.

An energy-based control design method for a three-stage cascaded multilevel SST was proposed in [164]; by selecting the total energy in the two dc link capacitors as the control objective, this approach resulted in a control design.

An advanced control methodology based on fuzzy logic controllers was proposed in [165–167]. References [65,168] analyzed up to four different control strategies for a 20 kVA SST with a seven-level cascaded rectifier stage, three output parallel dual active bridges (DAB), DC/DC stage and an inverter stage. References [169,170] studied soft-switching techniques for MV isolated bidirectional DC/DC NPC-based converters.

A control strategy for a cascaded H-bridge based converter was proposed [171]: the input-stage part was responsible for the power quality improvement and high-voltage DC link voltage balance; the DAB stage was responsible for maintaining the low-voltage DC link voltage; the output-stage part was responsible for the output terminal voltage regulation and parallel module current sharing control. Power synchronization and interleaving modulation were adopted in the output-stage part.

Reference [68] presented a cascaded H-bridge converter-based SST to interface a 7.2 kV AC grid and a 400 V DC distribution; a single-phase dq vector control was used. A new voltage balance control method was proposed to resolve the voltage unbalance of the dc links in H-bridges; see also [85,172].

A hierarchical power management strategy, including primary, secondary, and tertiary control, for a DC microgrid was proposed in [70,173].

Reference [174] investigated the concept of convertible static transmission controller (CSTC) using modular converter. Algebraic models of the CSTC were derived in two different configurations (series–shunt and shunt–shunt).

References [175–177] analyzed the black start operation of a single phase SST with the master-slave control mode and using dual loop structures.

Reference [178] proposed a current sensorless controller for balancing the power in the DC-DC stage of a cascaded multilevel converter-based SST; the equalization of the active power component of duty cycles in the cascaded multilevel rectifier stage can be a good indicator of power balance. Additionally, the power balance of the DC-DC stage can guarantee the voltage balance in the rectifier stage if the differences among the power devices are negligible. Reference [179] proposed a trapezoid current modulated discontinuous conduction mode AC-DC DAB converter for a two-stage SST; the soft switching converter exhibited a high efficiency, and could be operated in open loop control without current sensors.

Reference [116] presented a power and voltage balance control scheme of a cascaded H-bridge modular inverter for microgrid applications operating under unbalanced conditions; the control method was designed to address the presence of power and voltage unbalance.

A control architecture with two communication networks aimed at improving the communication modularity among power modules of a SST was presented in [180]. The communication structure was based on a two full-duplex RS-485 networks (one for each SST side) from which the central unit communicates and controls the local units using of a custom protocol.

Reference [181] proposed a sliding mode control scheme for the rectifier stage with constant power load. This approach can stabilize the dc-link voltage and guarantee the input current sinusoidal in the presence of significant variations in the load power.

References [93,94,182,183] presented a predictive control for a matrix converter-based SST; the goal was to reduce the complexity of the traditional modulation strategy and improve its performance.

Other works related to control strategies of SST were presented in [128,150,152,158,184–218].

3. High-frequency transformer

The high-frequency transformer (HFT) is a fundamental component of the SST, and a requirement to achieve a reduction of size with respect to conventional transformers. To fulfill high-voltage, high-power, and high-frequency operation requirements, several issues and challenges need to be addressed [9]: (i) the selection of the magnetic material is critical to achieve high power density and low losses; (ii) the winding configuration can significantly affect the efficiency at high frequency; (iii) thermal behavior is a challenge to consider in order to avoid breakdown for a high-voltage and high-power designs; (iv) a high-voltage operation makes the insulation requirement another challenge, especially when oil is eliminated and a compact design is required.

It is important to keep in mind that a higher frequency causes extra losses in the magnetic core (as a result of eddy currents) and in windings (due to skin and proximity effects); that is, a volume reduction at higher frequencies is at the expense of increased (core and winding) losses. A thermal management strategy is also important to more accurately evaluate the power losses.

Consequently, multiple degrees of freedom exist when optimizing the design of a HFT; they can be categorized in electric, geometric, and material parameters. For instance, two electric parameters to be accounted for are the number of turns (it determines the ratio of flux density and current density, so it has to be set such that the sum of core and winding losses is minimized) and the operating frequency (when increased it reduces the core losses but increases the proximity losses in the winding and the switching losses of the semiconductors). Two very important specifications of a HFT are the power rating and the operating frequency since, when combined with the magnetic core and conductor materials, they strongly influence the efficiency and power-density.

Reference [219] proposed a transformer for 12 and 24-pulse rectifier systems; a size reduction of 1/3 with respect to a conventional 60 Hz design could be achieved by operating the transformer core at 990 Hz and utilizing conventional grain oriented steel.

Reference [61] proposed optimum size and weight reduction of a 7 kVA dry-type HFT, whose high-voltage side insulation should withstand 15 kV. The design required a relatively high leakage inductance (lack of leakage inductance may lead to additional inductors which results in higher size, weight, and cost), so meeting the leakage inductance requirement became an important issue. Since the leakage inductance depends on the winding arrangement and the number of turns, to adjust the leakage flux, a two-winding arrangement (the windings on both sides are totally separated and one winding is totally covered up with the other winding) and several core materials were considered. Metglas amorphous alloy cores turned out to be the best choice.

Reference [220] presented an accurate equivalent circuit of a MV coaxial winding by comparing results from a finite element method (FEM) and lab measurements. The design provides uniformly and symmetrically distributed electromagnetic flux with good electric and magnetic shielding. An overall efficiency of 99.5% remaining below 100 °C under oil-free and natural convection was achieved for a 30 kVA transfer.

Reference [221] presented the optimization of a HFT with different targets (i.e., weight, volume, and cost) under certain constraints like a given cooling performance, insulation requirements, and selected semiconductors. By means of a detailed transformer model it was possible to find the optimum frequency with respect to size, weight or efficiency for different designs, and systematically investigate improvements arising from different core materials, wire structures, geometries and cooling designs.

A design and optimization method for HFTs was proposed in [9]. The authors carried out a comparison of different magnetic materials; the main conclusion was that a nanocrystalline core is the option that better satisfies both power density and efficiency requirements.

Reference [222] presented the optimization of a water-cooled HFT prototype for maximum power density and efficiency; the electric and thermal specifications, as well as certain dimensions that define clearance space and cooling system, were specified.

Reference [223] presented an optimization methodology applied to a 50 kW, 5 kHz HFT, and aimed at finding the highest power density while the efficiency, isolation, thermal and leakage inductance requirements are meet taking into account thermal management.

Reference [224] proposed a toroidal design for a Metglas core using a procedure aimed at optimizing the number of turns and minimizing (core plus winding) losses.

Reference [225] detailed an optimization procedure for a 166 kW/20 kHz prototype. The authors could achieve 99.4% efficiency at a power density of 44 kW/dm^3. As a consequence of the relatively high-power rating, the cooling system became a major challenge. The work also provided analytic solutions for high-frequency losses, which were separated into skin and proximity losses.

Reference [226] presented the optimal design of 20/0.4 kV HFT; the goal was to maximize efficiency and power density, and minimize weight. The maximum allowable temperature rise was considered as an inequality constraint while desired values of leakage and magnetizing inductances were considered as equality constraints. Results did show that an efficiency and power density above 99.70% and 13 kW/dm^3 can be achieved using a nanocrystalline-based core.

For more details on the design of HFTs for SST applications, see references [202, 227–236].

4. Solid state transformer efficiency

It is widely accepted that the efficiency of the current SST designs is lower than that of their conventional iron-and-copper counterpart; see, for instance, [6,8,237]. To accurately evaluate SST efficiency, the losses of all SST components must be taken into account; these losses include conduction and switching losses in power electronics converters, filter and HFT losses. The losses can be estimated through experimental test setups or by detailed modelling in simulation tools such as Matlab/Simulink or EMTP-like tools. The lower SST efficiency is basically due to the high losses of power electronic converters and the need of filters at both SST sides.

There are some great challenges regarding to the design of efficient converter topologies. Standardized voltages equal or above 10 kV are considered for MV side applications by most utilities. Since the maximum operating voltage of Si semiconductors is about 3.6 kV, multilevel topologies are necessary for the MV side of actual SST configurations. An alternative is to use SiC semiconductors. Although, this technology is not mature enough at the time this work is prepared, it appears as one of the best options for highly efficient and compact SST designs.

Basically, the improvement of SST efficiency can be accomplished by means of the following approaches: (i) advanced converter configurations combined with optimized control strategies for all SST stages; (ii) optimized design of the HFT; (iii) use of wide band gap semiconductors (i.e., SiC), since they can provide lower losses even when working at higher switching frequencies. This latter improvement is due, among other things, to the significant reduction of the number of semiconductors that can be accomplished with this technology.

Although not many works have been dedicated to the estimation and reduction of SST losses, some experience is already available.

An interesting conclusion of a study presented in [238] was that soft switching control might be a good option for enhancing SST efficiency.

Reference [239] analyzed the efficiency of five different topologies of SST with considering commercially available Si semiconductors. Reference [240] studied the efficiency of three modular SSTs under daily loading profile. A computer model for representing semiconductor losses was proposed in [241].

Reference [242] analyzed the behavior of a SST model implemented in OpenDSS for power flow calculations; the SST efficiency was estimated as a function of load level and power factor.

The efficiency of the SST was also analyzed in [100,137,140,232,243–259].

5. Computer modeling and simulation

The SST is a versatile device that can provide new power quality solutions to the future smart grid. Given the complexity of the actual designs and the difficulties that arise when its performance as component of an actual power system has to be analyzed, computer-based simulation appears as a reasonable alternative.

A significant experience is already available in SST modeling. Several types of SST models have been developed and tested; they depend of the application to be analyzed and can be categorized into three main types: switching (detailed) models, average models, and steady-state models. Other approaches (i.e., state-variable model) have been considered. Table 2 provides a selection of SST models implemented to date; the table provides the main characteristics of the implemented models, the simulation tool and the modeling approach.

Table 2. SST simulation tools.

Simulation tool	Modeling approach	Configuration	References
Matlab/Simulink	Switching model	Single phase, three-stage, multilevel	[59,262]
Matlab/Simulink	Switching model	Single phase, three stage, multilevel, bidirectional	[68]
Matlab/Simulink	Average model	Three phase, three stage, multilevel, bidirectional	[84]
Matlab/Simulink	Switching model	Three phase, single stage, multilevel, bidirectional	[97]
Matlab/Simulink	Switching model	Single phase, three stage, multilevel, bidirectional	[106,107,109,281]
Matlab/Simulink	Average model	Three phase, three stage, two level, bidirectional	[111,112,279,280]
Matlab/Simulink	Switching model	Three phase, three stage, multilevel, bidirectional	[117,241]
Matlab/Simulink	Average model	Three phase, three stage, two level, bidirectional	[159]
Matlab/Simulink	Switching model	Three-phase, single-stage, two level, bidirectional	[160,162,263–267]
Matlab/Simulink	Average model	Single phase, three stage, multilevel, bidirectional	[164,270]
Matlab/Simulink	Average model	Single phase, three stage, multilevel, bidirectional	[168]
Matlab/Simulink	Average model	Single phase, three stage, multilevel, bidirectional	[173]
Matlab/Simulink	Switching model	Three phase, single stage, two level, bidirectional	[260]
Matlab/Simulink	Average model	Three phase, three stage, two level, bidirectional	[261]
Matlab/Simulink	Average model	Three phase, three stage, multilevel, bidirectional	[268]
Matlab/Simulink	Average model	Single phase, three stage, multilevel, bidirectional	[269]
Matlab/Simulink	Switching model	Three phase, three stage, multilevel, bidirectional	[271]
Matlab/Simulink	Average model	Three phase, three stage, two level, bidirectional	[272]
Matlab/Simulink	Switching model	Three phase, three stage, two level, bidirectional	[273,274]
Matlab/Simulink	Switching model	Single phase, three stage, multilevel, bidirectional	[275–277]
Matlab/Simulink	Switching model	Three phase, three stage, two level	[278]
Matlab/Simulink	Switching model	Three phase, three stage, multilevel, bidirectional	[282]
Matlab/Simulink	Switching model	Three phase, three stage, two level, bidirectional	[283]
Matlab/Simulink	Average model	Single phase, three stage, two level, bidirectional	[284]
Matlab/Simulink	Switching model	Three phase, three stage, two level , bidirectional	[285]
Matlab/Simulink	Average model	Three phase, two stage, multilevel, bidirectional	[286]
Matlab/Simulink	Average model	Single phase, two and three stage, two level, bidirectional	[287]
Matlab/Simulink	Switching model	Single phase, two stage, two level, bidirectional	[288]
Matlab/Simulink	Switching model	Three-phase, three-stage, multilevel, bidirectional	[289]
Matlab/Simulink	Switching model	Three phase, three stage, multilevel, bidirectional	[290]
Matlab/Simulink	Switching model	Three phase, three stage, two level, bidirectional	[291]
Matlab/Simulink	Average model	Single phase, three stage, two level, bidirectional	[292]
Matlab/Simulink	Switching model	Three phase, three stage, two level, bidirectional	[293]
Matlab/Simulink	Switching model	Three phase, three stage, multilevel, bidirectional	[294]
Matlab/Simulink	Switching model	Single phase, two stage, multilevel, bidirectional	[295]

Continued on next page

Simulation tool	Modeling approach	Configuration	References
Matlab/Simulink	Switching model	Single phase, three stages, multilevel, bidirectional	[296]
Matlab/Simulink	Average model	Single phase, two stage, two level, bidirectional	[297]
Matlab/Simulink	Switching model	Three phase, three stage, two level, bidirectional	[298]
Matlab/PLECS	Average model	Single phase, two stage, multilevel, bidirectional	[62]
Matlab/PLECS	Switching model	Single phase, three stage, two level, bidirectional	[177]
Matlab/PLECS	Switching model	Single phase, single stage, multilevel, bidirectional	[299]
Matlab/ PLECS	Average model	Single phase, three stage, two level, bidirectional	[300,301]
Matlab/PLECS	Switching/Average models	Single phase, three stage, two level, bidirectional	[302]
Matlab/PLECS	Switching model	Three phase, three stage, multilevel	[303]
Matlab/PLECS	Switching model	Single phase, two stage, two level, bidirectional	[304]
Matlab/PLECS/RTDS	Average model	Single-phase, three-stage, two level, bidirectional	[305]
PLECS	Switching model	Three phase, single stage, two level, bidirectional	[306]
PLECS	Average model	Single phase, three stage, two level, bidirectional	[307]
PLECS	Average model	Single phase, three stage, multilevel, bidirectional	[308]
PLECS/SPICE	Phasor-based model	Three phase, three stage, multilevel, bidirectional	[309]
PSPICE	Switching model	Single phase, three stage, two level, bidirectional	[80]
SPICE	Switching model	Single phase, two stage, two level	[310,311]
SPICE/SABER	Switching model	Single phase, single stage, multilevel	[118]
SABER	Switching model	Single phase, single stage, two level, bidirectional	[312]
PSCAD/EMTDC	Switching model	Single phase, three stage, multilevel, bidirectional	[76]
PSCAD/EMTDC	Average model	Single phase, three stage, multilevel, bidirectional	[176]
PSCAD/EMTDC	Switching model	Single phase, single stage, two level, bidirectional	[313,314]
PSCAD/EMTDC	Switching model	Three phase, three stage, multilevel	[315]
PSCAD/EMTDC	Switching model	Three phase, three stage, two level, bidirectional	[316]
PSCAD/EMTDC	Switching model	Three phase, three stage, multilevel, bidirectional	[317]
PSCAD/EMTDC	Average model	Three phase, three stage, two level, bidirectional	[318]
PSCAD/EMTDC	Average model	Three phase, three stage, two level, bidirectional	[319]
PSCAD/EMTDC	Switching model	Single phase, three stage, multilevel, bidirectional	[320,321]
PSCAD/EMTDC	Switching model	Three phase, three stage, two level, bidirectional	[322]
PSCAD/EMTDC	Switching model	Three phase, three stage, two level, bidirectional	[323]
PSCAD/EMTDC	Switching model	Three phase, three stage, multilevel, bidirectional	[324]
PSCAD/EMTDC	Average model	Single phase, three stage, two level, bidirectional	[325]
PSCAD/EMTDC	Switching model	Three phase, three stage, two level, bidirectional	[326]
PSCAD/EMTDC	Average model	Three phase, three stage, multilevel, bidirectional	[327]
PSCAD/EMTDC	Switching model	Three phase, three stage, multilevel, bidirectional	[328]
PSCAD/EMTDC	Switching model	Single stage, DC/DC, multilevel, bidirectional	[329]
EMTP/ATP	Switching model	Three phase, three stage, multilevel, bidirectional	[115]
EMTP/ATP	Switching model	Three phase, three stage, two level, bidirectional	[330]
EMTP/ATP	Switching model	Three phase, three stage, multilevel, bidirectional	[331]
PSIM	Switching model	Single-phase, two-stage, multilevel, bidirectional	[85]
PSIM	Switching model	Single stage, DC/DC, multilevel, bidirectional	[170,333]

Continued on next page

Simulation tool	Modeling approach	Configuration	References
PSIM	Switching model	Single-phase, two-stage, multilevel, bidirectional	[179]
PSIM	Switching model	Single stage, DC/DC, two level, bidirectional	[332]
PSIM	Switching model	Three phase, three stage, multilevel, bidirectional	[334,335]
PSIM	Average model	Single phase, single stage, two level, bidirectional	[336]
Digsilent	Average model	Three phase, three stage, two level, bidirectional	[337]
Simplorer	Switching model	Three phase, three stage, multilevel	[338]
RTDS	Average model	Single phase, three stage, multilevel, bidirectional	[339]
RTDS	Average model	Single phase, three stage, two level, bidirectional	[340]
OPAL-RT	Switching model	Single phase, three stage, two level, bidirectional	[341]
Multisim/Labview	Switching model	Single phase, single stage, two level, bidirectional	[342,343]
OPENDSS	Steady-state model	Three phases, three stage, two level, bidirectional	[242]

Not much experience on SST models is currently available for steady-state power flow calculations. Reference [242] presented a SST model for OpenDSS implementation. The model can be used to explore and assess the impact of the SST on distribution system performance (i.e., in steady-state power flow calculations) considering either a short- or long-term evaluation.

On the other hand, a significant experience is already available with switching (detailed) models. The use of these models generally requires the use of very short time-step sizes (i.e., equal or shorter than 1 μs), which implies long simulation times and limits the size of the system that can be practically analyzed. The most popular tool for this type of models is MATLAB/Simulink, although some important experience is also available with EMTP-like tools.

Real-time simulation platforms are widely used for transient simulation of power systems, testing of protection devices, or rapid control prototyping. The need of a very short time-step for switching models of SSTs can be a drawback with many of these simulation platforms. To mitigate or circumvent this limitation the so-called *dynamic average models* (DAM) can be developed: a DAM approximates the behavior of a converter by applying the moving average operator at the switching frequency to the detailed switching model; the switching effects are removed from the model, but the dynamic behavior is preserved. DAMs, named *average models* hereinafter, can reproduce with a high accuracy the transient behavior of the original detailed switching model but using a larger time step size, facilitating the implementation of transient models in real-time simulation platforms. Figure 3 shows three different average models proposed in the literature.

Simulation results with an average model of the three-stage SST for the Future Renewable Electric Energy Delivery and Management (FREEDM) green hub system were presented in [270,344].

Reference [269] presented a SST average model that was validated by comparing results from those with the detailed switching model; see also [345,346].

Reference [347] proposed a SST average model aimed at analyzing the transient performance of a distribution network.

References [111] and [348] presented an average model of a bidirectional SST for feasibility studies and real-time implementation; several cases were studied to evaluate the behavior of the model under different operating conditions, check its feasibility for power quality improvements, and explore the implementation in a real-time simulation platform; see also [112,279].

Reference [349] presented a SST average model for studying renewable energy integration. A three-phase test system model including substation and loads was implemented in PSCAD.

Reference [261] proposed a model for transient stability studies; the dynamic model neglects high-frequency transients. To verify the accuracy of the proposed model, a comparison between results from such simplified dynamic model and a detailed model implemented in Matlab/Simulink were carried out; see also [350]. Reference [309] proposed a modular dynamic-phasor SST model for stability analysis; the model provided a significant reduction of simulation time. Reference [337] detailed the implementation of a SST model in Digsilent Power Factory; the model was based on the dynamic average technique and is compatible with LV-side three-phase, four-wire configuration.

The SST performance was analyzed in [351] by means of a 70th-order state-space model. The system model included renewable generation and storage systems and their corresponding interface circuits to DC and AC buses. The model was used to evaluate the performance of a distribution system under grid connected and islanded conditions.

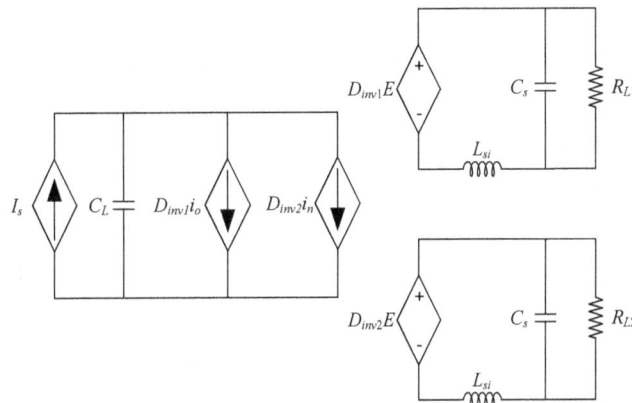

a) Average model for a DC/AC inverter [269]

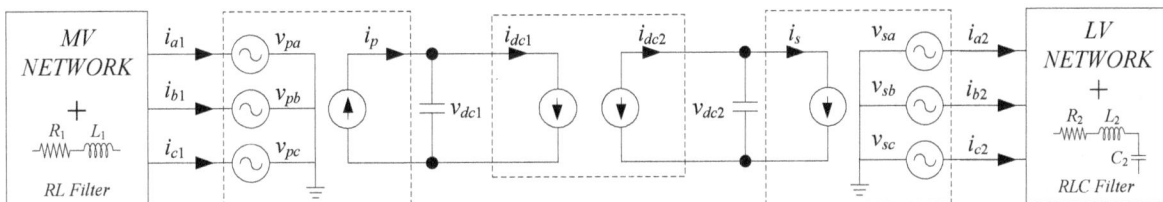

b) Average model for a three-stage bidirectional SST [111]

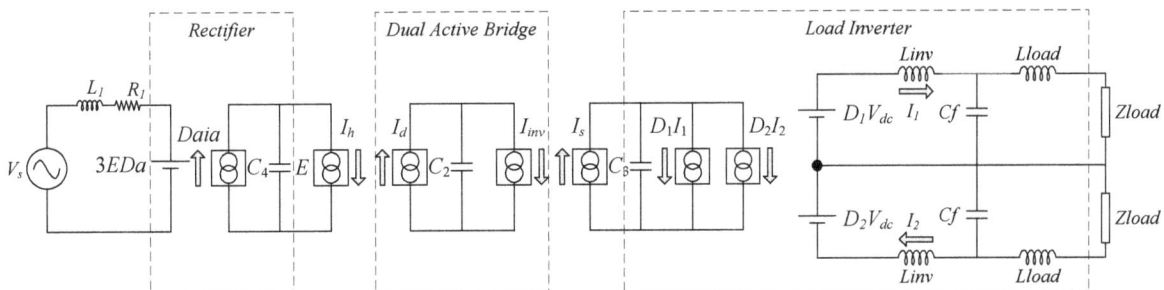

c) Average model for a three-stage SST [349]

Figure 3. Some average models for the solid state transformer.

Reference [331] presented a model of a MV/LV bidirectional SST. A multilevel converter configuration of the MV side is obtained by cascading a single-phase cell made of the series connection of an H bridge and a dual active bridge (DC/DC converter); the aim was to configure a realistic SST design suitable for MV levels. The SST model, including the corresponding controllers, was built and encapsulated as a custom-made model in the ATP version of the EMTP for application in distribution system studies [115,352].

Reference [117] presented a model of a bidirectional MV/LV SST for distribution system studies. A modular multilevel converter configuration is used in the MV side of the STT. The LV side uses a three-phase four-wire configuration that can be connected to both load and generation. The model developed was implemented in MATLAB/Simulink, and its behavior was tested by carrying out several case studies under different operating conditions.

Detailed dynamic analysis of buck-type SST under general load condition was presented in [353]; the study showed that under general inductive load condition, the open-loop system is marginally stable and serious output voltage oscillation can be provoked by any random disturbance upon input voltage, control signal or output current.

A SST model and its mapping to IEC 61850 was presented in [354].

A significant experience is also available on SST modeling for implementation in real-time simulation platforms. Reference [302] proposed an average model suitable for simulation with RTDS; the model was tested and validated by comparing its performance to that of detailed full switching model and a cycle-by-cycle average model built in Matlab and PLECS [305].

Reference [63] introduced the development of a platform intended as a distributed controller for grid intelligence system at FREEDM Systems Center. This platform can serve as a real-time local converter controller and a communication node for distributed deployment of energy management schemes. One device it controls is the SST, a key element to interface renewable energy sources to distribution systems in FREEDM Center. Both the hardware design and software structure for SST control were presented. The communication part used the Distributed Network Protocol 3.0.

Reference [340] proposed a new protection technique for SST; it was verified by hardware in the loop (HIL) testing.

Reference [95] proposed a platform based on the Xilinx Zynq®-7000 family to test SST performance; SST functions include: link to information and communication technologies, voltage transformation, integration of distributed renewable energy resources. The platform embedded a double-core ARM® Cortex™-A9 processor and Field Programmable Gates Array technology; all within a programmable system on a chip.

The implementation and simulation of the SST in either off- and on-line simulation platforms has been the subject of many works not listed in Table 2; see [126,178,355–415].

6. Applications of the solid state transformer

Future distribution grids will be characterized by a growing need for integration of renewable energy sources, energy storage devices and other smart grid technologies. The SST can perform as a universal interface for integrating distributed energy resources or as part of microgrids of any architecture. The potential applications of the SST have been analyzed in some references; see, for instance, [6,100,416–419]. In general, it is assumed that instead of using SST as a simple replacement of a conventional transformer, the SST will provide additional functionalities that could significantly improve power quality. Figure 4 schematizes some potential applications of the SST in a future distribution system. The SST might work as an interface of a DC-based fast charger (not shown in the figure), a traction system, a distributed energy source with or without energy storage capability [300], in microgrid architectures [420], or providing reactive power compensation and active harmonic filtering to any type of loads. A review of the SST applications proposed in the literature is provided in this section. The applications have been classified into three groups: the SST as a component of the distribution system, application in traction systems, and other applications (heating, lighting, smart house).

SST in the future grid: Reference [64] presented the next generation power distribution system architecture: the FREEDM system, which enables the plug-and-play of distributed renewable energy resources and storage devices. The FREEDM system is a highly attractive candidate for the future power distribution system; for more details, see references [67,71,292,421–423].

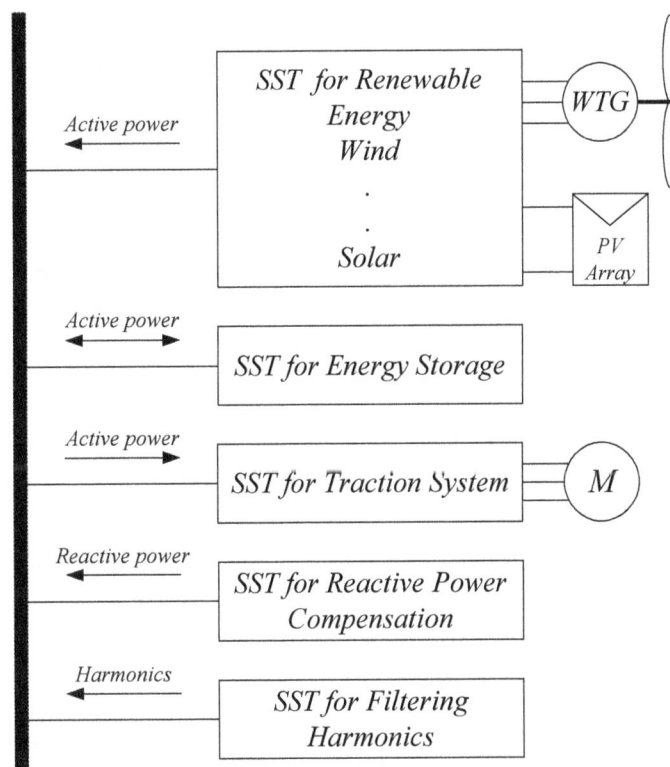

Figure 4. Potential applications of the SST.

Reference [86] proposed the application of the SST in hybrid microgrids: converters use SiC-based semiconductors commercially available; high-frequency magnetic components are made of iron-based nanocrystalline soft magnetic material; the high-voltage grid interface adopts multilevel cascade structure; the isolated bidirectional DC/DC converter employs dual-phase-shift control to decrease circulating current and increase efficiency; the isolation unit is responsible to control voltage and power balancing to avoid control coupling. Experimental results were presented in [424].

References [425,426] studied the application of the SST as an energy router in future smart grids.

A three-port converter based on the Cuk topology was presented in [427]: the converter interfaces one unidirectional input power port (envisioned to be a DC power source such as PV or fuel cell) and two bidirectional output ports, representing respectively a grid-tied inverter DC bus and a storage system.

Reference [286] presented the SST as interface of hybrid DC and AC microgrid systems. Reference [75] proposed a three-phase MMC-based SST that can be link a DC systems and LV renewable energy systems. Reference [428] presented a SST topology based on a quad-active-bridge converter that can provide isolation for load, generation and storage; see also [288].

Reference [429] proposed the application to grid connected PV systems; see also [430]. Reference [313] proposed the SST as interface of a system that combined a PV array and a battery storage: since the power can flow from the network to the PV-battery side and vice versa, the battery can be charged from either the network or the PV array. Reference [431] presented the SST as interface in a PV-assisted charging station with an autonomous energy management strategy. For application of the SST as interface of PV systems, see also [297,430], and [432–438].

References [69,439] studied a SST-interfaced wind energy conversion system with integrated active power transfer, reactive power compensation and voltage conversion functions; the proposed configuration can effectively suppress the voltage fluctuation caused by the transient nature of wind energy without additional reactive power compensation. For the application of the SST in wind energy conversion systems, see also [98,298,322,440,441].

Reference [442] described the design and performance of a bidirectional isolated DC/DC converter using a 20 kHz HFT for a 53.2 V, 2 kWh lithium-ion battery energy storage system.

Reference [443] analyzed a flywheel energy storage system for wind farms fed from a DC system via a SST. The application of the SST in wind energy systems was also studied in [204,438,444,445]. For more details on the application of SSTs in microgrids, see references [66,116,173,326,341,446–449].

The application of the SST in distribution system has also been analyzed in [206,215,253,256,259,337,382,404,405,450–466].

SST application in traction: The constraints of weight and size on the traction transformer are becoming stronger with the new generation of trains, in which reliability and efficiency become very important. The SST is considered a viable solution for the replacement of bulky low-frequency transformers in railway systems operating at 16⅔ Hz. Reference [467] presented an overview of SST technology for traction applications. The multilevel converter proposed in [468] exhibited reduced weight and size, and an improved global life cycle cost; the proposed multilevel topology consisted of sixteen bidirectional direct current converters (cycloconverters), fed from a 15 kV/16.7 Hz catenary through a choke inductor and connected to sixteen medium frequency transformers (400 Hz) supplied by sixteen four-quadrant converters connected in parallel to a 1.8 kV DC link. For more information on the application of the SST in traction, see references [469–472].

Other applications: A single-stage bidirectional SST for induction heating applications was proposed in [60,473]: the SST can simultaneously track the maximum power point and improve the output power factor by using an adjustable switching frequency controller. Reference [474] presented a single-stage bidirectional SST for lighting: the system supplies multi-lamp units that are controlled simultaneously by the SST using a PWM scheme; the SST contains a single-input multi-output HFT that provides galvanic isolation in each unit; the design exhibits good efficiency, low weight and small volume, and allows operation without any bulky storage elements; the control strategy may achieve fluorescent lamps operation free from voltage flicker and disturbances, improving illumination, protecting lamps, and increasing lifetime of lamps.

Reference [314] proposed the combination of a bidirectional SST and a dynamic voltage restorer to correct voltage distortions on a sensitive load.

Reference [475] presented the application of a dynamic power limiter at the PCC of a microgrid; the limiter was a high-frequency isolated power-converter system comprised of a HFT and three-phase to single-phase matrix converters.

Reference [476] presented the SST potential in integrating sources and appliances at the domestic level. Reference [477] proposed a smart house fed from a microgrid-supplied SST.

7. Conclusion

The solid state transformer (SST) offers several benefits for future smart grids: DC and high-frequency AC power supply, enhanced power quality performance, fast voltage control, reactive power compensation, reactive power control at both primary and secondary sides. The SST can also provide operational benefits, such as an efficient management of distribution resources by incorporating on-line monitoring. In a few words, the SST can be seen as a universal interface that can provide power quality improvements, efficient management of distributed resources, and a link between systems operating at different power frequencies.

This paper has presented a bibliographical survey on the work carried out to date on design, testing, modelling, and potential applications of the SST. The number of references included in this paper confirms the interest in the SST and the foreseen benefits that this device can offer.

The paper has been organized taking into account the aspects mentioned above; the various sections has been dedicated to summary the work made on SST configurations and control strategies, design of the HFT to be used in the isolation stage, efficiency, modelling and validation, and SST applications.

The two tables included in the paper provide a selected list of prototypes and computer models implemented in different simulations tools, including real-time simulation platforms. The high number of both lab prototypes and computer models already built prove that the SST technology is becoming mature. Actually, some SST designs are already working; for instance, in traction systems.

The most challenging issue is the low SST efficiency: the high number of semiconductors needed for any multilevel configuration, the amount of semiconductor losses and the need of filters at both SST sides are three important factors that have impact on efficiency. Not much field experience is currently available with actual designs and real costs (including operation and maintenance costs).

Reliability is another aspect for which some work is required: the high number of semiconductors that are needed to build a multilevel configuration increase the probability of failure and losses. Redundant designs that could increase reliability would also increase losses and costs, and in turn reduce efficiency. For more details on this subject see [53,54].

Although the field experience does not yet suffice to decide about the most convenient SST configuration, it seems that those configurations based on a three-stage offer the best operational benefits in most power system applications.

Some studies show that with the present technology the volume, weight and manufacturing cost of a SST could exceed those of a traditional iron-and-copper transformer [478]. The usage of SiC-based technologies could provide a solution to all these drawbacks and permit smaller, lighter, more efficient, and cheaper designs.

Conflict of interest

The authors declare no conflicts of interest in this paper.

References

1. Liu G, Polis MP, Wang B (1992) Solid-state power transformer circuit. Patent No.: 5119285.
2. Sudhoff SD (1999) Solid state transformer. Patent No.: 5943229.
3. Lai JS, Mansoor A, Maltra A, et al. (2005) Multifunction hybrid intelligent universal transformer. Patent No.: US 6954366 B2.
4. Lai JS, Mansoor A, Maltra A, et al. (2006) Multilevel converter based intelligent universal transformer. Patent No.: US 7050311 B2.
5. Gupta RK, Mohapatra KK, Castelino G, et al. (2011) Soft switching power electronic transformer. Patent No.: US 2011/0007534 A1.
6. She X, Huang AQ, Burgos R (2013) Review of solid-state transformer technologies and their application in power distribution systems. *IEEE J Em Sel Top P* 1: 186–198.
7. Huber JE, Kolar JW (2016) Solid-state transformers: On the origins and evolution of key concepts. *IEEE Ind Electron Mag* 10: 19–28.
8. Huang AQ (2016) Medium-voltage solid-state transformer. *IEEE Ind Electron Mag* 10: 29–42.
9. Ronanki D, Williamson S (2018) Evolution of power converter topologies and technical considerations of power electronic transformer-based rolling stock architectures. *IEEE T on Transport Electri* 4: 211–219.
10. Pena-Alzola R, Gohil G, Mathe L, et al. (2013) Review of modular power converters solutions for smart transformer in distribution system. IEEE Energy Conversion Congress and Exposition (ECCE). *IEEE*, 380–387.
11. Rathod DK (2014) Solid state transformer (SST) review of recent developments. *Adv Electron Electr Eng* 4: 45–50.
12. Gajowik T, Rafał K, Malinowski M (2017) Review of multilevel converters for application in solid state transformers. *Przegląd Elektrotechniczny* 93: 1–5.
13. Falcones S, Mao X, Ayyanar R (2010) Topology comparison for solid state transformer implementation. IEEE Power and Energy Society General Meeting. *IEEE*, 1–8.
14. Shri A (2013) A solid-state transformer for interconnection between the medium and the low voltage grid. Master Thesis, Delft University.
15. Harada K, Yamasaki F, Jinno K (1996) Intelligent transformer. In: 27th Annual IEEE Power Electronics Specialists Conference (PESC). Baveno, Italy, 1337–1341.
16. Kang M, Enjeti PN, Pitel IJ (1999) A simplified auto-connected electronic transformer (SACET) approach upgrades standard 6-pulse rectifier equipment with 12-pulse characteristics to facilitate harmonic compliance. IEEE Power Electronics Specialists Conference (PESC). *IEEE*, 199–204.
17. Kang M, Woo BO, Enjeti PN, et al. (1999) Open-delta auto-connected electronic transformer (OD-ACET) based multi-pulse rectifier systems. Applied Power Electronics Conference and Exposition (APEC). *IEEE*, 234–240.

18. Kang M, Woo BO, Enjeti P, et al. (1999) Autoconnected-electronic-transformer-based multipulse rectifiers for utility interface of power electronic systems. *IEEE T Ind Appl* 35: 646–656.

19. Kang M, Enjeti PN, Pitel IJ (1999) Analysis and design of electronic transformers for electric power distribution system. *IEEE T Power Electr* 14: 1133–1141.

20. Visser AJ, Enslin JHR, Mouton HDT (2002) Transformerless series sag compensation with a cascaded multilevel inverter. *IEEE T Ind Electron* 49: 824–831.

21. Tolbert LM, Peterson WA, White CP, et al. (2002) A bi-directional DC-DC converter with minimum energy storage elements. Industry Applications Conference. *IEEE*, 1572–1577.

22. EPRI (2006) EPRI intelligent universal transformer:risk appraisal and project plans. *EPRI Report* 1012434.

23. Aijuan J, Hangtian L, Shaolong L (2006) A new high-frequency AC link three-phase four-wire power electronic transformer. 1st IEEE Conference on Industrial Electronics and Applications (ICIEA). *IEEE*, 1–6.

24. Haibo L, Chengxiong M, Jiming L, et al. (2008) Parallel operation of electronic power transformer and conventional transformer. 3rd International Conference on Electric Utility Deregulation and Restructuring and Power Technologies (DRPT). *IEEE*, 1802–1808.

25. Falcones S, Ayyanar R, Mao X (2012) A DC–DC multiport-converter-based solid-state transformer integrating distributed generation and storage. *IEEE T Power Electr* 28: 2192–2203.

26. Ling C, Ge B, Bi D, et al. (2011) An effective power electronic transformer applied to distribution system. International Conference on Electrical Machines and Systems (ICEMS). *IEEE*, 1–6.

27. Wang G, Baek S, Elliott J, et al. (2011) Design and hardware implementation of Gen-1 silicon based solid state transformer. 26th Annual IEEE Applied Power Electronics Conference and Exposition (APEC). *IEEE*, 1344–1349.

28. She X, Huang AQ, Wang G (2011) 3-D space modulation with voltage balancing capability for a cascaded seven-level converter in a solid-state transformer. *IEEE T Power Electr* 26: 3778–3789.

29. Liu X, Liu L, Li H, et al. (2012) Study on the start-up schemes for the three-stage solid state transformer applications. IEEE Energy Conversion Congress and Exposition, (ECCE). *IEEE*, 3528–3532.

30. Kadavelugu A, Wang G, Bhattacharya S, et al. (2012) Auxiliary power supply for solid state transformers. IEEE Energy Conversion Congress and Exposition (ECCE). *IEEE*, 1426–1432.

31. Tripathi AK, Mainali K, Patel D, et al. (2013) Closed loop D-Q control of high-voltage high-power three-phase dual active bridge converter in presence of real transformer parasitic parameters. IEEE Energy Conversion Congress and Exposition (ECCE). *IEEE*, 5488–5495.

32. Nan C, Ayyanar R (2013) Dual active bridge converter with PWM control for solid state transformer application. IEEE Energy Conversion Congress and Exposition (ECCE). *IEEE*, 4747–4753.

33. Madhusoodhanan S, Patel D, Bhattacharya S, et al. (2013) Protection of a transformerless intelligent power substation. 4th IEEE International Symposium on Power Electronics for Distributed Generation Systems (PEDG). *IEEE*, 1–8.

34. Toit D, Mouton HDT, Kennel R, et al. (2013) Predictive control of series stacked flying-capacitor active rectifiers. *IEEE T Ind Inform* 9: 697–707.

35. Madhusoodhanan S, Tripathi A, Kadavelugu A, et al. (2014) Experimental validation of the steady state and transient behavior of a transformerless intelligent power substation. 29th Annual IEEE Applied Power Electronics Conference and Exposition (APEC). *IEEE*, 3477–3484.

36. Roy S, De A, Bhattacharya S (2014) Current source inverter based cascaded solid state transformer for AC to DC power conversion. International Power Electronics Conference (IPEC). *IEEE*, 651–655.

37. Yang T, Meere R, Feely O, et al. (2014) Performance of 3-phase 4-wire solid state transformer under imbalanced loads. IEEE PES T&D Conference and Exposition. *IEEE*, 1–5.

38. Yu X, She X, Huang A, et al. (2014) distributed power balance strategy for DC/DC converters in solid state transformer. 29th Annual IEEE Applied Power Electronics Conference and Exposition (APEC). *IEEE*, 989–994.

39. Sidorov AV, Zinoviev GS (2015) power electronic transformer based on AC voltage regulator. 16th International Conference of Young Specialists on Micro/Nanotechnologies and Electron Devices (EDM). *IEEE*, 477–481.

40. Ouyang S, Liu J, Wang X, et al. (2015) Comparison of four power electronic transformer topologies on unbalanced load correction capacity. IEEE Energy Conversion Congress and Exposition (ECCE). *IEEE*, 3702–3709.

41. Roasto I, Miñambres-Marcos V, Romero-Cadaval E, et al. (2015) Design and evaluation of a base module of active power electronic transformer. 9th International Conference on Compatibility and Power Electronics (CPE). *IEEE*, 384–389.

42. Zong S, Zhu Q, Yu W, et al. (2015) Auxiliary power supply for solid state transformer with ultra high voltage capacitive driving. IEEE Applied Power Electronics Conference and Exposition-APEC. *IEEE*, 1008–1013.

43. Reddy BD, Sahoo SK (2015) Testing of solid state transformer. *Int J Nov Res Electr Mech Eng* 2: 1–7.

44. Ramos-Ruiz J, Krishnamoorthy H, Enjeti P (2015) Adding capacity to an existing electric power distribution network using a solid state transformer system. IEEE Energy Conversion Congress and Exposition (ECCE). *IEEE*, 6059–6066.

45. Hambridge S, Huang AQ, Yu R (2015) Solid state transformer (SST) as an energy router: Economic dispatch based energy routing strategy. IEEE Energy Conversion Congress and Exposition (ECCE). *IEEE*, 2355–2360.

46. Mainali K, Madhusoodhanan S, Tripathi A, et al. (2015) Start-up scheme for solid state transformers connected to medium voltage grids. IEEE Applied Power Electronics Conference and Exposition (APEC). *IEEE*, 1014–1021.

47. Cazakevicius FE, Quevedo JO, Beltrame RC, et al. (2015) High insulation voltage gate-driver applied to a solid state transformer. 13th IEEE Brazilian Power Electronics Conference and 1st Southern Power Electronics Conference (COBEP/SPEC). *IEEE*, 1–6.

48. Zhu Q, Wang L, Zhang L, et al. (2016) Improved medium voltage AC-DC rectifier based on 10k kV SiC MOSFET for solid state transformer (SST) application. IEEE Applied Power Electronics Conference and Exposition (APEC). *IEEE*, 2365–2369.

49. Yang T, Meere R, O'Loughlin C, et al. (2016) Performance of solid state transformers under imbalanced loads in distribution systems. In: IEEE Applied Power Electronics Conference and Exposition (APEC). *IEEE*, 2629–2636.

50. Kadavelugu A, Wang G, Bhattacharya S, et al. (2016) Auxiliary power supply for Solid State Transformers. IEEE International Conference on Electronics, Circuits and Systems (ICECS). *IEEE*, 193–196.

51. Ronan ER, Sudhoff SD, Glover SF, et al. (2000) Application of power electronics to the distribution transformer. IEEE Applied Power Electronics Conference and Exposition (APEC). *IEEE*, 861–867.

52. Ronan ER, Sudhoff SD, Glover SF, et al. (2002) A power electronic-based distribution transformer. *IEEE T Power Deliver* 17: 537–543.

53. EPRI (2004) Feasibility study for the development of high-voltage, low-current power semiconductor devices. *EPRI Report* 1001698.

54. EPRI (2004) Development of a new multilevel converter-based intelligent universal transformer : design analysis. *EPRI Report* 1002159.

55. EPRI (2005) Bench model development of a new multilevel converter-based intelligent universal transformer. *EPRI Report* 1010549.

56. Lai JS, Maitra A, Goodman F (2006) Performance of a distribution intelligent universal transformer under source and load disturbances. Industry Applications Conference. *IEEE*, 719–725.

57. Jin AJ, Li HT, Li SL (2006) A new matrix type three-phase four-wire power electronic transformer. IEEE Annual Power Electronics Specialists Conference (PESC). *IEEE*, 1–6.

58. EPRI (2008) 100-kVA intelligent universal transformer development. *EPRI Report* 1016034.

59. Iman-Eini H, Farhangi S, Schanen JL, et al. (2009) A modular power electronic transformer based on a cascaded H-bridge multilevel converter. *Electr Power Syst Res* 79: 1625–1637.

60. Sabahi M, Hosseini SH, Sharifian MB, et al. (2010) Bi-directional power electronic transformer with maximum power-point tracking capability for induction heating applications. *IET Power Electron* 3: 724–731.

61. Yu Du SB, Wang G, Bhattacharya S (2010) Design considerations of high voltage and high frequency three phase transformer for Solid State Transformer application. Energy Conversion Congress and Exposition (ECCE). *IEEE*, 1551–1558.

62. Wang F, Huang A, Wang G, et al. (2012) Feed-forward control of solid state transformer. 27th Annual IEEE Applied Power Electronics Conference and Exposition (APEC). *IEEE*, 1153–1158.

63. Wang F, Lu X, Wang W, et al. (2012) Development of distributed grid intelligence platform for solid state transformer. 3rd IEEE International Conference on Smart Grid Communications (SmartGridComm). *IEEE*, 481–485.

64. Huang A, She X, Yu X, et al. (2013) Next generation power distribution system architecture: the future renewable electric energy delivery and management (freedm) system. The Third International Conference on Smart Grids, Green Communications and IT Energy-aware Technologies.

65. She X, Huang AQ, Ni X (2013) A cost effective power sharing strategy for a cascaded multilevel converter based Solid state transformer. IEEE Energy Conversion Congress and Exposition (ECCE). *IEEE*, 372–379.

66. Yu X, She X, Huang A (2013) Hierarchical power management for DC microgrid in islanding mode and Solid State transformer enabled mode. 39th Annual Conference of the IEEE Industrial Electronics Society (IECON). *IEEE*, 1656–1661.

67. Yu X, She X, Ni X, et al. (2013) Power management strategy for DC microgrid interfaced to distribution system based on solid state transformer. IEEE Energy Conversion Congress and Exposition (ECCE). *IEEE*, 5131–5136.

68. Zhao T, Wang G, Bhattacharya S, et al. (2013) Voltage and power balance control for a cascaded h-bridge converter-based solid-state Transformer. *IEEE T Power Electr* 28: 1523–1532.

69. She X, Huang AQ, Wang F, et al. (2013) Wind energy system with integrated functions of active power transfer , reactive power compensation , and voltage conversion. *IEEE T Ind Electron* 60: 4512–4524.

70. Yu X, She X, Ni X, et al. (2014) System integration and hierarchical power management strategy for a solid-state transformer interfaced microgrid system. *IEEE T Power Electr* 29: 4414–4425.

71. She X, Yu X, Wang F, et al. (2014) Design and demonstration of a 3.6-kV-120-V/10-kVA solid-state transformer for smart grid application. *IEEE T Power Electr* 29: 3982–3996.

72. Dujic D, Mester A, Chaudhuri T, et al. (2011) Laboratory scale prototype of a power electronic transformer for traction applications. 14th European Conference on Power Electronics and Applications (EPE). *IEEE*, 1–10.

73. Besselmann T, Mester A, Dujic D (2014) Power electronic traction transformer: Efficiency improvements under light-load conditions. *IEEE T Power Electr* 29: 3971–3981.

74. Liu X, Li H, Wang Z (2012) A start-up scheme for a three-stage solid-state transformer with minimized transformer current response. *IEEE T Power Electr* 27: 4832–4836.

75. Li Z, Wang P, Chu Z, et al. (2013) A three-phase 10 kVAC-750 VDC power electronic transformer for smart distribution grid. 15th European Conference on Power Electronics and Applications (EPE). *IEEE*, 1–9.

76. Han BM, Choi NS, Lee JY (2014) New bidirectional intelligent semiconductor transformer for smart grid application. *IEEE T Power Electr* 29: 4058–4066.

77. Lee JY, Yoon YD, Han BM (2014) New intelligent semiconductor transformer with bidirectional power-flow capability. *IEEE T Power Deliver* 29: 299–301.

78. Wang F, Wang G, Huang A, et al. (2014) Design and operation of a 3.6 kV high performance solid state transformer based on 13kV SiC MOSFET and JBS diode. IEEE Energy Conversion Congress and Exposition (ECCE). *IEEE*, 4553–4560.

79. Wang F, Wang G, Huang A, et al. (2014) Standalone operation of a single phase medium voltage solid state transformer in distribution grid. 40th Annual Conference of the IEEE Industrial Electronics Society (IECON). *IEEE*, 5221–5226.

80. Tan K, Yu R, Guo S, et al. (2014) Optimal design methodology of bidirectional LLC resonant DC/DC converter for solid state transformer application. 40th Annual Conference of the IEEE Industrial Electronics Society (IECON). *IEEE*, 1657–1664.

81. Wang F, Wang G, Huang A, et al. (2014) Rectifier stage operation and controller design for a medium voltage solid state transformer with LCL filter. IEEE Energy Conversion Congress and Exposition (ECCE). *IEEE*, 3642–3649.

82. Wang F, Wang G, Huang A, et al. (2014) A 3.6kV high performance solid state transformer based on 13kV SiC MOSFET. 5th IEEE International Symposium on Power Electronics for Distributed Generation Systems (PEDG). *IEEE*, 1–8.

83. Oliveira SVG, Castellani DG, De Novaes Y, et al. (2014) AC-AC modular multilevel converter applied to solid-state transformers. 40th Annual Conference of the IEEE Industrial Electronics Society (IECON). *IEEE*, 1174–1180.

84. Morawiec M, Lewicki A, Krzemiński Z (2015) Power electronic transformer for smart grid application. First Workshop on Smart Grid and Renewable Energy (SGRE). *IEEE*, 1–6.

85. Yun HJ, Kim HS, Ryu MH, et al. (2015) A simple and practical voltage balance method for a solid-state transformer using cascaded H-bridge converters. 9th International Conference on Power Electronics (ICPE). *IEEE*, 2415–2420.

86. Zhao B, Song Q, Liu W (2015) A practical solution of high-frequency-link bidirectional solid-state transformer based on advanced components in hybrid microgrid. *IEEE T Ind Electron* 62: 4587–4597.

87. Madhusoodhanan S, Tripathi A, Patel D, et al. (2015) Solid-state transformer and MV grid Tie applications enabled by 15 kV SiC IGBTs and 10 kV SiC MOSFETs based multilevel converters. *IEEE T Ind Appl* 51: 3343–3360.

88. Ge J, Zhao Z, Yuan L, et al. (2015) Energy feed-forward and direct feed-forward control for solid-state transformer. *IEEE T Power Electr* 30: 4042–4047.

89. Chen H, Prasai A, Moghe R, et al. (2016) A 50-kVA three-phase solid-state transformer based on the minimal topology: Dyna-C. *IEEE T Power Electr* 31: 8126–8137.

90. Chen H, Prasai A, Divan D (2017) Dyna-C: A minimal topology for bidirectional solid-state transformers. *IEEE T Power Electr* 32: 995–1005.

91. Garcia P, Saeed S, Navarro-Rodriguez A, et al. (2016) Switching frequency optimization for a solid state transformer with energy storage capabilities. IEEE Energy Conversion Congress and Exposition, Proceedings (ECCE). *IEEE*, 1–8.

92. Wang Z, Castellazzi A (2016) Impact of SiC technology in a three-port active bridge converter for energy storage integrated solid state transformer applications. 4th IEEE Workshop on Wide Bandgap Power Devices and Applications (WiPDA). *IEEE*, 84–89.

93. Liu Y, Liu Y, Abu-Rub H, et al. (2016) Model predictive control of matrix converter based solid state transformer. IEEE International Conference on Industrial Technology (ICIT). *IEEE*, 1248–1253.

94. Liu Y, Liu Y, Ge B, et al. (2017) Interactive grid interfacing system by matrix-converter based solid state transformer with model predictive control. *IEEE T Ind Informatics*.

95. Nila-Olmedo N, Mendoza-Mondragon F, Espinosa-Calderon A, et al. (2016) ARM + FPGA platform to manage solid-state-smart transformer in smart grid application. International Conference on ReConFigurable Computing and FPGAs. *IEEE*, 1–6.

96. Chen H, Divan D (2016) Soft-switching solid state transformer (S4T). IEEE Energy Conversion Congress and Exposition (ECCE). *IEEE*, 1–10.

97. Zhang J, Wang Z, Shao S (2017) A three-phase modular multilevel DC–DC converter for power electronic transformer applications. *IEEE J Em Sel Top Pn* 5: 140–150.

98. Gao R, She X, Husain I, et al. (2017) Solid-state transformer interfaced permanent magnet wind turbine distributed generation system with power management functions. *IEEE T Ind Appl* 53: 3849–3861.

99. Aggeler D, Biela J, Kolar JW (2008) Solid-state transformer based on SiC JFETs for future energy distribution systems. ETH Power Electronic Systems Laboratory. *IEEE*.

100. Merwe JWD, Mouton HDT (2009) The solid-state transformer concept: A new era in power distribution. IEEE Africon Conference. *IEEE*, 1–6.

101. Merwe WVD, Mouton T (2009) Solid-state transformer topology selection. IEEE International Conference on Industrial Technology. *IEEE*, 1–6.

102. Stefanski K, Qin H, Chowdhury BH, et al. (2010) Identifying techniques, topologies and features for maximizing the efficiency of a distribution grid with solid state power devices. North American Power Symposium (NAPS). *IEEE*, 1–7.

103. Sabahi M, Goharrizi AY, Hosseini SH, et al. (2010) Flexible power electronic transformer. *IEEE Trans Power Electron* 25: 2159–2169.

104. Wang X, Liu J, Xu T, et al. (2012) Comparisons of different three-stage three-phase cascaded modular topologies for power electronic transformer. IEEE Energy Conversion Congress and Exposition (ECCE). *IEEE*, 1420–1425.

105. Cheng L, Xie Y, Lu X, et al. (2012) The topology analysis and compare of high-frequency power electronic transformer. Asia-Pacific Power and Energy Engineering Conference (APPEEC). *IEEE*, 1–6.

106. Shojaei A, Joos G (2013) A topology for three-stage solid state transformer. IEEE Power and Energy Society General Meeting. *IEEE*, 1–5.

107. Shojaei A, Joos G (2013) A modular solid state transformer with a single-phase medium-frequency transformer. IEEE Electrical Power & Energy Conference (EPEC). *IEEE*, 1–5.

108. Qin H, Kimball JW (2013) Solid-state transformer architecture using AC-AC dual-active-bridge converter. *IEEE T Ind Electron* 60: 3720–3730.

109. Shojaei A, Joos G (2013) A modular multilevel converter-based power electronic transformer. IEEE Energy Conversion Congress and Exposition, (ECCE). *IEEE*, 367–371.

110. Viktor B, Indrek R, Tõnu L (2011) Intelligent Transformer: Possibilities and Challenges. *Sci J Riga Tech Univ Power Electr Eng* 29: 95–100.

111. Martinez-Velasco JA, Alepuz S, Gonzalez-Molina F, et al. (2014) Dynamic average modeling of a bidirectional solid state transformer for feasibility studies and real-time implementation. *Electr Power Syst Res* 117: 143–153.

112. Alepuz S, Gonzalez-Molina F, Martin-Arnedo J, et al. (2014) Development and testing of a bidirectional distribution electronic power transformer model. *Electr Power Syst Res* 107: 230–239.

113. Hatua K, Dutta S, Tripathi A, et al. (2011) Transformer less intelligent power substation design with 15kV SiC IGBT for grid interconnection. IEEE Energy Conversion Congress and Exposition (ECCE). *IEEE*, 4225–4232.

114. Madhusoodhanan S, Tripathi A, Patel D, et al. (2014) Solid-state transformer and MV grid Tie applications enabled by 15 kV SiC IGBTs and 10 kV SiC MOSFETs based multilevel converters. International Power Electronics Conference (IPEC). *IEEE*, 1626–1633.

115. González-Molina F, Martin-Arnedo J, Alepuz S, et al. (2015) EMTP model of a bidirectional multilevel solid state transformer for distribution system studies. Power & Energy Society General Meeting. *IEEE*, 1–5.

116. Wang L, Zhang D, Wang Y, et al. (2016) Power and voltage balance control of a novel three-phase solid-state transformer using multilevel cascaded H-bridge inverters for microgrid applications. *IEEE T Power Electr* 31: 3289–3301.

117. Adabi ME, Martinez-Velasco JA, Alepuz S (2017) Modeling and simulation of a MMC-based solid-state transformer. *Electr Eng*, 1–13.

118. Yang L, Zhao T, Wang J, et al. (2007) Design and analysis of a 270 kW five-level DC/DC converter for solid state transformer using 10 kV SiC power devices. IEEE Power Electronics Specialists Conference (PESC). *IEEE*, 245–251.

119. Zhao T, Yang L, Wang J, et al. (2007) 270 kVA solid state transformer based on 10 kV SiC power devices. IEEE Electric Ship Technologies Symposium (ESTS). *IEEE*, 145–149.

120. Abedini A, Lipo T (2010) A novel topology of solid state transformer. 1st Power Electronics and Drive Systems and Technologies Conference (PEDSTC). *IEEE*, 101–105.

121. Aijuan J, Hangtian L, Shaolong L (2005) A three-phase four-wire high-frequency AC link matrix converter for power electronic transformer. 8th International Conference on Electrical Machines and Systems (ICEMS). *IEEE*, 1295–1300.

122. Merwe WVD, Mouton T (2009) Natural balancing of the two-cell back-to-back multilevel converter with specific application to the solid-state transformer concept. 4th IEEE Conference on Industrial Electronics and Applications (ICIEA). *IEEE*, 2955–2960.

123. Gupta RK, Mohapatra KK, Mohan N (2009) A novel three-phase switched multi-winding power electronic transformer. IEEE Energy Conversion Congress and Exposition (ECCE). *IEEE*, 2696–2703.

124. Basu K, Gupta RK, Nath S, et al. (2010) Research in matrix-converter based three-phase power-electronic transformers. International Power Electronics Conference (IPEC). *IEEE*, 2799–2803.

125. Zhu H, Li Y, Wang P, et al. (2012) Design of power electronic transformer based on modular multilevel converter. Asia-Pacific Power and Energy Engineering Conference (APPEEC). *IEEE*, 1–4.

126. Reddy SY, Bharat S (2012) A modular power electronic transformer for medium voltage application. 34th International Conference on Software Engineering (ICSE). *IEEE*, 1–6.

127. Xinyu W, Jinjun L, Taotao X, et al. (2012) Research of three-phase single-stage matrix converter for power electronic transformer. 7th International Power Electronics and Motion Control Conference (IPEMC). *IEEE*, 599–603.

128. Dannier A, Rizzo R (2012) An overview of power electronic transformer: control strategies and topologies. 21st International Symposium on Power Electronics, Electrical Drives, Automation and Motion (SPEEDAM). *IEEE*, 1552–1557.

129. Rajesh C, Kishor M, Chandra NP (2012) reduced switch topology of power electronic transformer. *Int J Eng Res Appl* 2: 2786–2792.

130. Hwang SH, Liu X, Kim JM, et al. (2013) Distributed digital control of modular-based solid-state transformer using DSP+FPGA. *IEEE Trans Ind Electron* 60: 670–680.

131. Kirsten AL, Oliveira THD, Roncalio JGP, et al. (2013) Performance analysis of modular converter for solid state transformers. Brazilian Power Electronics Conference. *IEEE*, 1060–1066.

132. Dutta S, Roy S, Bhattacharya S (2014) A mode switching, multiterminal converter topology with integrated fluctuating renewable energy source without energy storage. 29th Annual IEEE Applied Power Electronics Conference and Exposition (APEC). *IEEE*, 419–426.

133. Chen H, Prasai A, Divan D (2014) Stacked modular isolated dynamic current source converters for medium voltage applications. 29th Annual IEEE Applied Power Electronics Conference and Exposition (APEC). *IEEE*, 2278–2285.

134. Hosseini SH, Sedaghati F, Sabahi M, et al. (2014) A new configuration of modular isolated bidirectional DC-DC converter. 27th Canadian Conference on Electrical and Computer Engineering (CCECE). *IEEE*, 1–6.

135. Sepahvand H, Corzine SMK, Bhattacharya S, et al. (2014) Topology selection for medium-voltage three-phase SiC solid-state transformer. 3rd International Conference on Renewable Energy Research and Applications (ICRERA). *IEEE*, 485–489.

136. Kunov G (2014) Matlab-Simulink model of solid-state transformer realized with matrix converters. 18th International Symposium on Electrical Apparatus and Technologies (SIELA). *IEEE*, 1–4.

137. Yang T, O'Loughlin C, Meere R, et al. (2014) Investigation of modularity in DC-DC converters for solid state transformers. 5th IEEE International Symposium on Power Electronics for Distributed Generation Systems (PEDG). *IEEE*, 1–8.

138. Roy S, De A, Bhattacharya S (2014) Multi-port solid state transformer for inter-grid power flow control. International Power Electronics Conference (IPEC). *IEEE*, 3286–3291.

139. Prasai A, Chen H, Divan D (2014) Dyna-C: A topology for a bi-directional solid-state transformer. 29th Annual IEEE Applied Power Electronics Conference and Exposition (APEC). *IEEE*, 1219–1226.

140. Prasai A, Chen H, Moghe R, et al. (2014) Dyna-C: Experimental results for a 50 kVA 3-phase to 3-phase solid state transformer. 29th Annual IEEE Applied Power Electronics Conference and Exposition (APEC). *IEEE*, 2271–2277.

141. Vasiladiotis M, Member S, Rufer A (2015) A modular multiport power electronic transformer with integrated split battery energy storage for versatile ultrafast EV charging stations. *IEEE T Ind Electron* 62: 3213–3222.

142. Wang Z, Zhang J, Sheng K (2015) Modular multilevel power electronic transformer. 9th International Conference on Power Electronics and ECCE Asia (ICPE-ECCE Asia). *IEEE*, 315–321.

143. Wang Z, Wang T, Zhang J (2015) Three phase modular multilevel DC/DC converter for power electronic transformer application. IEEE Energy Conversion Congress and Exposition (ECCE). *IEEE*, 6277–6284.

144. De A, Bhattacharya S (2015) Design, analysis and implementation of discontinuous mode Dyna-C AC/AC converter for solid state transformer applications. IEEE Energy Conversion Congress and Exposition (ECCE). *IEEE*, 5030–5037.

145. Lopez M, Rodriguez A, Blanco E, et al. (2015) Design and implementation of the control of an MMC-based solid state transformer. 13th IEEE International Conference on Industrial Informatics (INDIN). *IEEE*, 1583–1590.

146. Huber JE, Kolar JW (2015) Analysis and design of fixed voltage transfer ratio DC/DC converter cells for phase-modular solid-state transformers. IEEE Energy Conversion Congress and Exposition (ECCE). *IEEE*, 5021–5029.

147. Mamede HR, Dos Santos WM, Martins DC (2015) Interconnection of DAB converters for application in solid-state transformers with redundancy. 6th IEEE International Symposium on Power Electronics for Distributed Generation Systems (PEDG). *IEEE*, 1–6.

148. Mario L, Briz F, Saeed M, et al. (2016) Comparative analysis of modular multiport power electronic transformer topologies. IEEE Energy Conversion Congress and Exposition (ECCE). *IEEE*, 1–8.

149. Briz F, López M, Rodríguez A, et al. (2016) Modular power electronic transformers:modular multilevel converter versus cascaded H-bridge solutions. *IEEE Ind Electron Mag* 10: 6–19.

150. Lin L, Lin Z, Haijun L (2016) Research on the topology and control strategy of bidirectional DC-DC converter used in the power electronic transformer. 12th IET International Conference on AC and DC Power Transmission (ACDC). *IEEE*, 1–6.

151. Wang X, Zhang G (2016) Research on power electronic transformer with bidirectional power flow. 8th IEEE International Power Electronics and Motion Control Conference (IPEMC-ECCE Asia). *IEEE*, 1212–1216.

152. Liu Y, Liu Y, Abu-Rub H, et al. (2016) Model predictive control of matrix converter based solid state transformer. IEEE International Conference on Industrial Technology (ICIT). *IEEE*, 1248–1253.

153. Qingshan W, Liang D (2016) Research on loss reduction of dual active bridge converter over wide load range for solid state transformer application. 11th International Conference on Ecological Vehicles and Renewable Energies (EVER). *IEEE*, 1–9.

154. Al-Hafri A, Ali H, Ghias A, et al. (2016) Transformer-less based solid state transformer for intelligent power management. 5th International Conference on Electronic Devices, Systems and Applications (ICEDSA). *IEEE*, 1–4.

155. He L, Zhang J, Chen C, et al. (2017) A bidirectional bridge modular switched-capacitor-based Power electronics transformer. *IEEE T Ind Electron* 65: 718–726.

156. Costa LF, Buticchi G, Liserre M (2017) Quad-active-bridge dc-dc converter as cross-link for medium-voltage modular inverters. *IEEE T Ind Appl* 53: 1243–1253.

157. Zhao B, Song Q, Jianguo L, et al. (2017) A modular-multilevel-DC-link front-to-front DC solid state transformer based on high-frequency dual active phase-shift for HVDC grid integration. *IEEE T Ind Electron* 64: 8919–8927.

158. Shri A, Popovic J, Ferreira JA, et al. (2013) Design and control of a three-phase four-leg inverter for solid-state transformer applications. 15th European Conference on Power Electronics and Applications (EPE). *IEEE*, 1–9.

159. Liu H, Mao C, Lu J, et al. (2009) Optimal regulator-based control of electronic power transformer for distribution systems. *Electr Power Syst Res* 79: 863–870.

160. Basu K, Mohan N (2010) A power electronic transformer for PWM AC drive with lossless commutation and common-mode voltage suppression. Joint International Conference on Power Electronics, Drives and Energy Systems (PEDES) & Power India. *IEEE*, 1–6.

161. Basu K, Somani A, Mohapatra KK, et al. (2010) A three-phase AC/AC power electronic transformer-based PWM AC drive with lossless commutation of leakage energy. International Symposium on Power Electronics, Electrical Drives, Automation and Motion (SPEEDAM). *IEEE*, 1693–1699.

162. Nath S, Mohapatra KK, Basu K, et al. (2010) Source based commutation in matrix converter fed power electronic transformer for power systems application. International Symposium on Power Electronics, Electrical Drives, Automation and Motion (SPEEDAM). *IEEE*, 1682–1686.

163. Brando G, Dannier A, Del Pizzo A (2010) A simple predictive control technique of power electronic transformers with high dynamic features. 5th IET International Conference on Power Electronics, Machines and Drives (PEMD).

164. Mao X, Falcones S, Ayyanar R (2010) Energy-based control design for a solid state transformer. IEEE Power and Energy Society General Meeting. *IEEE*, 1–7.

165. Sadeghi M, Gholami M (2011) Advanced control methodology for intelligent universal transformers based on fuzzy logic controllers. 10th WSEAS International Conference on Communications, Electrical & Computer Engineering. *IEEE*, 58–62.

166. Açıkgöz H, Keçecioğlu ÖF, Gani A, et al. (2015) Optimal control and analysis of three phase electronic power transformers. *Procedia Soc Behav Sci* 195: 2412–2420.

167. Liu B, Zha Y, Zhang T, et al. (2016) Fuzzy logic control of dual active bridge in solid state transformer applications. Tsinghua University-IET Electrical Engineering Academic Forum. *IEEE*, 1–4.

168. Wang G, She X, Wang F, et al. (2011) Comparisons of different control strategies for 20kVA solid state transformer. IEEE Energy Conversion Congress and Exposition (ECCE). *IEEE*, 3173–3178.

169. Ortiz G, Bortis D, Kolar JW, et al. (2012) Soft-switching techniques for medium-voltage isolated bidirectional DC/DC converters in solid state transformers. 38th Annual Conference on IEEE Industrial Electronics Society (IECON). *IEEE*, 5233–5240.

170. Moonem MA, Krishnaswami H (2014) Control and configuration of three-level dual-active bridge DC-DC converter as a front-end interface for photovoltaic system. 29th IEEE Applied Power Electronics Conference and Exposition (APEC). *IEEE*, 3017–3020.

171. Wang X, Liu J, Xu T, et al. (2013) Control of a three-stage three-phase cascaded modular power electronic transformer. 28th Annual IEEE Applied Power Electronics Conference and Exposition (APEC). *IEEE*, 1309–1315.

172. Cheng H, Gong Y, Gao Q (2015) The research of coordination control strategy in cascaded multilevel solid state transformer. IEEE International Conference on Mechatronics and Automation (ICMA). *IEEE*, 222–226.

173. Yu X, Ni X, Huang A (2014) Multiple objectives tertiary control strategy for solid state transformer interfaced DC microgrid. IEEE Energy Conversion Congress and Exposition (ECCE). *IEEE*, 4537–4544.

174. Yousefpoor N, Parkhideh B, Azidehak A, et al. (2014) Modular transformer converter-based convertible static transmission controller for transmission grid management. *IEEE T Power Electr* 29: 6293–6306.

175. Parks N, Dutta S, Ramachandram V, et al. (2012) Black start control of a solid state transformer for emergency power restoration. IEEE Energy Conversion Congress and Exposition (ECCE). *IEEE*, 188–195.

176. Dutta S, Ramachandar V, Bhattacharya S (2014) Black start operation for the solid state transformer created micro-grid under islanding with storage. IEEE Energy Conversion Congress and Exposition (ECCE). *IEEE*, 3934–3941.

177. Cho Y, Han Y, Beddingfield RB, et al. (2016) Seamless black start and reconnection of LCL-filtered solid state transformer based on droop control. IEEE Energy Conversion Congress and Exposition (ECCE). *IEEE*, 1–7.

178. She X, Huang AQ, Ni X (2014) Current sensorless power balance strategy for DC/DC converters in a cascaded multilevel converter based solid state transformer. *IEEE T Power Electron* 29: 17–22.

179. Zengin S, Boztepe M (2015) Trapezoid current modulated DCM AC/DC DAB converter for two-stage solid state transformer. 9th International Conference on Electrical and Electronics Engineering (ELECO). *IEEE*, 634–638.

180. Vargas T, Toebe A, Rech C (2015) Double network control architecture for a modular solid state transformer. 13th IEEE Brazilian Power Electronics Conference and 1st Southern Power Electronics Conference (COBEP/SPEC). *IEEE*, 1–6.

181. Liu B, Zha Y, Zhang T, et al. (2016) Sliding mode control for rectifier stage of solid state transformer. 14th International Workshop on Variable Structure Systems (VSS). *IEEE*, 286–289.

182. Liu B, Zha Y, Zhang T, et al. (2016) Predictive direct power control for rectifier stage of solid state transformer. 12th World Congress on Intelligent Control and Automation (WCICA). *IEEE*, 201–204.

183. Liu B, Zha Y, Zhang T (2017) D-Q frame predictive current control methods for inverter stage of solid state transformer. *IET Power Electr* 10: 687–696.

184. Jianfeng Z (2004) Simulation research on instantaneous control-based power electronics transformer. 4th International Power Electronics and Motion Control Conference (IPEMC), 1722–1725.

185. Krishnaswami H, Ramanarayanan V (2005) Control of high-frequency AC link electronic transformer. IEE Proceedings Electric Power Appl. *IEEE*, 509–516.

186. Fan S, Mao C, Chen L (2006) Optimal coordinated PET and generator excitation control for power systems. *Int J Elec Power* 28: 158–165.

187. Liu H, Mao C, Lu J, et al. (2007) Parallel operation of electronic power transformer based on distributed logic control. 42nd International Universities Power Engineering Conference (UPEC). *IEEE*, 93–101.

188. Wang D, Mao C, Lu J (2008) Coordinated control of EPT and generator excitation system for multidouble-circuit transmission-lines system. *IEEE T Power Deliver* 23: 371–379.

189. Wang D, Mao C, Lu J (2009) Operation and control mode of electronic power transformer. IEEE PES/IAS Conference on Sustainable Alternative Energy (SAE). *IEEE*, 1–5.

190. Brando G, Dannier A, Del Pizzo A, et al. (2010) A high performance control technique of power electronic transformers in medium voltage grid-connected PV plants. International Conference on Electrical Machines (ICEM). *IEEE*, 1–6.

191. Shah J, Wollenberg BF, Mohan N (2011) Decentralized power flow control for a smart micro-grid. IEEE Power and Energy Society General Meeting. *IEEE*, 1–6.

192. Schietekat LM (2011) Design and Implementation of the Main Controller of a Solid-State Transformer. Master Thesis, Stellenbosch University.

193. Fan H, Li H (2011) A distributed control of input-series-output-parallel bidirectional DC-DC converter modules applied for 20 kVA solid state transformer. 26th IEEE Applied Power Electronics Conference and Exposition (APEC). *IEEE*, 939–945.

194. Tajfar A, Mazumder SK (2011) A transformer-flux-balance controller for a high-frequency-link inverter with applications for solid-state transformer, renewable/alternative energy sources, energy storage, and electric vehicles. IEEE Electric Ship Technologies Symposium (ESTS). *IEEE*, 121–126.

195. Zhao T, She X, Bhattacharya S, et al. (2011) Power synchronization control for capacitor minimization in solid state transformers (SST). IEEE Energy Conversion Congress and Exposition (ECCE). *IEEE*, 2812–2818.

196. Parks NB (2012) Black start control of a solid state transformer for emergency distribution power restoration. Master Thesis, North Carolina State University.

197. Qin H, Kimball JW (2012) Closed-loop control of DC-DC dual active bridge converters driving single-phase inverters. IEEE Energy Conversion Congress and Exposition (ECCE). *IEEE*, 173–179.

198. Sadeghi M, Gholami M (2012) Neural predictive model control for intelligent universal transformers in advanced distribution automation of tomorrow. 11th international conference on Applications of Electrical and Computer Engineering. *IEEE*, 40–45.

199. Shen Z, Baran ME (2013) Gradient based centralized optimal Volt/Var control strategy for smart distribution system. IEEE PES Innovative Smart Grid Technologies Conference (ISGT). *IEEE*, 1–6.

200. Ardeshir JF, Ajami A, Jalilvand A (2013) Flexible power electronic transformer for power flow control applications. *J Oper Autom Power Eng* 1: 147–155.

201. Kim SG (2014) A comprehensive study of dual active bridge converter and deep belief network controller for bi-directional solid state transformer. Master Thesis, Grad Sch UNIST.

202. Zheng Z, Gao Z, Gu C, et al. (2014) Stability and voltage balance control of a modular converter with multiwinding high-frequency transformer. *IEEE T Power Electr* 29: 4183–4194.

203. Pu Y, Miao H, Zeng CB, et al. (2014) A solid-state transformer with controllable input power factor and output voltage. International Conference on Power System Technology (POWERCON). *IEEE*, 2940–2944.

204. Huang H, Mao C, Lu J, et al. (2014) Electronic power transformer control strategy in wind energy conversion systems for low voltage ride-through capability enhancement of directly driven wind turbines with permanent magnet synchronous generators (D-PMSGs). *Energies* 7: 7330–7347.

205. Shanshan W, Yubin W, Yifei L (2015) Sensorless power balance control for cascaded multilevel converter based solid state transformer. 18th International Conference on Electrical Machines and Systems (ICEMS). *IEEE*, 925–1929.

206. Gu C, Zheng Z, Li Y (2015) Control strategies of a multiport power electronic transformer (PET) for DC distribution applications. IEEE Electric Ship Technologies Symposium (ESTS). *IEEE*, 135–139.

207. Xu K, Fu C, Wang Y, et al. (2015) Voltage and current balance control for the ISOP converter-based power electronic transformer. 18th International Conference onElectrical Machines and Systems (ICEMS). *IEEE*, 378–382.

208. Wu Q, Wang G, Feng J, et al. (2015) A novel comprehensive control scheme of modular multilevel converter-based power electronic transformer. 5th International Conference on Electric Utility Deregulation and Restructuring and Power Technologies (DRPT). *IEEE*, 2253–2258.

209. Ouyang S, Liu J, Song S, et al. (2016) Reactive component injection control of the modular multi-output power electronic transformer. IEEE Annual Southern Power Electronics Conference (SPEC). *IEEE*, 1–7.

210. Wang X, Liu J, Ouyang S, et al. (2016) Control and exprimental of an H-bridge-based three-phase three-stage modular power electronic transformer. *IEEE T Power Electr* 31: 2002–2011.

211. Shah D, Crow ML (2016) Online volt-var control for distribution systems with solid-state transformers. *IEEE T Power Deliver* 31: 343–350.

212. Liu Y, Wang W, Liu Y, et al. (2016) Control of single-stage AC-AC solid state transformer for power exchange between grids. 11th IEEE Conference on Industrial Electronics and Applications (ICIEA). *IEEE*, 892–896.

213. Li R, Xu L, Yao L, et al. (2016) Active control of dc fault currents in dc solid-state transformers during ride-through operation of multi-terminal HVDC systems. *IEEE T Energy Convers* 31: 1336–1346.

214. Ye Q, Li H (2016) Stability analysis and improvement of solid state transformer (SST)-paralleled inverters System using negative impedance feedback control. IEEE Applied Power Electronics Conference and Exposition (APEC). *IEEE*, 2237–2244.

215. Liu J, Yang J, Zhang J, et al. (2017) Voltage balance control based on dual active bridge DC/DC converters in a power electronic traction transformer. *IEEE T Power Electr* 33: 1696–1714.

216. Zhao B, Song Q, Li J, et al. (2017) Full-process operation, control, and experiments of modular high-frequency-link DC transformer based on dual active bridge for flexible MVDC distribution: a practical tutorial. *IEEE T Power Electr* 32: 6751–6766.

217. Zhang C, Zhao Z (2017) Dual-timescale control for power electronic zigzag transformer. *CES T Electr Mach Syst* 1: 315–321.

218. Almaguer J, Cardenas V, Aganza-Torres A, et al. (2017) Power control of a multi-cell solid-state transformer with extended functions. IEEE International Conference on Industrial Technology (ICIT). *IEEE*, 183–188.

219. Kang M, Woo BO, Enjeti PN, et al. (1998) Auto-connected electronic transformer (ACET) based multi-pulse rectifiers for utility interface of power electronic systems. IEEE Industry Applications Conference (IAS). *IEEE*, 1554–1561.

220. Baek S, Cougo B, Bhattacharya S, et al. (2012) Accurate equivalent circuit modeling of a medium-voltage and high-frequency coaxial winding DC-link transformer for solid state transformer applications. IEEE Energy Conversion Congress and Exposition (ECCE). *IEEE*, 1439–1446.

221. Drofenik U (2012) A 150kW medium frequency transformer optimized for maximum power density. 7th International Conference on Integrated Power Electronics Systems (CIPS). *IEEE*, 1–6.

222. Ortiz G, Leibl M, Kolar JW, et al. (2013) Medium frequency transformers for solid-state-transformer applications-Design and experimental verification. IEEE International Conference on Power Electronics and Drive Systems (PEDS). *IEEE*, 1285–1290.

223. Bahmani MA, Thiringer T, Kharezy M (2016) Optimization and experimental validation of medium-frequency high power transformers in solid-state transformer applications. IEEE Applied Power Electronics Conference and Exposition (APEC). *IEEE*, 3043–3050.

224. Alam KS, Tria LAR, Zhang D, et al. (2016) Design of a bi-directional DC-DC converter for solid-state transformer (SST) application by exploiting the shoot through mode. IEEE International Conference on Sustainable Energy Technologies (ICSET). *IEEE*, 276–281.

225. Leibl M, Ortiz G, Kolar JW (2017) Design and experimental analysis of a medium-frequency transformer for solid-state transformer applications. *IEEE J Em Sel Top P* 5: 110–123.

226. Beiranvand H, Rokrok E, Rezaeealam B, et al. (2017) Optimal design of medium-frequency transformers for solid-state transformer applications. 8th Power Electronics, Drive Systems & Technologies Conference (PEDSTC). *IEEE*, 154–159.

227. Duarte J, Hendrix M, Simoes MG (2004) A three-port bi-directional converter for hybrid fuel cell systems. 35th IEEE Annual Power Electronics Specialists Conference (PESC). *IEEE*, 1–22.

228. Baek S (2009) Design consideations of high voltage and high frequency transformer for solid state transformer application. PhD Thesis, North Carolina State University.

229. Bhattacharya S, Zhao T, Wang G, et al. (2010) Design and development of generation-I silicon based solid state transformer. 25th Annual IEEE Applied Power Electronics Conference and Exposition. *IEEE*, 1666–1673.

230. Baek S, Bhattacharya S (2011) Analytical modeling of a medium-voltage and high-frequency resonant coaxial-type power transformer for a solid state transformer application. IEEE Energy Conversion Congress and Exposition (ECCE). *IEEE*, 1873–1880.

231. Baek S, Roy S, Bhattacharya S, et al. (2013) Power flow analysis for 3-port 3-phase dual active bridge dc/dc converter and design validation using high frequency planar transformer. IEEE Energy Conversion Congress and Exposition (ECCE). *IEEE*, 388–395.

232. Rothmund D, Ortiz G, Guillod T, et al. (2015) 10kV SiC-based isolated DC-DC converter for medium voltage-connected solid-state transformers. IEEE Applied Power Electronics Conference and Exposition (APEC). *IEEE*, 1096–1103.

233. Reddy BD, Sahoo SK (2015) Design of solid state transformer. *Int J Adv Res Electr Electron Instrum Eng* 4: 357–364.

234. Alam KS, Tria LAR, Zhang D, et al. (2016) Design and comprehensive modelling of solid-state transformer (SST) based substation. IEEE International Conference on Power System Technology (POWERCON). *IEEE*, 1–6.

235. Huang P, Mao C, Wang D, et al. (2017) Optimal design and implementation of high-voltage high-power silicon steel core medium frequency transformer. *IEEE T Ind Electron* 64: 4391–4401.

236. Ortiz G, Leibl MG, Huber JE, et al. (2017) Design and experimental testing of a resonant DC-DC converter for solid-state transformers. *IEEE T Power Electr* 32: 7534–7542.

237. Huang AQ, Wang L, Tian Q, et al. (2016) Medium voltage solid state transformers based on 15 kV SiC MOSFET and JBS diode. 42nd Annual Conference of the IEEE Industrial Electronics Society (IECON). *IEEE*, 6996–7003.

238. Qin H, Kimball JW (2010) A comparative efficiency study of silicon-based solid state transformers. IEEE Energy Conversion Congress and Exposition (ECCE). *IEEE*, 1458–1463.

239. Pena-Alzola R, Gohil G, Mathe L, et al. (2013) Review of modular power converters solutions for smart transformer in distribution system. IEEE Energy Conversion Congress and Exposition (ECCE). *IEEE*, 380–387.

240. Yang T, Meere R, McKenna K, et al. (2015) The evaluation of a modular solid state transformer and low-frequency distribution transformer under daily loading profile. 17th European Conference on Power Electronics and Applications (EPE-ECCE). *IEEE*, 1–10.

241. Adabi ME, Martinez-Velasco JA (2017) MMC-based solid-state transformer model including semiconductor losses. *Electr Eng*, 1–18.

242. Guerra G, Martinez-Velasco JA (2017) A solid state transformer model for power flow calculations. *Int J Electr Power Energy Syst* 89: 40–51.

243. Heinemann L, Mauthe G (2001) The universal power electronics based distribution transformer, an unified approach. 32nd IEEE Annual Power Electronics Specialists Conference (PESC). *IEEE*, 504–509.

244. Sabahi M, Hosseini SH, Sharifian MB, et al. (2010) Zero-voltage switching bi-directional power electronic transformer. *IET Power Electron* 3: 818–828.

245. Fan H, Li H (2010) High frequency high efficiency bidirectional DC-DC converter module design for 10 kVA solid state transformer. 25th Annual IEEE Applied Power Electronics Conference and Exposition (APEC). *IEEE*, 210–215.

246. Fan H, Li H (2010) A novel phase-shift bidirectional dc-dc converter with an extended high-efficiency range for 20 kVA solid state transformer. IEEE Energy Conversion Congress and Exposition (ECCE). *IEEE*, 3870–3876.

247. Fan H, Li H (2011) High-frequency transformer isolated bidirectional DC-DC converter modules with high efficiency over wide load range for 20 kVA solid-state transformer. *IEEE T Power Electr* 26: 3599–3608.

248. Wang Z, Xu J, Hatua K, et al. (2012) Solid state transformer specification via feeder modeling and simulation. IEEE Power and Energy Society General Meeting. *IEEE*, 1–5.

249. Sastry J, Bala S (2013) Considerations for the design of power electronic modules for hybrid distribution transformers. IEEE Energy Conversion Congress and Exposition (ECCE). *IEEE*, 1422–1428.

250. Beldjajev V, Lehtla T, Zakis J (2013) Impact of component losses on the efficiency of the LC-filter based dual active bridge for the isolation stage of power electronic transformer. 8th International Conference-Workshop Compatibility in Power Electronics (CPE). *IEEE*, 132–137.

251. Ortiz G, Uemura H, Bortis D, et al. (2013) Modeling of soft-switching losses of IGBTs in high-power high-efficiency dual-active-bridge DC/DC converters. *IEEE T Electron Dev* 60: 587–597.

252. Ortiz G, Gammeter C, Kolar JW, et al. (2013) Mixed MOSFET-IGBT bridge for high-efficient medium-frequency dual-active-bridge converter in solid state transformers. 14th IEEE Workshop on Control and Modeling for Power Electronics (COMPEL). *IEEE*, 1–8.

253. Kolar JW, Ortiz G (2014) Solid-state-transformers: key components of future traction and smart grid systems. International Power Electronics Conference (IPEC). *IEEE*, 1–15.

254. Oggier GG, Ordonez M (2014) High efficiency switching sequence and enhanced dynamic regulation for DAB converters in solid-state transformers. 29th Annual IEEE Applied Power Electronics Conference and Exposition (APEC). *IEEE*, 326–333.

255. FREEDM (2014) Solid-state transformer research and development at North Carolina State university's FREEDM systems center. *IEEE Power Electron Mag* 1: 10–11.

256. She X, Yu X, Wang F, et al. (2014) Design and demonstration of a 3.6kV-120V/10KVA solid state transformer for smart grid application. 29th Annual IEEE Applied Power Electronics Conference and Exposition (APEC). *IEEE*, 3429–3436.

257. Ouyang S, Liu J, Song S, et al. (2015) Operation and efficiency analysis of an MAB based three-phase three-stage power electronic transformer. 2nd IEEE International Future Energy Electronics Conference (IFEEC). *IEEE*, 1–6.

258. Montoya RJG, Mallela A, Balda JC (2015) An evaluation of selected solid-state transformer topologies for electric distribution systems. IEEE Applied Power Electronics Conference and Exposition (APEC). *IEEE*, 1022–1029.

259. Choi H, Park H, Jung J (2015) Design methodology of dual active bridge converter for solid state transformer application in smart grid. 9th International Conference on Power Electronics (ICPE). *IEEE*, 196–201.

260. Mirmousa H, Zolghadri MR (2007) A novel circuit topology for three-phase four-wire Distribution Electronic Power Transformer. 7th International Conference on Power Electronics and Drive Systems (PEDS). *IEEE*, 1215–1222.

261. Wang D, Mao C, Lu J (2007) Modelling of electronic power transformer and its application to power system. *IET Gener Transm Dis* 1: 887–895.

262. Iman-Eini H, Schanen JL, Farhangi S, et al. (2008) A power electronic based transformer for feeding sensitive loads. IEEE Power Electronics Specialists Conference (PESC). *IEEE*, 2549–2555.

263. Mohapatra KK, Mohan N (2008) Matrix converter fed open-ended power electronic transformer for power system application. IEEE Power and Energy Society General Meeting. *IEEE*, 1–6.

264. Nath S, Mohapatra KK, Mohan N (2009) Output voltage regulation in matrix converter fed power electronic transformer for power systems application in electric ship. IEEE Electric Ship Technologies Symposium (ESTS). *IEEE*, 203–206.

265. Nath S, Mohan N (2011) A solid state power converter with sinusoidal currents in high frequency transformer for power system applications. IEEE International Conference on Industrial Technology. *IEEE*, 110–114.

266. Shahani A, Basu K, Mohan N (2012) A power electronic transformer based on indirect matrix converter for PWM AC drive with lossless commutation of leakage energy. 6th IET International Conference on Power Electronics, Machines and Drives (PEMD). *IEEE*, 1–6.

267. Basu K, Shahani A, Sahoo AK, et al. (2015) A single-stage solid-state transformer for PWM AC drive with source-based commutation of leakage energy. *IEEE T Power Electr* 30: 1734–1746.

268. Liu H, Mao C, Lu J, et al. (2009) Electronic power transformer with supercapacitors storage energy system. *Electr Power Syst Res* 79: 1200–1208.

269. Zhao T, Zeng J, Bhattacharya S, et al. (2009) An average model of solid state transformer for dynamic system simulation. IEEE Power and Energy Society General Meeting. *IEEE*, 1–8.

270. Falcones S, Mao X, Ayyanar R (2010) Simulation of the FREEDM green hub with solid state transformers and distributed control. Future Renewable Electric Energy Distribution. *IEEE*, 1–5.

271. Wang D, Mao C, Lu J, et al. (2010) Auto-balancing transformer based on power electronics. *Electr Power Syst Res* 80: 28–36.

272. Liu H, Yang J, Mao C, et al. (2011) Nonlinear control of electronic power transformer for distribution system using feedback linearization. IEEE Power Engineering and Automation Conference (PEAM). *IEEE*, 22–26.

273. Banaei MR, Salary E (2011) Power quality improvement based on novel power electronic transformer. 2nd Power Electronics, Drive Systems and Technologies Conference (PEDSTC). *IEEE*, 286–291.

274. Banaei MR, Salary E (2014) Mitigation of voltage sag, swell and power factor correction using solid-state transformer based matrix converter in output stage. *Alexandria Eng J* 53: 563–572.

275. Sadeghi M, Gholami M (2011) A novel distribution automation involving intelligent electronic devices as IUT. *Int J Circuits Syst Signal Process* 5: 443–450.

276. Sadeghi M, Gholami M (2012) Genetic algorithm optimization methodology for PWM Inverters of Intelligent universal transformer for the advanced distribution automation of future. *Indian J Sci Technol* 5: 2035–2040.

277. Sadeghi M (2013) Modern methodology introducing for three layers intelligent universal transformers in advanced distribution automation equipping pi voltage and current source controllers. *Int J Inf Electron Eng* 3: 258–261.

278. Mala RC, Tripathy S, Tadepalli S, et al. (2012) Performance analysis of three phase solid state transformers. International Conference on Devices, Circuits and Systems (ICDCS). *IEEE*, 486–490.

279. Alepuz S, González-Molina F, Martin-Arnedo J, et al. (2013) Time-domain model of a bidirectional distribution electronic power transformer. International Power System Transients Conference.

280. Alepuz S, Gonzalez F, Martin-Arnedo J, et al. (2013) Solid state transformer with low-voltage ride-through and current unbalance management capabilities. 39th Annual Conference of the IEEE Industrial Electronics Society (IECON). *IEEE*, 1278–1283.

281. Shojaei A (2014) Design of modular multilevel converter-based solid state transformers. Master Thesis, McGillUniversity.

282. Yang J, Mao C, Wang D, et al. (2013) Fast and continuous on-load voltage regulator based on electronic power transformer. *IET Electr Power Appl* 7: 499–508.

283. Kalyan BT, Prasad PR (2013) Analysis and design of power electronic transformer based power quality improvement. *IOSR J Electr Electron Eng* 5: 61–69.

284. Roasto I, Romero-Cadaval E, Martins J (2013) Active power electronic transformer based on modular building blocks. 39th Annual Conference of the IEEE Industrial Electronics Society (IECON). *IEEE*, 5922–5927.

285. Hagh MT (2013) Dual output power electronic transformer. 21st Iranian Conference on Electrical Engineering (ICEE).

286. Liu Y, Escobar-Mejia A, Farnell C, et al. (2014) Modular multilevel converter with high-frequency transformers for interfacing hybrid DC and AC microgrid systems. 5th IEEE International Symposium on Power Electronics for Distributed Generation Systems (PEDG). *IEEE*, 1–6.

287. Shanmugam D, Indiradevi K (2014) Implementation of multiport dc-dc converter- based solid state transformer in smart grid system. International Conference on Computer Communication and Informatics (ICCCI). *IEEE*, 1–6.

288. Ponraj T, George A (2014) A solid state transformer integrating distributed generation and storage. *Int J Innov Res Comput Commun Eng* 2: 4029–4035.

289. Sahoo AK, Mohan N (2014) A power electronic transformer with sinusoidal voltages and currents using modular multilevel converter. International Power Electronics Conference (IPEC). *IEEE*, 3750–3757.

290. Ahmed KY, Yahaya NZ, Asirvadam VS (2014) Optimal analysis and design of power electronic distribution transformer. *Res J Appl Sci Eng Technol* 7: 1734–1743.

291. Contreras JP, Ramirez JM (2014) Multi-fed power electronic transformer for use in modern distribution systems. *IEEE T Smart Grid* 5: 1532–1541.

292. Yu X, She X, Zhou X, et al. (2014) Power management for DC microgrid enabled by solid-state transformer. *IEEE T Smart Grid* 5: 954–965.

293. Zhabelova G, Yavarian A, Vyatkin V, et al. (2015) Data center energy efficiency and power quality : an alternative approach with solid state transformer. 41st Annual Conference of the IEEE Industrial Electronics Society (IECON). *IEEE*, 1294–1300.

294. Tarisciotti L, Zanchetta P, Watson A, et al. (2015) Multiobjective modulated model predictive control for a multilevel solid-state transformer. *IEEE T Ind Appl* 51: 4051–4060.

295. Sun Y, Liu J, Li Y, et al. (2016) Research on voltage and switching balance control for cascaded power electronic transformer under hybrid PWM modulation. 8th IEEE International Power Electronics and Motion Control Conference (IPEMC-ECCE Asia). *IEEE*, 2033–2038.

296. Gu C, Zheng Z, Xu L, et al. (2016) Modeling and control of a multiport power electronic transformer (PET) for electric traction applications. *IEEE T Power Electron* 31: 915–927.

297. Liu B, Zha Y, Zhang T, et al. (2016) Solid state transformer application to grid connected photovoltaic inverters. International Conference on Smart Grid and Clean Energy Technologies (ICSGCE). *IEEE*, 248–251.

298. Syed I, Khadkikar V (2017) Replacing the grid interface transformer in wind energy conversion system with solid-state transformer. *IEEE T Power Syst* 32: 2152–2160.

299. Qin H, Kimball JW (2009) AC-AC dual active bridge converter for solid state transformer. In: IEEE Energy Conversion Congress and Exposition (ECCE). *IEEE*, 3039–3044.

300. Shanmugam D, Balakrishnan D, Indiradevi K (2012) Solid state transformer integration in smart grid system. *Int J Sci Technol* 3: 8–14.

301. Shanmugam D, Balakrishnan D, Indiradevi K (2013) A multiport dc-dc converter-based solid state transformer in smart grid system. *Int J Recent Trends Electr Electron Eng* 3: 1–9.

302. Jiang Y, Breazeale L, Ayyanar R, et al. (2012) Simplified solid state transformer modeling for real time digital simulator (RTDS). IEEE Energy Conversion Congress and Exposition (ECCE). *IEEE*, 1447–1452.

303. Ghias AMYM, Ciobotaru M, Agelidis VG, et al. (2012) Solid state transformer based on the flying capacitor multilevel converter for intelligent power management. IEEE Power Engineering Society Conference and Exposition in Africa (PowerAfrica). *IEEE*, 1–7.

304. Malan WL, Vilathgamuwa DM, Walker GR, et al. (2016) A three port resonant solid state transformer with minimized circulating reactive currents in the high frequency link. IEEE Annual Southern Power Electronics Conference (SPEC). *IEEE*, 1–6.

305. Mo R, Mao C, Lu J, et al. (2012) Three-stage solid state transformer modeling through real time digital simulation with controller hardware-in-the-loop. 7th International Power Electronics and Motion Control Conference (IPEMC). *IEEE*, 1116–1119.

306. Castelino G, Mohan N (2013) Modulation and commutation of matrix converter based power electronic transformer using a single FPGA.39th Annual Conference of the IEEE Industrial Electronics Society (IECON). *IEEE*, 4892–4898.

307. Qin H, Kimball JW (2014) Closed-loop control of DC-DC dual-active-bridge converters driving single-phase inverters. *IEEE T Power Electr* 29: 1006–1017.

308. Ouyang S, Liu J, Wang X, et al. (2015) A single phase power electronic transformer considering harmonic compensation in scott traction system. 9th International Conference on Power Electronics and ECCE Asia (ICPE-ECCE Asia). *IEEE*, 2620–2627.

309. Parimi M, Monika M, Rane M, et al. (2016) Dynamic phasor-based small-signal stability analysis and control of solid state transformer. 6th IEEE International Conference on Power Systems (ICPS). *IEEE*, 1–6.

310. Saghaleini M, Farhangi S (2007) Distributed power supply with power factor correction: A solution to feed all modules in power electronic transformers. 7th International Conference on Power Electronics (ICPE). *IEEE*, 1225–1229.

311. Saghaleini M, Hekmati A, Farhangi S (2007) An advanced distributed power supply for power electronic transformers. 33rd Annual Conference of the IEEE Industrial Electronics Society (IECON). *IEEE*, 2038–2043.

312. Xue J (2010) Single-phase vs. three-phase high power high frequency transformers. Master Thesis, Virginia Polytech Inst State University.

313. Hosseini SH, Sharifian MBB, Sabahi M, et al. (2009) A tri-directional power electronic transformer for photo voltaic based distributed generation application. IEEE Power and Energy Society General Meeting. *IEEE*, 1–5.

314. Hosseini SH, Sharifian MBB, Sabahi M, et al. (2009) Bi-directional power electronic transformer based compact dynamic voltage restorer. IEEE Power and Energy Society General Meeting. *IEEE*, 1–5.

315. Hu W, Cheng J, Chen M, et al. (2012) Research on distribution IUT based on three voltage level topology. China International Conference on Electricity Distribution (CICED). *IEEE*, 1–5.

316. Mazgar FN, Hagh MT, Babaei E (2012) Distribution electronic power transformer with reduced number of power switches. 3rd Power Electronics and Drive Systems Technology (PEDSTC). *IEEE*, 324–329.

317. Venkat J, Shukla A, Kulkarni SV (2014) Operation of a three phase solid state transformer under unbalanced load conditions. IEEE International Conference on Power Electronics, Drives and Energy Systems (PEDES). *IEEE*, 1–6.

318. Khazraei M, Prabhala VAK, Ahmadi R, et al. (2014) Solid state transformer stability and control considerations. 29th Annual IEEE Applied Power Electronics Conference and Exposition (APEC). *IEEE*, 2237–2244.

319. Shah DG, Crow ML (2014) Stability design criteria for distribution systems with solid-state transformers. *IEEE T Power Deliver* 29: 2588–2595.

320. Zhou K, Jin Q, Lan Z, et al. (2015) The study of power electronic transformer on power flow control and voltage regulation in DC micro-grid. 5th International Conference on Electric Utility Deregulation and Restructuring and Power Technologies (DRPT). *IEEE*, 2166–2172.

321. Zhou K, Chu H (2015) Study on the optimal DC voltage control for power electronic transformer. 5th International Conference on Electric Utility Deregulation and Restructuring and Power Technologies (DRPT). *IEEE*, 2270–2275.

322. Paladhi S, Ashok S (2015) Solid state transformer application in wind based DG system. IEEE International Conference on Signal Processing, Informatics, Communication and Energy Systems (SPICES). *IEEE*, 1–5.

323. Jakka VNSR, Shukla A (2016) A triple port active bridge converter based multi- fed power electronic transformer. IEEE Energy Conversion Congress and Exposition (ECCE). *IEEE*, 1–8.

324. Jakka VNSR, Shukla A (2016) Integration of AC and DC sources using multi-source fed power electronic transformer (MSF-PET) for modern power distribution system applications. 18th European Conference on Power Electronics and Applications. *IEEE*, 1–9.

325. Shah D, Crow M (2016) Online volt-var control for distribution systems with solid state transformers. *IEEE T Power Deliver* 31: 343–350.

326. Wen H, Yang R (2016) Power management of solid state transformer in microgrids. IEEE PES Asia-Pacific Power and Energy Engineering Conference (APPEEC). *IEEE*, 1399–1404.

327. Liu C, Zhi Y, Zhang Y, et al. (2016) New breed of solid-state transformer distribution tail system for flexible power conversion between medium-voltage distribution and low-voltage customer side. 1EEE PES Asia-Pacific Power and Energy Conference (APPEEC). *IEEE*, 403–407.

328. Zhou T, Xu Y (2017) Fault characteristic analysis and simulation of power electronic transformer based on MMC in distribution network. First IEEE International Conference on Energy Internet. *IEEE*, 332–337.

329. Fan B, Li Y, Wang K, et al. (2017) Hierarchical system design and control of an MMC-based power-electronic transformer. *IEEE T Ind Inform* 13: 238–247.

330. Martin-Arnedo J, Gonzalez-Molina F, Martinez-Velasco JA, et al. (2012) Development and testing of a distribution electronic power transformer model. IEEE Power and Energy Society General Meeting. *IEEE*, 1–6.

331. Martin-Arnedo J, González-Molina F, Martinez-Velasco JA, et al. (2017) EMTP model of a bidirectional cascaded multilevel solid state transformer for distribution system studies. *Energies* 10: 521–539.

332. Moonem MA, Krishnaswami H (2012) Analysis of dual active bridge based power electronic transformer as a three-phase inverter. 38th Annual Conference on IEEE Industrial Electronics Society (IECON). *IEEE*, 238–243.

333. Moonem MA, Krishnaswami H (2012) Analysis and control of multi-level dual active bridge DC-DC converter. IEEE Energy Conversion Congress and Exposition (ECCE). *IEEE*, 1556–1561.

334. Li H, Wang Y, Yu C (2016) Control of three-phase cascaded multilevel converter based power electronic transformer under unbalanced input voltages. 42nd Annual Conference of the IEEE Industrial Electronics Society (IECON). *IEEE*, 3299–3304.

335. Li H, Wang Y, Yu C (2016) Research on voltage balance and power balance control for three-phase cascaded multilevel converter based power electronic transformer. 42nd Annual Conference of the IEEE Industrial Electronics Society (IECON). *IEEE*, 3588–3593.

336. Facchinello GG, Mamede H, Brighenti LL, et al. (2016) AC-AC hybrid dual active bridge converter for solid state transformer. 7th IEEE International Symposium on Power Electronics for Distributed Generation Systems (PEDG). *IEEE*, 1–8.

337. Hunziker NS (2016) Solid-state transformer modeling for analyzing its application in distribution grids. PCIM Europe; International Exhibition and Conference for Power Electronics, Intelligent Motion, Renewable Energy and Energy Management. *IEEE*, 2167–2174.

338. Lai JS, Maitra A, Mansoor A, et al. (2005) Multilevel intelligent universal transformer for medium voltage applications. Industry Applications Conference. *IEEE*, 1893–1899.

339. Tatcho P, Jiang Y, Li H (2011) A novel line section protection for the FREEDM system based on the solid state transformer. IEEE Power and Energy Society General Meeting. *IEEE*, 1–8.

340. Tatcho P, Li H, Jiang Y, et al. (2013) A novel hierarchical section protection based on the solid state transformer for the future renewable electric energy delivery and management (FREEDM) system. *IEEE T Smart Grid* 4: 1096–1104.

341. Ye Q, Mo R, Li H (2017) Multiple resonances mitigation of paralleled inverters in a solid-state transformer (SST) enabled ac microgrid. *IEEE T Smart Grid*.

342. Maheswari M, Kumar NS (2015) Design and control of power electronic transformer with power factor correction. International Conference on Circuit, Power and Computing Technologies (ICCPCT). *IEEE*, 1–6.

343. Devi SV, Kumar NS (2016) Design of power electronic transformer based variable speed wind energy conversion system. International Conference on Circuit, Power and Computing Technologies (ICCPCT). *IEEE*, 1–7.

344. Falcones S, Mao X, Ayyanar R (2011) Simulink block-set for modeling distribution systems with solid state transformer. *IEEE*, 1–4.

345. Gonzalez-Agudelo D, Escobar-Mejia A, Ramirez-Murrillo H (2016) Dynamic model of a dual active bridge suitable for solid state transformers. 13th International Conference on Power Electronics (CIEP). *IEEE*, 350–355.

346. Busada C, Chiacchiarini H, Jorge SG, et al. (2017) Modeling and control of a medium voltage three-phase solid-state transformer. 11th IEEE International Conference on Compatibility, Power Electronics and Power Engineering (CPE-POWERENG). *IEEE*, 556–561.

347. Posada CJ, Ramirez JM, Correa RE (2012) Modeling and simulation of a solid state transformer for distribution systems. IEEE Power and Energy Society General Meeting. *IEEE*, 1–6.

348. He L, Zhang J, Chen C, et al. (2018) Bidirectional bridge modular switched-capacitor-based power electronics transformer. *IEEE T on Ind Electron* 65: 718–726.

349. Ramachandran V, Kuvar A, Singh U, et al. (2014) A system level study employing improved solid state transformer average models with renewable energy integration. IEEE Power and Energy Society General Meeting. *IEEE*, 1–5.

350. Yu Z, Ayyanar R, Husain I (2015) A detailed analytical model of a solid state transformer. IEEE Energy Conversion Congress and Exposition (ECCE). *IEEE*, 723–729.

351. Arafat-Khan MT, Milani AA, Chakrabortty A, et al. (2016) Comprehensive dynamic modeling of a solid-state transformer based power distribution system. IEEE Energy Conversion Congress and Exposition (ECCE). *IEEE*, 1–8.

352. Martin-Arnedo J, González-Molina F, Martínez-Velasco JA, et al. (2016) Implementation of a custom-made model for a multilevel solid state transformer. *EEUG*.

353. Huasheng M, Bo Z, Jianchao Z, et al. (2005) Dynamic characteristics analysis and instantaneous value control design for buck-type power electronic transformer (PET). 31st Annual Conference of IEEE Industrial Electronics Society (IECON). *IEEE*, 1043–1047.

354. Bahadornejad M, Nair NK, Zhabelova G, et al. (2011) Modeling solid state transformer in IEC 61850. 37th Annual Conference on IEEE Industrial Electronics Society (IECON). *IEEE*, 2706–2710.

355. Kieferndorf R, Venkataramanan G, Manjrekar MD (2000) A power electronic transformer (PET) fed nine-level H-bridge inverter for large induction motor drives. IEEE Industry Applications Conference. *IEEE*, 2489–2495.

356. Wrede H, Staudt V, Steimel A (2002) Design of an electronic power transformer. 28th IEEE Annual Conference of the Industrial Electronics Society (IECON). *IEEE*, 1380–1385.

357. Wang D, Mao C, Lu J, et al. (2005) The research on characteristics of electronic power transformer for distribution system. IEEE/PES Transmission and Distribution Conference and Exhibition: Asia and Pacific. *IEEE*, 1–5.

358. Iman-Eini H, Farhangi S (2006) Analysis and design of power electronic transformer for medium voltage levels. 37th IEEE Power Electronics Specialists Conference (PESC). *IEEE*, 1–5.

359. Iman-Eini H, Farhangi S, Schanen JL, et al. (2007) Design of power electronic transformer based on cascaded H-bridge multilevel converter. IEEE International Symposium on Industrial Electronics (ISIE). *IEEE*, 877–882.

360. Jovcic D (2009) Bidirectional, high-power DC transformer. *IEEE T Power Deliver* 24: 2276–2283.

361. Castelino G, Basu K, Mohan N (2010) Power electronic transformer with reduced number of switches: Analysis of clamp circuit for leakage energy commutation. In: Joint International Conference on Power Electronics, Drives and Energy Systems (PEDES) & Power India. *IEEE*, 656–661.

362. Wang J, Wang G, Bhattacharya S, et al. (2010) Comparison of 10-kV SiC power devices in solid-state transformer. IEEE Energy Conversion Congress and Exposition (ECCE). *IEEE*, 3284–3289.

363. Basu K, Mohan N (2010) A power electronic transformer for three phase PWM AC/AC drive with loss less commutation and common-mode voltage suppression. 36th Annual Conference on IEEE Industrial Electronics Society (IECON). *IEEE*, 315–320.

364. Castelino G, Basu K, Mohan N (2010) Power electronic transformer with reduced number of switches: Analysis of clamp circuit for leakage energy commutation. Joint International Conference on Power Electronics, Drives and Energy Systems (PEDES) &Power India. *IEEE*, 1–8.

365. Basu K, Somani A, Mohapatra KK, et al. (2010) A power electronic transformer-based three-phase PWM AC drive with lossless commutation of leakage energy. International Symposium on Power Electronics, Electrical Drives, Automation and Motion (SPEEDAM). *IEEE*, 1194–1200.

366. Wang Z, Yu K (2010) The research of power electronic transformer (PET) in smart distribution network. International Conference on Power System Technology (POWERCON). *IEEE*, 1–7.

367. Banaei MR, Salary E (2010) Mitigation of current harmonics and unbalances using power electronic transformer. 25th International Power System Conference (PSC). *IEEE*, 1–9.

368. Wang H, Guo TX, Li QM, et al. (2011) Development and applicability analysis of intelligent solid state transformer. 4th International Conference on Electric Utility Deregulation and Restructuring and Power Technologies (DRPT). *IEEE*, 1150–1154.

369. Beldjajev V, Roasto I (2011) Analysis of new bidirectional DC-DC converter based on current doubler rectifier. 10th International Symposium onTopical Problems in the Field of Electrical and Power Engineering. *IEEE*, 234–237.

370. Zhang J, Wang W, Bhattacharya S (2012) Architecture of solid state transformer-based energy router and models of energy traffic. IEEE PES Innovative Smart Grid Technologies (ISGT). *IEEE*, 1–8.

371. Banaei MR, Salary E (2012) Four-wire solid state transformer to improve current quality. *Gazi Univ J Sci* 25: 887–899.

372. Rodríguez JR, Moreno-Goytia EL, Venegas V (2012) State of the art, modeling and simulation of an advanced power electronics transformer. North American Power Symposium (NAPS). *IEEE*, 1–6.

373. Wang F, She X, Wang G, et al. (2012) Parallel operation of solid state transformer. IEEE Energy Conversion Congress and Exposition (ECCE). *IEEE*, 1433–1438.

374. Qin H (2012) Dual active bridge converters in solid state transformers. PhD Thesis, Missouri University Sci Technol.

375. Beldjajev V, Roasto I, Zakis J (2012) Isolation stage for power electronic transformer: dual active bridge based isolation stage for power electronic transformer. International Symposium on Power Electronics, Electrical Drives, Automation and Motion (SPEEDAM). *IEEE*, 849–854.

376. Zhu M, Zhao R, Zhang H, et al. (2012) A novel solution using two-port network models for transient analysis of full-bridge DC-DC converter in solid state transformer. 15th IEEE International Conference on Electrical Machines and Systems (ICEMS). *IEEE*, 1–4.

377. Wani M, Kurundkar K, Bhawalkar MP (2012) Use of power electronic converters to suppress transformer inrush current. IEEE International Conference on Power Electronics, Drives and Energy Systems (PEDES). *IEEE*.

378. Ji Z, Sun Y, Wang S, et al. (2012) Design of a three-phase cascaded power electronic transformer based on energy internet. International Conference on Sustainable Power Generation and Supply (SUPERGEN). *IEEE*, 1–6.

379. Beldjajev V, Roasto I, Zakis J (2012) Isolation stage for power electronic transformer: Dual active bridge vs bi-directional current doubler rectifier. 21st International Symposium on Power Electronics, Electrical Drives, Automation and Motion (SPEEDAM). *IEEE*, 849–854.

380. Pavlović Z (2013) Multiple input-output bidirectional solid state transformer based on a series resonant converter. PhD Thesis, University Politécnica Madrid.

381. Madhusoodhanan S, Cho Y, Kadavelugu A, et al. (2013) Comparative evaluation of SiC devices for PWM buck rectifier based active front end converter for MV grid interface. IEEE Energy Conversion Congress and Exposition (ECCE). *IEEE*, 3034–3041.

382. Carr J, Wang Z, Bhattacharya S, et al. (2013) Transient overvoltage rating and BIL of the transformerless intelligent power substation. IEEE Power and Energy Society General Meeting. *IEEE*, 1–5.

383. Carr J, Wang Z, Bhattacharya S, et al. (2013) Overloading and overvoltage evaluation of a Transformerless Intelligent Power Substation. IEEE Power and Energy Society General Meeting. *IEEE*, 1–5.

384. Wang X, Ouyang S, Liu J, et al. (2013) Comparison on unbalanced-load handling capability of two power electronic transformer topologies. IEEE Energy Conversion Congress and Exposition (ECCE). *IEEE*, 5266–5272.

385. Sang Z, Mao C, Lu J, et al. (2013) Analysis and simulation of fault characteristics of power switch failures in distribution electronic power transformers. *Energies* 6: 4246–4268.

386. Pena-Alzola R, Mathe L, Liserre M, et al. (2013) DC-bias cancellation for phase shift controlled dual active bridge. 39th Annual Conference of the IEEE Industrial Electronics Society (IECON). *IEEE*, 596–600.

387. Roasto I, Romero-Cadaval E, Martins J, et al. (2013) Active power electronic transformer as a power conditioner for nonlinear loads. 8th International Conference-Workshop Compatibility in Power Electronics (CPE). *IEEE*, 63–68.

388. Beldjajev V, Rang T, Zakis J (2013) Steady state analysis of the commutating LC filter based dual active bridge for the isolation stage of power electronic transformer. 8th International Conference-Workshop Compatibility in Power Electronics (CPE). *IEEE*, 138–143.

389. Reddy SM, Khandrika B (2013) A comparison study of solid state transformers using different switching techniques. *Int J Sci Res* 2: 314–318.

390. Huber J, Ortiz G, Krismer F, et al. (2013) η-ρ Pareto optimization of bidirectional half-cycle discontinuous-conduction-mode series-resonant DC/DC converter with fixed voltage transfer ratio. 28th Annual IEEE Applied Power Electronics Conference and Exposition (APEC). *IEEE*, 1413–1420.

391. Huber JE, Kolar JW (2013) Optimum number of cascaded cells for high-power medium-voltage AC–DC converters. IEEE Energy Conversion Congress and Exposition (ECCE). *IEEE*, 359–366.

392. Wang X, Ouyang S, Liu J, et al. (2014) Research of the voltage and current sharing issue of an H-bridge based power electronic transformer. 29th Annual IEEE Applied Power Electronics Conference and Exposition (APEC). *IEEE*, 2216–2222.

393. Guillod T, Huber JE, Ortiz G, et al. (2014) Characterization of the voltage and electric field stresses in multi-cell solid-state transformers. IEEE Energy Conversion Congress and Exposition (ECCE). *IEEE*, 4726–4734.

394. Ji IH, Wang S, Lee B, et al. (2014) Design and fabrication of high current AlGaN/GaN HFET for Gen III solid state transformer. 2nd IEEE Workshop on Wide Bandgap Power Devices and Applications (WiPDA). *IEEE*, 63–65.

395. Wang Q, Liang D, Du J (2014) Performance study of solid state transformer applying BP artificial neural network PID regulator. 17th International Conference on Electrical Machines and Systems (ICEMS). *IEEE*, 2440–2444.

396. Raju R (2014) Silicon carbide-high voltage , high frequency conversion. NIST High Megawatt Variable Speed Drive Technology Workshop.

397. Huber JE, Kolar JW (2014) Common-mode currents in multi-cell solid-state transformers. International Power Electronics Conference (IPEC). *IEEE*, 766–773.

398. Hooshyar H, Baran ME (2014) Fault analysis on distribution feeders employing solid state transformers. IEEE Power and Energy Society General Meeting. *IEEE*, 1–5.

399. Sailaja M, Wahab S, Reddy ML (2014) A cascaded H-bridge converter instigated for a solid-state transformer to restrain voltage and power. *Int J Ind Electron Electr Eng* 2: 25–29.

400. Tashackori A, Hosseini SH, Sabahi M (2015) Power quality improvement using a power electronic transformer basde DVR. 23rd Iranian Conference on Electrical Engineering (ICEE). *IEEE*, 1597–1601.

401. Zhou H, Zhang W, Wu X (2015) Optimization of reactive power for active distribution network with power electronic transformer. 12th International Conference on the European Energy Market (EEM). *IEEE*, 1–5.

402. Roy RB, Rokonuzzaman M, Hossam-E-Haider M (2015) Design and analysis of the power electronic transformer for power quality improvement. International Conference on Electrical Engineering and Information Communication Technology (ICEEICT). *IEEE*, 1–5.

403. Miñambres-Marcos V, Roasto I, Romero-Cadaval E, et al. (2015) Single-phase power electronics transformer with active functions for smart grid. 9th International Conference on Compatibility and Power Electronics (CPE). *IEEE*, 528–533.

404. Guillod T, Krismer F, Färber R, et al. (2015) Protection of MV/LV solid-state transformers in the distribution grid. 41th Annual Conference of the IEEE Industrial Electronics Society (IECON). *IEEE*, 3531–3538.

405. Shah DG, Crow ML (2015) Stability assessment extensions for single-phase distribution solid-state transformers. *IEEE T Power Deliver* 30: 1636–1638.

406. Wang D, Tian J, Mao C, et al. (2016) A 10-kV/400-V 500-kVA electronic power transformer. *IEEE T Ind Electron* 63: 6653–6663.

407. Mamede HR, Dos Santos WM, Martins DC (2016) A DAB-based solid-state transformer with high reliability as to the power supply. 13th IEEE Brazilian Power Electronics Conference and 1st Southern Power Electronics Conference (COBEP/SPEC). *IEEE*, 1–6.

408. Sundaramoorthy RS, Udhayakumar G (2016) Power management using modified solid state transformer for AC distribution system. International Conference on Emerging Trends in Engineering, Technology and Science (ICETETS). *IEEE*, 1–6.

409. Yang Z, Xiaopin Y (2016) Fast equivalent modeling of input series output parallel HVDC solid state transformer with LLC structure. IEEE International Conference on Power and Renewable Energy. *IEEE*, 160–164.

410. Huber JE, Rothmund D, Wang L, et al. (2016) Full-ZVS modulation for all-SiC ISOP-type isolated front end (IFE) solid-state transformer. IEEE Energy Conversion Congress and Exposition (ECCE). *IEEE*, 1–8.

411. Rana MM, Rahman R, Rahman M (2016) Solid state transformer based on cascade multilevel converter for distribution network. 9th International Conference on Electrical and Computer Engineering (ICECE). *IEEE*, 527–529.

412. Huang AQ, Zhu Q, Wang L, et al. (2017) 15 kV SiC MOSFET: an enabling technology for medium voltage solid state transformers. *CPSS T Power Electron Appl* 2: 118–130.

413. Costa LF, Carne GD, Buticchi G, et al. (2017) The smart transformer: a solid-state transformer tailored to provide ancillary services to the distribution grid. *IEEE Ind Electron Mag* 4: 56–67.

414. Huber JE, Böhler J, Rothmund D, et al. (2017) Analysis and cell-level experimental verification of a 25 kW all-SiC isolated front end 6.6 kV/400 V AC-DC solid-state transformer. *CPSS T Power Electron Appl* 2: 140–148.

415. Baranwal R, Castelino GF, Iyer K, et al. (2018) A dual active bridge based single phase AC to DC power electronic transformer with advanced features. *IEEE T Power Electr* 33: 313–331.

416. She X, Burgos R, Wang G, et al. (2012) Review of solid state transformer in the distribution system: From components to field application. IEEE Energy Conversion Congress and Exposition (ECCE). *IEEE*, 4077–4084.

417. She X, Huang A (2013) Solid state transformer in the future smart electrical system. IEEE Power and Energy Society General Meeting. *IEEE*, 1–5.

418. Contreras JP, Ramirez JM, Marin JV, et al. (2013) Distribution systems equipped with power electronic transformers. IEEE Grenoble PowerTech (POWERTECH). *IEEE*, 1–6.

419. Rothmund D, Ortiz G, Kolar JW (2014) SiC-based unidirectional solid-state transformer concepts for directly interfacing 400V DC to medium-voltage AC distribution systems. 36th IEEE International Telecommunications Energy Conference (INTELEC). *IEEE*, 1–9.

420. Roasto I, Romero-Cadaval E, Martins J, et al. (2012) State of the art of active power electronic transformers for smart grids. 38th Annual Conference on IEEE Industrial Electronics Society (IECON). *IEEE*, 5241–5246.

421. Dutta S, Bhattacharya SRS (2013) Integration of multi-terminal DC to DC hub architecture with solid state transformer for renewable energy integration. IEEE Energy Conversion Congress and Exposition (ECCE). *IEEE*, 4793–4800.

422. Doncker RWD (2014) Power electronic technologies for flexible DC distribution grids. International Power Electronics Conference Power (IPEC). *IEEE*, 736–743.

423. Huang A, Cheng L, Palmour JW, et al. (2014) Ultra high voltage SiC power devices and its impact on future power delivery system. International Exhibition and Conference for Power Electronics, Intelligent Motion, Renewable Energy and Energy Management. *IEEE*, 19–26.

424. Rodrigues WA, Santana RAS, Cota APL, et al. (2016) Integration of solid state transformer with DC microgrid system. IEEE Annual Southern Power Electronics Conference (SPEC). *IEEE*, 1–6.

425. Rashidi M, Nasiri A, Cuzner R (2016) Application of multi-port solid state transformers for microgrid-based distribution systems. IEEE International Conference on Renewable Energy Research and Applications (ICRERA). *IEEE*, 605–610.

426. Wang S, Zheng Z, Li Y, et al. (2016) A modular DC solid state transformer for future onboard DC grid. In: International Conference on Electrical Systems for Aircraft, Railway, Ship Propulsion and Road Vehicles and International Transportation Electrification Conference (ESARS-ITEC). *IEEE*, 1–6.

427. Biswas S, Dhople S, Mohan N (2013) A three-port bidirectional DC-DC converter with zero-ripple terminal currents for PV/microgrid applications. 39th Annual Conference of the IEEE Industrial Electronics Society (IECON). *IEEE*, 340–345.

428. Falcones S, Ayyanar R, Mao X (2013) A DC-DC Multiport-converter-based solid-state transformer integrating distributed generation and storage. *IEEE T Power Electr* 28: 2192–2203.

429. Brando G, Dannier A, Rizzo R (2009) Power electronic transformer application to grid connected photovoltaic systems. International Conference on Clean Electrical Power. *IEEE*, 685–690.

430. Zengin S, Boztepe M (2014) Modified dual active bridge photovoltaic inverter for solid state transformer applications. International Symposium on Fundamentals of Electrical Engineering (ISFEE). *IEEE*, 1–4.

431. Chen Q, Liu N, Hu C, et al. (2017) Autonomous energy management strategy for solid-state transformer to integrate PV-assisted EV charging station participating in ancillary service. *IEEE T Ind Inform* 13: 258–269.

432. Taghizadeh M, Sadeh J, Kamyab E (2011) Protection of grid connected photovoltaic system during voltage sag. International Conference on Advanced Power System Automation and Protection (APAP). *IEEE*, 2030–2035.

433. Xin H, Buhan Z (2011) The applications of the electronic power transformer in photovoltaic systems. International Conference on Electrical and Control Engineering (ICECE). *IEEE*, 3691–3694.

434. Foureaux NC, Adolpho L, Silva SM, et al. (2014) Application of solid state transformers in utility scale solar power plants. 40th IEEE Photovoltaic Specialist Conference (PVSC). *IEEE*, 3695–3700.

435. Mazza LC, Oliveira D, Antunes F, et al. (2016) Bidirectional converter with high frequency isolation feasible to solid state transformer applications. 18th European Conference on Power Electronics and Applications (EPE'16 ECCE Europe). *IEEE*, 1–9.

436. Ankita V, Vijayakumari A (2016) A reduced converter count solid state transformer for grid connected photovoltaic applications. International Conference on Emerging Technological Trends (ICETT). *IEEE*, 1–7.

437. Razmkhah M, Azizian MR, Madadi KH (2017) Photovoltaic systems based on power electronic transformer with maximum power tracking capability. 22nd Electrical Power Distribution Conference. *IEEE*, 74–79.

438. Zhu R, Carne GD, Deng F, et al. (2017) Integration of large photovoltaic and wind system by means of smart transformer. *IEEE T Ind Electron* 64: 8928–8938.

439. She X, Wang F, Burgos R, et al. (2012) Solid state transformer interfaced wind energy system with integrated active power transfer, reactive power compensation and voltage conversion functions. IEEE Energy Conversion Congress and Exposition (ECCE). *IEEE*, 3140–3147.

440. Prakash TRD, Kumar RS (2013) Design of single phase power electronic transformer for low voltage miniature synchronous wind electric generator. *Int J Adv Inf Sci Technol* 19: 1–8.

441. Gao R, Husain I, Wang F, et al. (2015) Solid-state transformer interfaced PMSG wind energy conversion system. IEEE Applied Power Electronics Conference and Exposition (APEC). *IEEE*, 1310–1317.

442. Tan NML, Abe T, Akagi H (2012) Design and performance of a bidirectional isolated DC-DC converter for a battery energy storage system. *IEEE T Power Electr* 27: 1237–1248.

443. Said RG, Abdel-Khalik AS, El Zawawi A, et al. (2014) Integrating flywheel energy storage system to wind farms-fed HVDC system via a solid state transformer. 3rd International Conference on Renewable Energy Research and Applications (ICRERA). *IEEE*, 375–380.

444. Gao R, Husain I, Huang AQ (2016) An autonomous power management strategy based on DC bus signaling for solid-state transformer interfaced PMSG wind energy conversion system. IEEE Applied Power Electronics Conference and Exposition (APEC). *IEEE*, 3383–3388.

445. Sandeep PV, Vidyapeetham A (2016) Grid connected wind driven permanent magnet synchronous generator with high frequency solid state transformer. International Conference on Emerging Technological Trends (ICETT). *IEEE*, 1–6.

446. She X, Huang AQ, Lukic S, et al. (2012) On integration of solid-state transformer with zonal DC microgrid. *IEEE T Smart Grid* 3: 975–985.

447. Yu J, Wu Z, Bhattacharya S (2013) Power dispatch strategy in microgrid integrated with solid state transformer. IEEE Power and Energy Society General Meeting. *IEEE*, 1–5.

448. Rodrigues WA, Morais LMF, Oliveira TR, et al. (2016) Analysis of solid state transformer based microgrid system. 12th IEEE International Conference on Industry Applications (INDUSCON). *IEEE*, 1–6.

449. Wei L, Hong F, Andong W, et al. (2017) Research of microgrid connecting interface based on multi-port power electronic transformer. 20th International Conference on Electrical Machines and Systems (ICEMS). *IEEE*, 1–6.

450. Kang M, Enjeti PN, Pitel IJ (1997) Analysis and design of electronic transformers for electric power distribution system. IEEE Industry Applications Conference (IAS). *IEEE*, 1689–1694.

451. Manjrekar MD, Kieferndorf R, Venkataramanan G (2000) Power electronic transformers for utility applications. IEEE Industry Applications Conference. *IEEE*, 2496–2502.

452. Chinthavali MS (2003) Silicon carbide GTO thyristor loss model for HVDC application. *Master Thesis*, University Tennessee.

453. Ratanapanachote S (2004) Applications of an electronic transformer in a power distribution system. *PhD Thesis*, Texas A&M University.

454. Dieckerhoff S, Bernet S, Krug D (2005) Power loss-oriented evaluation of high voltage IGBTs and multilevel converters in transformerless traction applications. *IEEE T Power Electr* 20: 1328–1336.

455. Wang D, Mao C, Lu J, et al. (2007) Theory and application of distribution electronic power transformer. *Electr Power Syst Res* 77: 219–226.

456. Krishnamurthy H, Ayyanar R (2008) Stability analysis of cascaded converters for bidirectional power flow applications. 30th IEEE InternationalTelecommunications Energy Conference (INTELEC). *IEEE*, 1–8.

457. Miller LE, Schoene J, Kunte R, et al. (2013) Smart grid opportunities in islanding detection. IEEE Power and Energy Society General Meeting. *IEEE*, 1–4.

458. Thamaraiselvi R, Ramesh P, Baskaran J, et al. (2013) A survey of PV based solid state transformer for storage and distribution applications. *Int J Sci Eng Res* 4: 137–141.

459. Carne GD, Liserre M, Christakou K, et al. (2014) Integrated voltage control and line congestion management in active distribution networks by means of smart transformers. 23rd IEEE International Symposium on Industrial Electronics (ISIE). *IEEE*, 2613–2619.

460. Bansode SG, Joshi PM (2014) Solid state transformers : new approach and new opportunity. 11th IRF International Conference.

461. Gu C, Zheng Z, Li YD (2015) A power electronic transformer (PET) with converters for electric traction applications. IEEE Transportation Electrification Conference and Expo (ITEC). *IEEE*, 1–6.

462. Evans NM, Lagier T, Pereira A (2016) A preliminary loss comparison of solid-state transformers in a rail application employing silicon carbide (SiC) MOSFET switches. 8th IET International Conference onPower Electronics, Machines and Drives (PEMD). *IEEE*, 1–6.

463. Joca DR, Barreto LHSC, Oliveira DDS, et al. (2016) Three-phase AC-DC solid-state transformer for low-voltage DC power distribution applications. 12th IEEE International Conference on Industry Applications (INDUSCON). *IEEE*, 1–8.

464. Rodriguez LAG, Jones V, Oliva A, et al. (2017) A new SST topology comprising boost three-level AC/DC converters for applications in electric power distribution systems. *IEEE J Em Sel Top P* 5: 735–746.

465. Feng J, Chu WQ, Zhang Z, et al. (2017) Power electronic transformer based railway traction systems: challenges and opportunities. *IEEE J Em Sel Top P* 5: 1237–1253.

466. Zhao B, Song Q, Li J, et al. (2017) Modular multilevel high-frequency-link dc transformer based on dual active phase-shift principle for medium-voltage DC power distribution application. *IEEE T Power Electr* 32: 1779–1791.

467. Dujic D, Kieferndorf F, Canales F (2012) Power electronic transformer technology for traction applications-An overview. *Electronics* 16: 50–56.

468. Hugo N, Stefanutti P, Pellerin M, et al. (2007) Power electronics traction transformer. European Conference on Power Electronics and Applications (EPE). *IEEE*, 1–10.

469. Claesens M, Dujic D, Canales F, et al. (2012) Traction transformation: A power-electronic traction transformer (PETT). *ABB Rev*, 11–17.

470. Dujic D, Kieferndorf F, Canales F, et al. (2012) Power electronic traction transformer technology. 7th International Power Electronics and Motion Control Conference (IPEMC). *IEEE*, 636–642.

471. Zhao C, Weiss M, Mester A, et al. (2012) Power electronic transformer (PET) converter: Design of a 1.2MW demonstrator for traction applications. 21st International Symposium on Power Electronics, Electrical Drives, Automation and Motion (SPEEDAM). *IEEE*, 855–860.

472. Dujic D, Zhao C, Mester A, et al. (2013) Power electronic traction transformer-low voltage prototype. *IEEE T Power Electr* 28: 5522–5534.

473. Hosseini SH, Sharifian MB, Sabahi M, et al. (2008) Bi-directional power electronic transformer for induction heating systems. Canadian Conference on Electrical and Computer Engineering (CCECE). *IEEE*, 347–350.

474. Sabahi M, Hosseini SH, Sharifian MBB, et al. (2009) A three-phase dimmable lighting system using a bidirectional power electronic transformer. *IEEE T Power Electr* 24: 830–837.

475. Shah J, Gupta RK, Mohapatra KK, et al. (2010) Power management with a dynamic power limit by a power electronic transformer for micro-grid. IEEE PES General Meeting. *IEEE*, 1–5.

476. Bignucolo F, Bertoluzzo M, Fontana C (2015) Applications of the solid state transformer concept in the electrical power system. AEIT International Annual Conference (AEIT). *IEEE*, 1–6.

477. Yan J, Zhu X, Lu N (2015) Smart hybrid house test systems in a solid-state transformer supplied microgrid. IEEE Power and Energy Society General Meeting. *IEEE*, 1–5.

478. Huber JE, Kolar JW (2014) Volume/weight/cost comparison of a 1 MVA 10 kV/400 V solid-state against a conventional low-frequency distribution transformer. IEEE Energy Conversion Congress and Exposition (ECCE). *IEEE*, 4545–4552.

12

Distribution network topology identification based on synchrophasor

Stefania Conti, Santi A. Rizzo*, Nunzio Salerno and Giuseppe M. Tina

Dipartimento di Ingegneria Elettrica Elettronica e Informatica (DIEEI), University of Catania, viale A. Doria, 6-95125 Catania, Italy

* **Correspondence:** Email: santi.rizzo@dieei.unict.it

Abstract: A distribution system upgrade moving towards Smart Grid implementation is necessary to face the proliferation of distributed generators and electric vehicles, in order to satisfy the increasing demand for high quality, efficient, secure, reliable energy supply. This perspective requires taking into account system vulnerability to cyber attacks. An effective attack could destroy stored information about network structure, historical data and so on. Countermeasures and network applications could be made impracticable since most of them are based on the knowledge of network topology. Usually, the location of each link between nodes in a network is known. Therefore, the methods used for topology identification determine if a link is open or closed. When no information on the location of the network links is available, these methods become totally unfeasible. This paper presents a method to identify the network topology using only nodal measures obtained by means of phasor measurement units.

Keywords: accuracy; distribution network; measurement errors; numerical methods; phasor measurement unit; power system state estimation; system of equations; topology identification

1. Introduction

In transmission networks, state estimation (SE) is a key function, since its results are used in fundamental network applications, such as optimal power flow, evaluation of available transfer capability, estimation of voltage and transient stability [1]. SE processes and filters real-time data (in terms of analogic measures, i.e., line flow, bus injection, voltage magnitude, or digital measures, i.e., switching devices status) available at the control centre of the transmission system to best estimate the current state of the network in order to provide dependable results to other network applications [2]. Typically, the aforementioned real-time applications are based on the knowledge of network topology, which is determined from the switching devices status [3]. Moreover, SE also uses topology to estimate the state and to identify bad data [1]. For these reasons, a great effort has been lavished to detect topology error in transmission systems [2–7].

Traditionally, power distribution networks are either not observable or only partially observable [8], since they were excluded, in the past, from the deployment of communication infrastructures, automation, monitoring, and control systems [9]. Recently, the distribution system is facing a deeply transformation due, on the one hand, to the load growth and the proliferation of new "actors" (distributed generators, prosumers, electric vehicles, etc.) and, on the other hand, to the increasing demand for high quality, efficiency, security, reliability of energy supply, leading towards Smart Grid (SG) implementation [10]. In this complex scenario, SG implementation asks for the introduction of innovative paradigms, network control and management techniques, as well as planning and maintenance strategies that, in turn, ask for the integration of additional infrastructures, computational resources and data storage, for information exchange and processing, electrical vehicles, intelligent devices, network optimization procedures, remote control, automation and so on [11–17]. Many benefits are expected from this distribution network upgrade [18]; on the other hand, it also involves new problems related to distribution system vulnerability to cyber-physical attacks [19–21].

Many countermeasures against these attacks [22] require the knowledge of the network topology [23]. Nevertheless, many of the aforementioned applications are based on network topology too [9,24,25]. As said before, topology identification topic has been well investigated for transmission networks, differently from the distribution system, for which a traditional topology identification technique is not available at present. Of course, the distribution system upgrade could enable the Distribution System Operator (DSO) to adopt techniques similar to the ones used in transmission networks or even to develop new ones [25,26]. It is worth to highlight that these techniques are based on the assumption that the location of each link in the network is known, then "topology identification" means to determine if a link is open or closed [2–7,25,26]. A well-suited cyber attack can destroy such information and make network topology identification impossible, as well as make countermeasures and network applications impracticable. New techniques based on phasor measurement units (PMUs), specifically devised for distribution networks, could cope with this issue since they can provide information about power injection and voltage at the network nodes (therefore, the current absorbed from/supplied to the nodes is also known). In [9] the problem of reconstructing the topology of a portion of the distribution network using a dataset of voltage measurements is investigated, but it assumes that all power lines in the grid have the same inductance/resistance ratio and it neglects the information about power injections.

On the other hand, [23] performs a blind topology identification only by using power injection data at each bus, neglecting information about voltages, and it considers a simplified power flow, that is the linear DC power flow. In [8] both nodal measurement types are considered, but the algorithm is based on a linear coupled approximation for lossless AC power flow and the results are conditioned on a given assumption regarding correlations in power injections at the non-substation buses. In [27] network connectivity verification is proposed. It is based on an algorithm that neglects the presence of dispersed generation along the distribution network, since it assumes that voltage magnitude always decreases downstream along the feeder as it usually occurs in traditional passive distribution networks only.

The method proposed in this paper is able to identify the network topology after a critical event such as a cyber-physical attack or, more in general, when there is no information about the connections among the network nodes. It is able to exploit the knowledge of both nodal current and voltage measured by means of PMUs, and it is based on a full AC power flow avoiding all the limitations of the aforementioned methods.

Section II describes the method proposed for identifying the network admittance matrix. Section III describes the technique adopted to analyze the real part of the matrix in order to identify the network topology accounting for PMUs accuracy. Finally, section IV reports some numerical results.

2. Nodal admittance matrix identification by means of synchrophasors

The one-line diagram of a feeder line of a three-phase, symmetrical, radial MV distribution network is considered. A graphic representation of the network is reported in Figure 1, where a node (filled point) is a network bus where customers and/or generators are connected, or a switching substation.

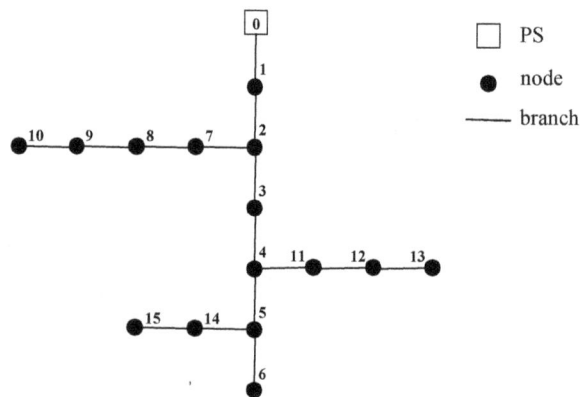

Figure 1. One-line diagram of a feeder line of a three-phase, symmetrical, radial distribution network.

A branch (solid line) is a link that represents the electrical equipment connecting two nodes. The primary substation (PS) bus takes the labeled 0, the other nodes are sequentially numbered such that the ones from the PS to node i take a number lower than i. These nodes are said "upstream" from node i while node i is said "downstream" from them. From the graph theory a tree, as the radial network shown in Figure 1, has n branches and $n+1$ nodes. So, each branch can be numbered as the downstream node which it is connected to. The Kirchhoff's current law can be written for each node and the following equation in matrix form can be written [28]:

$$[\mathbf{Y}][\mathbf{V}] = [\mathbf{J}] \tag{1}$$

where:

[**Y**] size (n+1) × (n+1)—is the nodal admittance matrix, whose elements can be directly obtained by topological inspection according to the following rule:

(1) y_{ii} (self-admittance) is the sum of the admittances related to the branches that are connected to node i;

(2) y_{ij} (mutual admittance) is the negative of the sum of the admittances related to the common branches between nodes i and j;

[**V**] size (n+1) × 1—node voltages array;

[**J**] size (n+1) × 1—load/generator currents array; the real part of the current is considered positive when it is injected into the node ("absorbed" load current).

The equations in (1) are linear and complex. In the considered scenario, the elements in [**Y**] are unknown and do not change; the elements in [**V**] and [**J**] are the synchrophasors (Figure 2).

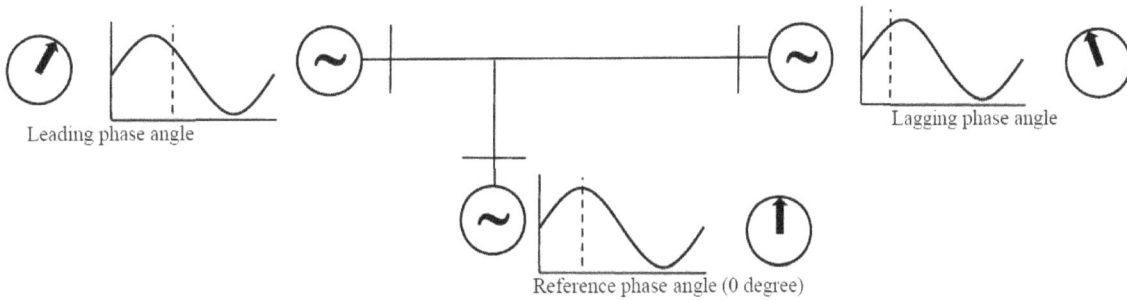

Figure 2. Synchrophasors [29].

After a cyber attack that destroyed the stored information about the network structure, firstly, [**Y**] is found by exploiting a set of equation systems like the one in (1) and the PMUs information about nodal voltages and currents. Subsequently, the network admittance matrix is used to identify the network topology, thus restoring the missing information. In fact, the network admittance matrix encloses information about topology by construction, since y_{ij} equal to zero implies that there is not a link between node i and j, otherwise the nodes are connected to each other. On the other hand, in a real world application, such a term could differ from zero even when there is not a link between node i and j due to voltage and current data accuracy and numerical issues.

At the first step, the nodes belonging to a given feeder line, fl, have to be found to identify the nodal admittance related to fl itself. To this aim, the sending-end switch of the feeder line is set to "closed status", while the sending-end switches of the other feeder lines are set to "open status", to search for the nodes belonging to fl. More specifically, the nodes whose voltage syncrophasor differs from zero are the ones belonging to fl. Then another switch is set to "closed status", and the new nodes whose voltage syncrophasor differs from zero to another feeder line. For a given PS, this procedure stops when the nodes of its feeders are identified.

After that, the following method could be adopted to identify matrix [**Y**] of a feeder. Note that, to convert [**Y**] of Eq1 into an equivalent vector, in order to explicit the unknowns, let us define the row string vector operator, rs with size $1 \times (n+1)^2$, the column string vector operator, cs with size $(n+1)^2 \times 1$, and the vectorization operator, v (transposed row string vector) with size $(n+1)^2 \times 1$, as follows [30]:

$$[\mathbf{rs}(\mathbf{Y})] \equiv \begin{bmatrix} y_{0,0} & y_{0,1} & \cdots & y_{0,n} & y_{1,0} & y_{1,1} & \cdots & y_{1,n} & \cdots & y_{n,0} & y_{n,1} & y_{n,n} \end{bmatrix} \qquad (2)$$

$$[\mathbf{cs(Y)}] \equiv [y_{0,0} \quad y_{1,0} \quad \cdots \quad y_{n,0} \quad y_{0,1} \quad y_{1,1} \quad \cdots \quad y_{n,1} \quad \cdots \quad y_{0,n} \quad y_{1,n} \qquad y_{n,n}]^T \tag{3}$$

$$[\mathbf{v(Y)}] \equiv [\mathbf{rs(Y)}]^T = [\mathbf{cs(Y}^T)] \tag{4}$$

Therefore, the following relations hold:

$$[\mathbf{rs(J)}] = [\mathbf{J}]^T \tag{5}$$

$$[\mathbf{cs(J)}] = [\mathbf{J}] \tag{6}$$

$$[\mathbf{v(J)}] = [\mathbf{J}] \tag{7}$$

Moreover, the following relation is obtained by considering the standard multiplication of three matrices M_1, M_2, M_3 [30]:

$$[\mathbf{v(M_1 M_2 M_3)}] = [\mathbf{M_1}] \otimes [\mathbf{M_3}]^T [v(\mathbf{M_2})] \tag{8}$$

in which symbol $\otimes$ denotes the Kronecker product.

Note that, Eq 1 does not change by pre-multiplying it by the identity matrix $[\mathbf{I}]$ (with size n+1):

$$[\mathbf{I}][\mathbf{Y}][\mathbf{V}] = [\mathbf{J}] \tag{9}$$

Considering relations (7) and (8), this system of equations can be rewritten as follows:

$$\overbrace{[\mathbf{I}] \otimes [\mathbf{V}]^T}^{(n+1) \times (n+1)^2} \; [\mathbf{v(Y)}] = [\mathbf{J}] \tag{10}$$

For example, when $n = 2$:

$$\begin{bmatrix} V_0 & V_1 & V_2 & 0 & 0 & 0 & 0 & 0 & 0 \\ 0 & 0 & 0 & V_0 & V_1 & V_2 & 0 & 0 & 0 \\ 0 & 0 & 0 & 0 & 0 & 0 & V_0 & V_1 & V_2 \end{bmatrix} \begin{bmatrix} y_{0,0} \\ y_{0,1} \\ y_{0,2} \\ y_{1,0} \\ y_{1,1} \\ y_{1,2} \\ y_{2,0} \\ y_{2,1} \\ y_{2,2} \end{bmatrix} = \begin{bmatrix} J_0 \\ J_1 \\ J_2 \end{bmatrix} \tag{11}$$

It is apparent that the number of equations is lower than the number of unknowns and, consequently, acquiring one set only of concurrent measures by means of the PMUs is not enough to calculate the elements of **[Y]**. Therefore, more equations are necessary, in other words many concurrent acquisitions at different instants must be performed. The first set of $(n+1)$ equations is obtained by using in (10) the nodal measures acquired at time t_1 by means of the PMUs. The second set of equations is obtained thanks to the subsequent concurrent measures, acquired at time t_2, and so on.

Therefore, the following system of equations is obtained by considering $(n+1)$ concurrent measures in $(n+1)$ different instants:

$$
\underbrace{\begin{bmatrix} [\mathbf{I}] \otimes [\mathbf{V}(t_1)]^T \\ [\mathbf{I}] \otimes [\mathbf{V}(t_2)]^T \\ \ldots \\ [\mathbf{I}] \otimes [\mathbf{V}(t_{n+1})]^T \end{bmatrix}}_{\| \quad [\mathbf{V'}]} [\mathbf{v}(\mathbf{Y})] = \underbrace{\begin{bmatrix} [\mathbf{J}(t_1)] \\ [\mathbf{J}(t_2)] \\ \ldots \\ [\mathbf{J}(t_{n+1})] \end{bmatrix}}_{\| \quad [\mathbf{J'}]}
\tag{12}
$$

where:

$$
\begin{aligned}
[\mathbf{V}(\tau)] &= [V_0(\tau) \quad V_1(\tau) \quad \ldots \quad V_k(\tau) \quad \ldots \quad V_n(\tau)]^T \\
[\mathbf{J}(\tau)] &= [J_0(\tau) \quad J_1(\tau) \quad \ldots \quad J_k(\tau) \quad \ldots \quad J_n(\tau)]^T \\
&with \quad \tau = t_1, t_2, \ldots, t_{n+1}
\end{aligned}
\tag{13}
$$

[V'], with size $(n+1)^2 \times (n+1)^2$, is the coefficient matrix and **[J']**, with size $(n+1)^2 \times 1$, is the known term array. System (12) is resolvable provided that the measures are not correlated among themselves, that is for each set of values measured at a given instant there must not exist a combination of the other sets that led to the same values in the considered set.

Note that, even though there are $(n+1)^2$ elements in **[Y]**, the number of admittances to be found is $(n+1)(n+2)/2$ for symmetry reasons. Let us apply the reduced form of operator v to account for this symmetric matrix [30]:

$$
[\overline{\mathbf{v}}(\mathbf{Y})] = \begin{bmatrix} \overbrace{y_{0,0} \quad y_{0,1} \quad \cdots \quad y_{0,n}}^{1 \times (n+1)} & \overbrace{y_{1,1} \quad y_{1,2} \quad \cdots \quad y_{1,n}}^{1 \times n} & \cdots & \overbrace{\cdots \quad \cdots}^{1 \times (n+1-k)} & \cdots & \overbrace{y_{n,n}}^{1 \times 1} \end{bmatrix}^T =
$$

$$
= [\ldots y_{rc} \ldots]^T \quad \text{where } r = 0,1,2,\ldots n \text{ and } c = r, r+1, \ldots n
\tag{14}
$$

The reduced operator in (14) can be obtained from the one in (4) by applying the row deletion matrix, **[D]** with size $0.5(n+1)(n+2) \times (n+1)^2$:

$$
[\overline{\mathbf{v}}(\mathbf{Y})] = [\mathbf{D}] \, [\mathbf{v}(\mathbf{Y})]
\tag{15}
$$

where **[D]** is obtained from a $(n+1)^2$-dimensioned identity matrix by deleting some rows. More specifically, the i-th row of this matrix is deleted if the i-th element of $[\mathbf{v}(\mathbf{Y})]$ is y_{rc}, with r greater than c. For example, when $n = 2$:

$$\begin{bmatrix} 1 & 0 & 0 & 0 & 0 & 0 & 0 & 0 & 0 \\ 0 & 1 & 0 & 0 & 0 & 0 & 0 & 0 & 0 \\ 0 & 0 & 1 & 0 & 0 & 0 & 0 & 0 & 0 \\ 0 & 0 & 0 & 1 & 0 & 0 & 0 & 0 & 0 \\ 0 & 0 & 0 & 0 & 1 & 0 & 0 & 0 & 0 \\ 0 & 0 & 0 & 0 & 0 & 1 & 0 & 0 & 0 \\ 0 & 0 & 0 & 0 & 0 & 0 & 1 & 0 & 0 \\ 0 & 0 & 0 & 0 & 0 & 0 & 0 & 1 & 0 \\ 0 & 0 & 0 & 0 & 0 & 0 & 0 & 0 & 1 \end{bmatrix} \leftrightarrow \begin{bmatrix} y_{0,0} \\ y_{0,1} \\ y_{0,2} \\ y_{1,0} \\ y_{1,1} \\ y_{1,2} \\ y_{2,0} \\ y_{2,1} \\ y_{2,2} \end{bmatrix} \begin{matrix} \\ \\ \\ \\ \\ \\ \leftrightarrow \\ \leftrightarrow \\ \end{matrix}$$

$$\Rightarrow [\mathbf{D}] = \begin{bmatrix} 1 & 0 & 0 & 0 & 0 & 0 & 0 & 0 & 0 \\ 0 & 1 & 0 & 0 & 0 & 0 & 0 & 0 & 0 \\ 0 & 0 & 1 & 0 & 0 & 0 & 0 & 0 & 0 \\ 0 & 0 & 0 & 0 & 1 & 0 & 0 & 0 & 0 \\ 0 & 0 & 0 & 0 & 0 & 1 & 0 & 0 & 0 \\ 0 & 0 & 0 & 0 & 0 & 0 & 0 & 0 & 1 \end{bmatrix} \tag{16}$$

On the other hand, the operator ν can be obtained from the reduced one by applying the row addition matrix, $[\mathbf{A}]$:

$$[\mathbf{v}(\mathbf{Y})] = [\mathbf{A}] \left| \overline{\mathbf{v}}(\mathbf{Y}) \right|$$

$$\text{size of } [\mathbf{A}] \quad (n+1)^2 \, x \frac{(n+1)(n+2)}{2} \tag{17}$$

where $[\mathbf{A}]$—size $(n+1)^2 \times 0.5(n+1)(n+2)$ —is obtained from a $(n+1)^2$-dimensioned identity matrix by deleting some columns and adding them to the others. More specifically, the i-th column of this matrix is deleted if the i-th element of $[\mathbf{v}(\mathbf{Y})]$ is y_{rc}, with r greater than c, and the column deleted is added to the j-th column provided that the j-th element of $[\mathbf{v}(\mathbf{Y})]$ is y_{cr}. For example, when $n = 2$:

$$\begin{bmatrix} 1 & 0 & 0 & 0 & 0 & 0 & 0 & 0 & 0 \\ 0 & 1 & 0 & 0 & 0 & 0 & 0 & 0 & 0 \\ 0 & 0 & 1 & 0 & 0 & 0 & 0 & 0 & 0 \\ 0 & 0 & 0 & 1 & 0 & 0 & 0 & 0 & 0 \\ 0 & 0 & 0 & 0 & 1 & 0 & 0 & 0 & 0 \\ 0 & 0 & 0 & 0 & 0 & 1 & 0 & 0 & 0 \\ 0 & 0 & 0 & 0 & 0 & 0 & 1 & 0 & 0 \\ 0 & 0 & 0 & 0 & 0 & 0 & 0 & 1 & 0 \\ 0 & 0 & 0 & 0 & 0 & 0 & 0 & 0 & 1 \end{bmatrix} \Rightarrow [\mathbf{A}] = \begin{bmatrix} 1 & 0 & 0 & 0 & 0 & 0 \\ 0 & 1 & 0 & 0 & 0 & 0 \\ 0 & 0 & 1 & 0 & 0 & 0 \\ 0 & 1 & 0 & 0 & 0 & 0 \\ 0 & 0 & 0 & 1 & 0 & 0 \\ 0 & 0 & 0 & 0 & 1 & 0 \\ 0 & 0 & 1 & 0 & 0 & 0 \\ 0 & 0 & 0 & 0 & 1 & 0 \\ 0 & 0 & 0 & 0 & 0 & 1 \end{bmatrix} \tag{18}$$

By applying to ν the row deletion and addition matrix in turn, the following relation holds:

$$[\mathbf{v}(\mathbf{Y})] = [\mathbf{A}] \left| \overline{\mathbf{v}}(\mathbf{Y}) \right| = [\mathbf{A}][\mathbf{D}] \, [\mathbf{v}(\mathbf{Y})] \tag{19}$$

Therefore, the equation system in (12) can be rewritten as follows to account for symmetries:

$$[\mathbf{V'}][\mathbf{A}] \left| \overline{\mathbf{v}}(\mathbf{Y}) \right| = [\mathbf{J'}] \tag{20}$$

and the sub-set of independent equations in the system is obtained by applying the row deletion matrix, provided that the measures are not correlated among themselves. Pre-multiplying the terms by the deletion matrix, the relations in (20) can be written as:

$$[\mathbf{D}][\mathbf{V'}][\mathbf{A}]\left|\overline{\mathbf{v}(\mathbf{Y})}\right| = [\mathbf{D}][\mathbf{J'}] \tag{21}$$

and the following reduced system of equations is obtained:

$$
\begin{aligned}
&\left[\overline{\mathbf{V'}}\right]\left[\overline{\mathbf{v}(\mathbf{Y})}\right] = \left[\overline{\Delta\mathbf{J'}}\right] \\
with \quad &\left[\overline{\mathbf{V'}}\right] = [\mathbf{D}][\mathbf{V'}][\mathbf{A}] \\
&\left[\overline{\Delta\mathbf{J'}}\right] = [\mathbf{D}][\Delta\mathbf{J'}]
\end{aligned} \tag{22}
$$

The reduced system of equations accounts for symmetries. On the one hand, it includes $(n+1)(n+2)$ unknowns, that are the real and imaginary parts of the admittances; on the other hand, there are $(n+1)(n+2)$ equations because each complex linear equation can be treated as two real linear equations.

3. Topology identification by using the nodal admittance matrix

The topology identification method proposed in this section is based on the use of the real part of $[\mathbf{Y}]$, i.e., the network conductance matrix $[\mathbf{G}]$. The resulting information about the links in the network are stored in an array, $[\mathbf{L}]$, where a link between node i and j is represented by a number, k, which is computed as follows:

$$
\begin{aligned}
&k = i(n+1) + j \\
&with \ i \in [0,n], \ j \in [i+1,n]
\end{aligned} \tag{23}
$$

For example, when the network depicted in Figure 1 is considered, the link array $[\mathbf{L}]$ is:

$$[\mathbf{L}] = \begin{Bmatrix} 1, 18, 35, 39, 52, 69, 75, 86, 94, \\ 120, 137, 154, 188, 205, 239 \end{Bmatrix} \tag{24}$$

Information about the topology can be easily derived from $[\mathbf{G}]$ since the element at position (i,j) is equal to zero when there is not a link between nodes i and j, otherwise the nodes are connected to each other. On the other hand, in practical applications, such an element could differ from zero even when nodes i and j are not directly connected to each other due to PMUs accuracy and numerical issues. Therefore, a method to properly analyse the conductance matrix is proposed in the following. More specifically, such a method checks the sign of the mutual conductance values and compares some of them with the self-conductance. Moreover, it imposes radiality and connectivity constraints. The main steps of the proposed method are reported in Figure 3, where:

a. *ascend_order_position* is a function that returns an array P such that:

$$G[i, P[p-1]] < G[i, P[p]]$$
$$with \quad p \in [1, n]$$

(25)

when it is applied to $(i, \mathbf{G})$;

b. α is a negative number with magnitude lower than one;

c. *apply_eq23* is a function that applies equation 23 (even when j is lower than $i+1$);

d. *remove_duplicated_links* is a function that removes from L each link k obtained for i higher than j;

e. *force_radiality* is a function that removes as many links as necessary to avoid loops;

f. *force_connectivity* is a function that adds as many links as it is necessary to avoid isolated set of nodes while keeping the radiality constraint satisfied.

The first link stored in [**L**] is the one connecting node 0 to the downstream node. Note that only one node is connected to node 0 when a specific feeder line is considered. Therefore, looking at the first row of [**G**], the column with the lowest value (that is the negative number with higher magnitude) indicates the node connected to 0 and, consequently, the related link is stored in [**L**] (see rows 3–4 in Figure 3).

When a generic node i is considered, the previous mechanism is firstly adopted. In other words, looking at the i-th row of [**G**], column j with the lowest value, i.e., $\mathbf{G}[i,P[1]]$, is found and the link between i and j is stored (see rows 7–8 in Figure 3). After that, the lowest one among the other elements in row i, i.e., $\mathbf{G}[i,P[2]]$, indicates another node that could be connected to i. The related link is stored in L provided that:

(1) the sum of the mutual-conductance magnitudes considered until that step does not exceed the self-conductance;

(2) the aforementioned lowest element is negative; and

(3) it has a magnitude comparable with the self-conductance (row 11 in Figure 3).

With reference to point 3, a coefficient α is used to establish if the element $\mathbf{G}[i,P[p]]$ can be considered exactly as "zero" although it actually differs from "zero". Such a difference is due to data accuracy and numerical issues.

After that, the next lowest value is considered and so on, as long as the three aforementioned conditions are satisfied (rows 11–15 in Figure 3). When the p-th lowest value is negative and it is comparable with the self-conductance (i.e., it satisfies conditions 2 and 3), but it hasn't been stored in [**L**] since the sum exceeds the self-conductance (i.e., it does not satisfy condition 1), the link is stored anyway if the addition of the mutual-conductance leads to a better approximation of the self-conductance (rows 16–19 in Figure 3).

A link between β and γ could be stored in [**L**] when i is equal to β, and the same link could be stored again in [**L**] when i is equal to γ. In other words, the same link is stored twice.

Conversely, a link between β and γ could be stored in [**L**] when i is equal to β, even though the link is supposed not to exist when i is equal to γ.

In both cases the link is added although in the first case it is more probable that the link actually exists. The function *remove_duplicated_links* erases the duplicates in the occurrence of the first case to avoid storing the same link twice (row 22 in Figure 3). Subsequently, two functions that, respectively, remove and add links to account for the topology constraints are executed (rows 23–24 in Figure 3).

```
1    L := ∅
2    i := 0
3    P := ascend_order_position (i. G)
4    L := L ∪ {apply_eq23(i.P[0])}
5    i := 1
6    while (i ≤ n)
7    P := ascend_order_position (i. G)
8    L := L ∪ {apply_eq23(i.P[0])}
9    sum := |G[i.P[0]]|
10   p := 1
11   while ( p < n & G[i.P[p]] < αG[i.i] & |G[i.P[p]]|+sum ≤ G[i.i] )
12   L := L ∪ {apply_eq23(i.P[p])}
13   sum := sum+|G[i.P[p]]|
14   p := p+1
15   endwhile
16   if ( p < n & G[i.P[p]] < αG[i.i] & |G[i.P[p]]| < 2(G[i.i]-sum) )
17   L := L ∪ {apply_eq23(i.P[p])}
18   sum := sum+|G[i.P[p]]|
19   endif
20   i:=i+1
21   endwhile
22   L := remove_duplicated_links (L)
23   L := force_radiality (L)
24   L := force_connectivity (L)
```

Figure 3. Main steps of the method adopted to analyze the network conductance matrix in order to figure out the topology.

The ability of the method to correctly identify each link is strongly dependent on the accuracy of **[G]**, which in turn depends on the accuracy of the adopted PMUs, on the method adopted to solve the reduced system of Eq 22, and on the precision data type. Assuming a steady-state condition, a simple method to limit the impact of the PMUs' accuracy on the network conductance matrix is to repeat the acquisition of the set of nodal measures. More specifically, the values of V_i and I_i at time τ in Eq 13 can be obtained by averaging their measured values in the neighbourhood of τ.

For example, when R consecutive acquisitions are performed in the neighbourhood of a given time $\tau = t_m$, the voltage at the i-th node is computed as:

$$V_i(t_m) = \frac{\sum_{s=1}^{R} V_i\left(t_m + \left(s - \frac{R+1}{2}\right)\Delta\right)}{r}$$

$$with \qquad t_{m-1} < t_m - \left(\frac{r-1}{2}\right)\Delta \qquad\qquad (26)$$

$$t_m + \left(\frac{r-1}{2}\right)\Delta < t_{m+1}$$

where Δ has to be a time interval small enough to ensure that the nodal voltages and currents do not change (accounting for R). The voltage (and current) values computed by using (26) are used in (22) to improve admittance matrix accuracy.

4. Case study

The method proposed to estimate the nodal admittance matrix and to identify the topology by using its real part has been implemented in a Java program. The program has been applied to the distribution network represented in Figure 1. The frequency has been assumed always equal to the nominal one and the coefficient α has been set to -0.05.

Firstly, it has been tested in case of no measurement error. To this aim, the line voltage magnitude at node 0 has been assumed always equal to 20 kV with phase 0, while 16 sets of simultaneous nodal voltages are randomly created for the other 15 nodes imposing that nodal voltage constraints are satisfied ($\pm 10\%$ of nominal voltage as established by the standard EN 50160). After that, the simultaneous values of nodal currents are obtained by applying Eq 1 for each voltage set, so that each node could appear as a load point, a generation point or both. These sets of voltage and current values are provided to the program implementing the proposed method in order to emulate the concurrent measures of both voltage and current acquired in 16 different time instants.

The program has correctly evaluated the values of both real and imaginary parts of the nodal admittance matrix, as well as it has correctly identified the network topology (i.e., the identified link array is equal to the one in 24).

Subsequently, the program has been tested considering different PMUs accuracies. ANSI C 12.20 is the American national standard for electricity meter. The meters should satisfy Accuracy Class 5 (i.e., $\pm 0.5\%$) at least, the best ones comply with Accuracy Class 2 (i.e., $\pm 0.2\%$). Usually, PMU are very precise meters because their accuracy is less or equal to 0.1% [31].

For a given accuracy, 100 tests have been performed and, for each test, the estimated link array has been compared with the correct one in order to compute the number of elements correctly identified. The average percentage of links correctly identified is reported in Figure 4 for different number of consecutive measures, with $R \in [1,10]$. Therefore, 50,000 tests have been performed to evaluate the performance of the proposed approach on the 16-node test network of Figure 1.

In Figure 4, for a given PMUs' accuracy and R, the height of the related rectangle is equal to the average number of links correctly identified expressed in percentage.

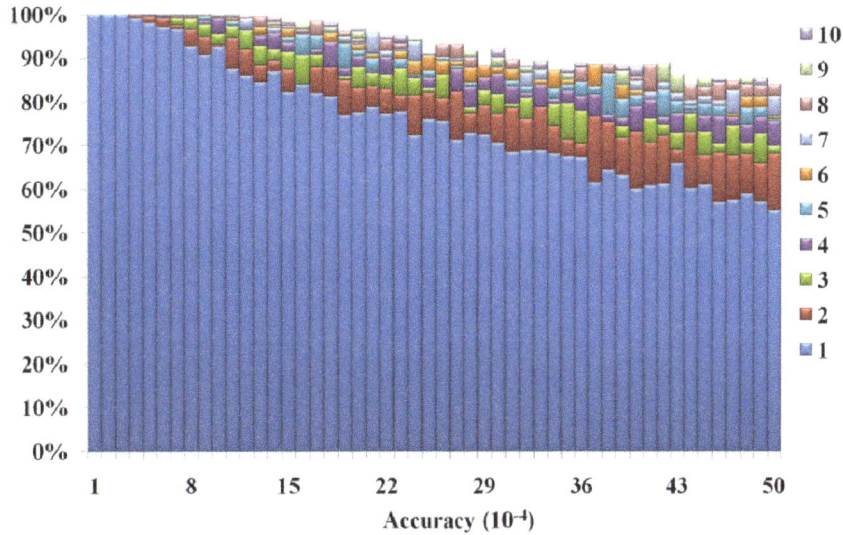

Figure 4. Average percentage of links correctly identified for the 16-node test system (Figure 1).

Moreover, for a given PMUs' accuracy the rectangles are stacked one behind the other for increasing values of R. More specifically, the first one, i.e., the one in front, refers to $R = 1$, the second refers to $R = 2$ and so on. This implies that, for a given accuracy, the rectangle related to $R = 2$ is visible only when its height is greater than the one related to $R = 1$. Similarly, for example, a rectangle related to $R = 5$ is visible only if its height is greater than the ones related to r equal to 1, 2, 3 and 4.

Figure 4 shows that the proposed topology identification method always enable to identify correctly the network without any new acquisition (i.e., $R = 1$) when the PMUs' accuracy is less than 0.04%. Moreover, more than the 90% of the links are correctly identified when only one acquisition is performed (i.e., $R = 1$) and the PMUs' accuracy is less or equal to 0.1%. When r is greater than 8, the method always enables to identify correctly the network if the PMUs' accuracy is less or equal to 0.1%. Moreover, even when poor quality PMUs are considered, for example with accuracy up to 0.5% (i.e., Accuracy Class 5), at least the 90% of the links are correctly identified if many re-acquisitions are executed.

The main fact arising from the results is that few re-acquisitions are necessary to obtain correct results with typical PMU's accuracy. Finally, the worse the PMUs' accuracy the greater the advantage of performing re-acquisitions.

The program has also been applied to the IEEE test feeder with 33 nodes [32] represented in Figure 5. The link array **[L]** related to the network is the following:

$$[\mathbf{L}] = \begin{cases} 1, 35, 51, 69, 88, 103, 137, 171, 190, 205, \\ 239, 273, 307, 341, 375, 409, 443, 477, \\ 511, 545, 613, 647, 681, 749, 783, 851, \\ 885, 919, 953, 987, 1021, 1055 \end{cases} \qquad (27)$$

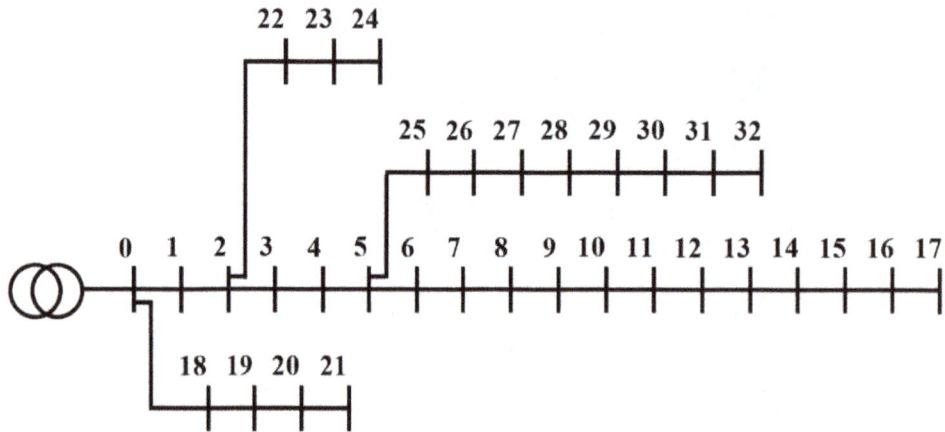

Figure 5. 33-node test system.

Similarly, to the previous test system, 33 sets of simultaneous nodal voltages are randomly created for node 1–32 then the related nodal currents are computed by applying Eq 1. These sets emulate the concurrent measures of both voltage and current acquired in 33 different time instants. When no measurement error is considered, that is the previous values are used without any modification, the proposed method has been able to find the correct value of all elements in the nodal admittance matrix. Moreover, the identified link array is equal to the one in 27 and hence the network topology has been correctly identified too. After that, 100 tests have been performed for each PMUs' accuracy as performed for the 16-node test system. More specifically, a slightly modification of the sets of concurrent measures has been done accordingly the given accuracy. The average percentage of links correctly identified is reported in Figure 6.

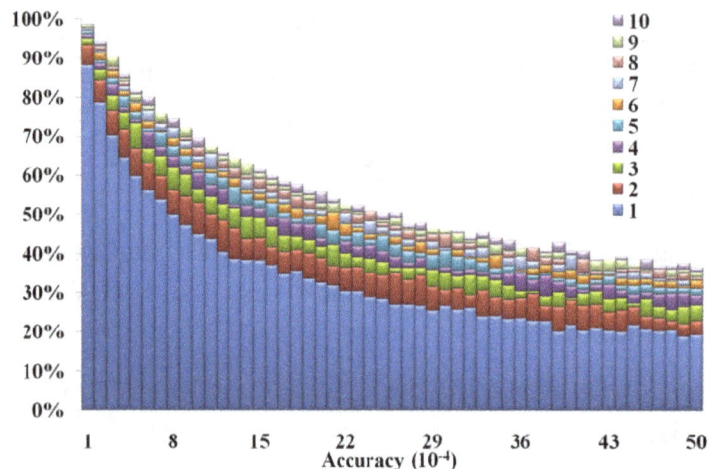

Figure 6. Average percentage of links correctly identified for the 33-node test system (Figure 5).

Comparing with the results obtained for the previous networks it is evident that the performance of the method degrades when the size of the network is doubled. On the other hand, the positive effect of performing more than one measure acquisition is very considerable even if the PMUs' accuracy is small, differently from the 16-node test system.

Therefore, the main fact arising from the tests performed on the 33-node test system is that the method proposed to estimate the nodal admittance matrix and to identify the topology still works but a large number of re-acquisitions are necessary to obtain sufficient or good results with typical PMU's accuracy.

5. Conclusion and future work

The basic theoretical aspects of the proposed method to identify the network topology using only voltage and current measures have been described. The method firstly uses these data to evaluate the elements of the nodal admittance matrix. It has been mathematically shown that to evaluate the value of all elements is necessary to concurrently measure both voltage and current at all nodes. Moreover, to this aim is also necessary to perform concurrent measures in different instants and the number of instants has to be equal to the number of nodes.

The second part of the method uses the real part of the nodal admittance matrix to identify the network topology. It checks the sign of the mutual conductance values and compares some of them with the self-conductance, and then it imposes radiality and connectivity constraints. The performance of the method to correctly identify each link strongly depends on the matrix accuracy and, in turn, on the accuracy of the adopted PMUs, on the numerical method adopted to solve the system of equations as well as on the precision of data storage.

The ability of the proposed method to identify the topology has been tested on two networks considering different PMUs' accuracy. When typical PMUs' accuracy is considered, the results show that the topology is always correctly identified in small networks. As the number of nodes increases, sometimes the topology is not correctly identified. However, a very great number of links in the network is correctly identified, and almost all if the measurements are very accurate.

Therefore, the proposed method is a promising starting point in the field of topology identification in distribution network subject to cyber attacks that have destroyed all information about network structure, historical data and so on. Further works will regard, by one hand, the development of very accurate PMUs, the research of the best method to solve the system of equations used for identify the nodal admittance matrix as well as the investigation of data storage precision on the performance. On the other hand, new features and improvement are necessary in the view of fully dependable topology identification in large networks.

Conflict of interest

The authors declare no conflict of interest in this paper.

References

1. Monticelli A (2004) Electric power system state estimation. *P IEEE* 88: 262–282.
2. Kumar DMV, Srivastava SC, Shah S, et al. (1996) Topology processing and static state estimation using artificial neural networks. *IEE P-Gener Transm D* 143: 99–105.
3. Clements KA, Costa AS (2002) Topology error identification using normalized Lagrange multipliers. *IEEE T Power Syst* 13: 347–353.

4.   Singh N, Glavitsch H (1991) Detection and identification of topological errors in on line power system analysis. *IEEE T Power Syst* 6: 324–331.

5.   Costa IS, Leao JA (1993) Identification of topology errors in power system state estimation. *IEEE T Power Syst* 8: 1531–1538.

6.   Silva APAD, Quintana VH, Silva APAD, et al. (1995) Pattern analysis in power system state estimation. *Int J Elec Power* 17: 51–60.

7.   Singh D, Pandey JP, Chauhan DS (2005) Topology identification, bad data processing, and state estimation using fuzzy pattern matching. *IEEE T Power Syst* 20: 1570–1579.

8.   Deka D, Backhaus S, Chertkov M (2015) Structure learning and statistical estimation in distribution networks-Part II. *Eprint Arxiv*, 1–1.

9.   Bolognani S, Bof N, Michelotti D, et al. (2013) Identification of power distribution network topology via voltage correlation analysis. IEEE Conference on Decision and Control. *IEEE*, 1659–1664.

10. Moslehi K, Kumar R (2010) A reliability perspective of the smart grid. *IEEE T Smart Grid* 1: 57–64.

11. Conti S, Rizzo SA, El-Saadany EF, et al. (2014) Reliability assessment of distribution systems considering telecontrolled switches and microgrids. *IEEE T Power Syst* 29: 598–607.

12. Rueda JL, Guaman WH, Cepeda JC, et al. (2013) Hybrid approach for power system operational planning with smart grid and small-signal stability enhancement considerations. *IEEE T Smart Grid* 4: 530–539.

13. Moeini-Aghtaie M, Farzin H, Fotuhi-Firuzabad M, et al. (2016) Generalized analytical approach to assess reliability of renewable-based energy hubs. *IEEE T Power Syst* 32: 368–377.

14. Minciardi R, Sacile R (2012) Optimal control in a cooperative network of smart power grids. *IEEE Syst J* 6: 126–133.

15. Zubo RHA, Mokryani G, Rajamani HS, et al. (2016) Operation and planning of distribution networks with integration of renewable distributed generators considering uncertainties: A review. *Renew Sust Energ Rev* 72: 1177–1198.

16. Soltani NY, Kim SJ, Giannakis GB (2015) Real-time load elasticity tracking and pricing for electric vehicle charging. *IEEE T Smart Grid* 6: 1303–1313.

17. Gungor VC, Sahin D, Kocak T, et al. (2011) Smart grid technologies: communication technologies and standards. *IEEE T Ind Inform* 7: 529–539.

18. Hamilton B, Summy M (2011) Benefits of the smart grid [In My View]. *IEEE Power Energy M* 9: 104–112.

19. Ansari N (2013) CONSUMER: A novel hybrid intrusion detection system for distribution networks in smart grid. *IEEE T Emerg Top Com* 1: 33–44.

20. Conti S, Corte AL, Nicolosi R, et al. (2016) Impact of cyber-physical system vulnerability, telecontrol system availability and islanding on distribution network reliability. *Sust Energ Grid Netw* 6: 143–151.

21. Davis KR, Davis CM, Zonouz SA, et al. (2015) A cyber-physical modeling and assessment framework for power grid infrastructures. *IEEE T Smart Grid* 6: 2464–2475.

22. Hug G, Giampapa JA (2012) Vulnerability assessment of ac state estimation with respect to false data injection cyber-attacks. *IEEE T Smart Grid* 3: 1362–1370.

23. Li X, Poor HV, Scaglione A (2016) Blind topology identification for power systems. IEEE International Conference on Smart Grid Communications. *IEEE*, 91–96.

24. Erseghe T, Tomasin S, Vigato A (2013) Topology estimation for smart micro grids via powerline communications. *IEEE T Signal Proces* 61: 3368–3377.

25. Zhao L, Song WZ, Tong L, et al. (2014) Topology identification in smart grid with limited measurements via convex optimization. Innovative Smart Grid Technologies-Asia. *IEEE*, 803–808.

26. Singh R, Manitsas E, Pal BC, et al. (2010) A recursive bayesian approach for identification of network configuration changes in distribution system state estimation. *IEEE T Power Syst* 25: 1329–1336.

27. Luan W, Peng J, Maras M, et al. (2015) Smart meter data analytics for distribution network connectivity verification. *IEEE T Smart Grid* 6: 1964–1971.

28. Chatzarakis GE (2009) Nodal analysis optimization based on the use of virtual current sources: A powerful new pedagogical method. *IEEE T Educ* 52: 144–150.

29. Phasor Advanced FAQ-Phasor-RTDMS: Available from: http://www.phasor-rtdms.com/phaserconcepts/phasor_adv_faq.html.

30. Vetter WJ (1975) Vector structures and solutions of linear matrix equations. *Linear Algebra & Its Applications* 10: 181–188.

31. Depablos J, Centeno V, Phadke AG, et al. (2004) Comparative testing of synchronized phasor measurement units. Power Engineering Society General Meeting. *IEEE* 1: 948–954.

32. Baran ME, Wu FF (1989) Network reconfiguration in distribution systems for loss reduction and load balancing. *IEEE T Power Deliver* 4: 1401–1407.

PERMISSIONS

The contributors of this book come from diverse backgrounds, making this book a truly international effort. This book will bring forth new frontiers with its revolutionizing research information and detailed analysis of the nascent developments around the world.

We would like to thank all the contributing authors for lending their expertise to make the book truly unique. They have played a crucial role in the development of this book. Without their invaluable contributions this book wouldn't have been possible. They have made vital efforts to compile up to date information on the varied aspects of this subject to make this book a valuable addition to the collection of many professionals and students.

This book was conceptualized with the vision of imparting up-to-date information and advanced data in this field. To ensure the same, a matchless editorial board was set up. Every individual on the board went through rigorous rounds of assessment to prove their worth. After which they invested a large part of their time researching and compiling the most relevant data for our readers.

The editorial board has been involved in producing this book since its inception. They have spent rigorous hours researching and exploring the diverse topics which have resulted in the successful publishing of this book. They have passed on their knowledge of decades through this book. To expedite this challenging task, the publisher supported the team at every step. A small team of assistant editors was also appointed to further simplify the editing procedure and attain best results for the readers.

Apart from the editorial board, the designing team has also invested a significant amount of their time in understanding the subject and creating the most relevant covers. They scrutinized every image to scout for the most suitable representation of the subject and create an appropriate cover for the book.

The publishing team has been an ardent support to the editorial, designing and production team. Their endless efforts to recruit the best for this project, has resulted in the accomplishment of this book. They are a veteran in the field of academics and their pool of knowledge is as vast as their experience in printing. Their expertise and guidance has proved useful at every step. Their uncompromising quality standards have made this book an exceptional effort. Their encouragement from time to time has been an inspiration for everyone.

The publisher and the editorial board hope that this book will prove to be a valuable piece of knowledge for researchers, students, practitioners and scholars across the globe.

LIST OF CONTRIBUTORS

Theocharis Tsoutsos, Stavroula Tournaki, Maria Frangou and Marianna Tsitoura
Renewable and Sustainable Energy Systems Lab, School of Environmental Engineering, Technical University of Crete, University Campus, Chania, Crete, Greece

Fiona Bénard-Sora and Jean-Philippe Praene
University of La Reunion, PIMENT Laboratory, 117 rue du Général Ailleret 97430 Le Tampon, France

Yatina Calixte
University of La Reunion, Department of Building Sciences and Environment, France

W. Prasadini, Kumudu S. Perera and Kamal P. Vidanapathirana
Polymer Electronics Research Laboratory, Department of Electronics, Wayamba University of Sri Lanka, Kuliyapitiya, Sri Lanka

W. A. D. S. S. Weerasinghe, K. P. Vidanapathirana and K. S. Perera
Department of Electronics, Faculty of Applied Sciences, Wayamba University of Sri Lanka, Kuliyapitiya, 60200 Sri Lanka

Saad S. Alrwashdeh, Falah M. Alsaraireh and Mohammad A. Saraireh
Mechanical Engineering Department, Faculty of Engineering, Mutah University, Al Karak 61710, Jordan

Henning Markötter, Nikolay Kardjilov and Ingo Manke
Helmholtz-Zentrum Berlin, Hahn-Meitner-Platz 1, 14109 Berlin, Germany

Merle Klages and Joachim Scholta
Zentrum für Sonnenenergie- und Wasserstoff-Forschung Baden Württemberg (ZSW), Helmholtzstraße 8, 89081 Ulm, Germany

Muhammad Rashed Al Mamun
Department of Farm Power and Machinery, Faculty of Agricultural Engineering and Technology, Sylhet Agricultural University, Sylhet, Bangladesh

Anika Tasnim and Shahidul Bashar
Department of Farm Power and Machinery, Sylhet Agricultural University, Sylhet, Bangladesh

Md. Jasim Uddin
Department of Animal Nutrition, Faculty of Veterinary, Animal and Biomedical Sciences, Sylhet Agricultural University, Sylhet, Bangladesh

Nagaraj C
Research Scholar, Department of Electrical & Electronics Engineering, National Institute of Technology Karnataka (NITK), Surathkal, India

K Manjunatha Sharma
Associate Professor, Department of Electrical & Electronics Engineering, National Institute of Technology Karnataka (NITK), Surathkal, India

Steven B. Sherman, Zachary P. Cano, Michael Fowler and Zhongwei Chen
Department of Chemical Engineering, University of Waterloo, 200 University Avenue West, Waterloo, ON, N2L 3G1, Canada

Li Bin, Muhammad Shahzad, Qi Bing, Hafiz MA Khan and Nabeel AM Fahal
School of Electrical and Electronic Engineering, North China Electric Power University, Beijing 102206, P.R. China

Muhammad Ahsan
Department of Electrical Engineering, The Superior University Lahore, Lahore 54000, Pakistan

Muhammad U Shoukat
Department of Electrical Engineering, Government College University Faisalabad Sahiwal Campus, Sahiwal 57000, Pakistan

James D. McLellan and Richard E Blanchard
Centre for Renewable Energy Systems Technology, Loughborough University, UK

M. Ebrahim Adabi and Juan A. Martinez-Velasco
Universitat Politecnica de Catalunya, Barcelona, Spain

Stefania Conti, Santi A. Rizzo, Nunzio Salerno and Giuseppe M. Tina
Dipartimento di Ingegneria Elettrica Elettronica e Informatica (DIEEI), University of Catania, viale A. Doria, 6-95125 Catania, Italy

Index

www.ingramcontent.com/pod-product-compliance
Lightning Source LLC
Chambersburg PA
CBHW080526200326
41458CB00012B/4350